A Railroad Atlas of the United States in 1946

VOLUME 2: NEW YORK & NEW ENGLAND

CREATING THE NORTH AMERICAN LANDSCAPE
Gregory Conniff, Edward K. Muller, David Schuyler, Consulting Editors
George F. Thompson, Series Founder and Director

Published in cooperation with the
CENTER FOR AMERICAN PLACES
Santa Fe, New Mexico, and Staunton, Virginia

A RAILROAD ATLAS of the United States in 1946

VOLUME 2: NEW YORK & NEW ENGLAND

Richard C. Carpenter

THE JOHNS HOPKINS UNIVERSITY PRESS   BALTIMORE AND LONDON

© 2005 Richard C. Carpenter
All rights reserved. Published 2005
Printed in Singapore
2 4 6 8 9 7 5 3 1

The Johns Hopkins University Press
2715 North Charles Street
Baltimore, Maryland 21218-4363
www. press.jhu.edu

Library of Congress Cataloging-in-Publication Data

Carpenter, Richard C.
  A railroad atlas of the United States in 1946 /
Richard C. Carpenter
    p. cm. — (Creating the North American landscape)
    Includes bibliographical references and index.
    Contents: v. 2. New York & New England.
    ISBN 0-8018-8078-5 (alk. paper)
    1. Railroads—New York & New England—Maps.
2. Railroads—New York & New England—History. I. Title.
II. Series.

G1246.P3 C3 2003
385'.0973'0223—dc21   2002031038 2002040581

A catalog record for this book is available
from the British Library.

Cover photograph taken by Philip M. Carpenter, the author's father, on August 21, 1947, while looking north from the Route 138 overhead bridge as a westbound New Haven Railroad passenger train, powered by an I-4 "pacific" type steam locomotive, passes interlocking tower 133 at Kingston, Rhode Island. As a young visitor to this tower, the author first learned about railroad operations and witnessed the final years of steam operations on the New Haven.

The sketches on the following pages were inspired by the photographs contained in the publications referenced in each instance: page ix, photograph by Arthur H. Mitchell, from Peter E. Lynch, *The New Haven Railroad* (St. Paul, Minn.: MBI Publishing Co., 2003), p. 124; page xi, photograph by Robert Chamberlain, and page xv, photograph by Al Holtz, both from Geoffrey H. Doughty, *New York Central Facilities in Color* (Scotch Plains, N.J.: Morning Sun Books, Inc., 2002), pp. 114 and 64, respectively; page xvi, photograph by Phillip R. Hastings, from David P. Morgan, *The Mohawk That Refused to Abdicate and Other Tales* (Milwaukee: Kalmbach Publishing Co., 1975), p. 191.

Contents

Introduction ... vii

How to Use This Atlas ... xix

Acknowledgments ... xxiii

THE ATLAS

Key Map 3

Map Symbols and Abbreviations 4

The Maps 5

Appendix: List of Railroads in the Atlas ... 195

Notes on the Maps ... 197   References ... 209

Indexes

Coaling Stations 213
Interlocking Stations and Former
   Interlocking Stations 214
Passenger and Non-passenger Stations 223
Track Pans 254
Tunnels 255
Viaducts 256

Introduction

I was driving across northern Indiana along U.S. 224, east of Huntington, on a trip home from Chicago in 1990. Looking out across the fields, I could see the distinctly visible high fill that had once been the roadbed for the double-track Erie Railroad, by then abandoned. I was following it out of the corner of my eye, as I am wont to do with old railroad lines, when suddenly it disappeared. The fill, which had been cutting a 15-foot-high swath across the verdant fields for miles, was completely gone. I could now look across flat sections of corn and soy as far as my eyes could see. There was absolutely no sign that this farmland had ever been interrupted in any way, never mind that it had watched, for more than one hundred years, the passage of thousands upon thousands of trains, from long, lumbering freights to such passenger trains as the New York–to–Chicago Erie Limited, the Atlantic Express, and the Lake Cities.

As I drove on, I spotted one bit of evidence. It was almost paved over in a farmhouse driveway next to a fully grown cornfield. Four pieces of steel rail sat in the pavement, beginning and ending on each side. These four shining rails, now retired to a tranquil life on a picturesque Indiana farm, were all that remained—in these parts, at least—of the once mighty Erie Railroad. They bore silent witness to the golden age of American railroading, when lines like this criss-crossed the country carrying goods and people and when sublime stretches of summer-evening silence on the prairie were marked by the unmistakable throaty moan of the steam engine's whistle.

I saw many other abandoned railroad rights-of-way as I drove across the historic, complex railscape of northern Indiana, Ohio, and on eastward into Pennsylvania and New Jersey. Somehow it occurred to me that I should not leave

unrecorded or forgotten the existence of these many links in what had once been a United States railroad system totaling some 254,037 route miles. The story of American railroading in its heyday was, and is, a story worth telling. And I became determined to tell it with a clear, easy-to-read atlas.

It seems particularly appropriate to document the railroad network that provided the steel pathway over which President Franklin D. Roosevelt's "arsenal of democracy" had transported the results of its overwhelming war machine. Mr. Roosevelt himself loved to ride the American rails in his private car, the Ferdinand Magellan.

It is my hope that, by producing a graphic record of this transportation network, present and future generations may learn valuable lessons from one of the most glorious episodes of our transportation history. And what better year than 1946, when the impressive system of American railroads emerged from its magnificent performance during World War II to face the challenges of highway and airline competition in the post-war era. This year marked a historic turning point in American rail transportation. In 1946, railroads were still the primary mode of intercity transportation. Indeed, the previous four years of World War II had increased their modal share of passengers and freight, due to gasoline rationing and its severe restrictions on nonessential travel as well as massive wartime troop and materiel shipments. But, with the lifting of wartime restrictions, the steady prewar trend from rail to highway transport and the growing demand for the speed afforded by the airplane would challenge and fundamentally change the American railroad network.

The year 1946 would also mark the beginning of the end of a long period of relative stability in the number of United States railroad companies. For more than a quarter century, between 1920 and 1946, only four national, significant railroad mergers and consolidations occurred. In 1923, the Nickel Plate Road absorbed both the Toledo, St. Louis and Western and the Lake Erie & Western; in 1938, the Duluth, Missabe, and Iron Range absorbed through consolidation the Duluth, Missabe & Northern, Duluth & Iron Range, Interstate Transfer, and Spirit Lake Transfer; in 1939, the Gulf, Mobile & Ohio absorbed through consolidation the Gulf, Mobile & Northern, Mobile & Ohio, and the New Orleans Great Northern; and 1945 saw the Gulf, Mobile & Ohio absorb the Alton through merger.

Four more mergers would take place before 1950, the start of what Frank Wilner, in his 1997 book *Railroad Mergers: History, Analysis, Insight,* has called "the modern merger movement of 1950–1979." However, none of these mergers would directly involve the Mid-Atlantic states or the New York and New England states. This modern merger movement would, by 1990, remove more than 80,000 route miles from the railroad map of the United States. In 1946, there were 137

Class I railroads in the country. By 2001, there were only seven Class I railroad systems in the United States and Canada. Many new "regional" and "shortline" railroads would subsequently be created from those lines not wanted by the merged Class I railroad systems. Some of those would choose to use their former historic "fallen flag" names, such as Wheeling and Lake Erie, Ohio Central, and Wisconsin Central.

In 1946, nearly every rail line, both main and branch, was served by at least one passenger train per day, which also picked up and delivered the mail. The east-west main line of the Pennsylvania Railroad, through the center of the state of Pennsylvania, had some sixty-three daily scheduled passenger trains operating over its four tracks. Often these trains operated with several following separate train sections to accommodate the demand. Nearly all of them had names, some now famous, such as the Broadway Limited between New York City and Chicago and the Spirit of St. Louis between New York City and St. Louis. These particular trains were composed of Pullman cars only.

In 1946, all railroad lines carried freight, both car-load and less-than-car-load, and they moved everything from massive electric generators and giant steel bridge girders on heavy multiwheeled flat cars, to a child's bicycle in a Railway Express car. Nearly every town of any size had its own freight agent, coal unloading trestle, and bulk delivery track. At nearly every junction between two or more railroad companies, specially designated tracks existed to permit the interchange of freight cars. The loaded car interline routing between different railroads could be specifically chosen by the shipper from a variety of published routes, while the return of empty cars usually followed a long-standing "home route" pattern devised by the railroads. If shippers or receivers delayed cars beyond the normal two-day allowance, they were charged "demurrage." And each railroad was required to pay the car owner for every day one of the owner's cars was on line.

More than 1.3 million railroad workers were needed to manage this complex transportation delivery system. In addition to five-person train crews (often with a sixth, or "swing," crewman added in some states when cars exceeded certain limits per train—a holdover from applying freight car brakes by hand!), there were station agents, signal tower operators and signal maintainers, ticket collectors, track laborers, roundhouse and shop workers, yard clerks, and trainmasters. All of these employees, and many more as well, functioned and reported through a military-style chain of command to an all-powerful division superintendent.

My own memories of a typical 1946 railroad station scene may provide the reader with a more meaningful understanding of the feel of American railroading fifty-five years ago. It is important to remember that much of American daily life revolved around the coming and going of trains. If a train was late, people took note. The "train whistle in the night" caused generations of small-town young people to dream of moving to a better life—far away—often in the big city.

Looking down the track, you would see that the rails were jointed, causing the "clickety-clack" sound when trains ran over them. The railroad ties were all made of wood, oozing with creosote preservative on hot summer days. Nowadays, modern trains glide over uninterrupted welded rail, sometimes supported by concrete ties. Overhead bridges were blackened with coal soot, unmistakable accumulated evidence of the passage of countless steam locomotives. These bridges were guarded with "tell-tales," which brushed the heads of brakemen on car roofs, warning them to lie flat, lest they hit the bridge.

Along the edge of the track, parallel with a neat, hand-raked line delineating ballast from cinders, stood a stately procession of telegraph poles. Each pole often supported five or more cross arms, each carrying from eight to ten separate wires. Many of those wires were owned by the Western Union Telegraph Company, and they provided the visual identification mark of a railroad line. Some were for railroad company dispatching and message circuits, others for railroad signal power supply.

And then there were the signals. Many were still of the upper- or lower-quadrant semaphore type, with red or yellow arms and white and black stripes. Many types of buildings—stations, signal towers, freight houses, water tanks, section houses, and tool sheds—most of them wooden, stood along the railroad tracks, each painted in the unique color style of its railroad. Often this was a particular shade of gray, brown, yellow, or green.

At the typical small passenger station, the platform might be wooden, with a yellow stripe painted parallel to the rails to warn waiting passengers to stand back. The station would almost always have a bay window, from which the station agent could easily observe passing trains. There would likely be a train order signal on a high mast near this bay window, with brightly painted mechanical levers controlled by the agent. Train orders would be written on thin paper, nicknamed "flimsies," in either three or five duplicate copies, depending upon the nature of the order. Communication at some stations was still by Morse code telegraph, with its clatter of dots and dashes, and radio was just starting to be used. But mostly, the telephone was in use, with its distinctive headset and mouthpiece on a pantograph-type extension.

If there was a signal tower near the station, the metallic sound of interlocking machine levers being pushed or pulled could also be heard. The imminent approach of a train would be announced within the tower by a distinctive bell, gong, or buzzer, depending upon the direction or origin of the train. At night, colored lanterns—red, yellow, green, and white—stood ready for instant use, but normally they were not visible through a window, so as to not give a false signal to trains. Both stations and towers were ideal places from which to watch trains.

A visitor to a passenger station would see venerable wooden baggage carts, braced with cast iron straps, waiting to be pulled alongside an open baggage car door to allow the transfer of baggage, mail, and express. On the waiting room wall would be colorful posters and advertisements. Near the ticket window would be racks of multicolored railroad public timetables, all free for the taking. Air conditioning was provided by a slow-turning, electric ceiling fan. In winter, warmth was furnished by the pot-bellied stove. Refreshment was available from Chiclets®, gum machines and a porcelain or shiny metal water fountain. Hard wooden waiting room seats were available, from which passengers could listen to the stately ticking of a standard railroad pendulum clock.

Most important of all were the trains. In 1946, nearly everyone watched passenger trains. There were fast passenger "limiteds," often speeding through, nonstop—baggage cars followed by coaches, then the dining car, with perhaps a brief whiff of charcoal smoke from the kitchen stove. Finally came olive-green Pullman cars, with maybe even an open-platform observation car, carrying heavy colored-glass marker lanterns to indicate the end of the train. But the real variety was provided by the long, sometimes slow freight trains—box cars, tank cars, cattle cars, yellow "reefer" (refrigerator) cars, gondola cars, hopper cars, flat cars, and even a few flat cars with highway truck trailers. And there was always a caboose, usually bright red, at the end of the train. Now there is only the winking flash of the "end-of-train device" fastened to the rear coupler of the last freight car.

Especially fascinating was the seemingly infinite variety of railroad names and distinctive railroad company logos and markings—the keystone of the Pennsylvania, the oval of the New York Central, the U.S. Capitol dome of the Baltimore & Ohio, the gracefully interlacing scroll of the New Haven, the black diamond of the Lehigh Valley, the shield of the Union Pacific, the "Heart of Dixie" of the Seaboard Air Line Railway, the "Chessie" cat of the Chesapeake & Ohio, and many more. Then there was the unforgettable noise—the rolling, clattering, squealing, and straining of a long freight train.

As you examine the multi-colored maps, read the station names, and trace the course of the active and abandoned lines of this railroad atlas, try to imagine yourself in this railroad age of fifty-nine years ago. Soon a steady procession of railroad mergers, as well as highway and airline competition, would cause much of the 1946 American railroad scene to disappear forever.

NEW YORK AND NEW ENGLAND STATES

Volume 2 of *A Railroad Atlas of the United States in 1946* includes an area with some of the oldest railroad lines in America. When the era of new railroad construction began in the mid- to late nineteenth century, there were already many towns and cities in the seven states of this volume. Canals had been built in Connecticut, Rhode Island, and Massachusetts, and, in New York State, the magnificent and bold public work of the Erie Canal had connected the Atlantic Ocean with Lake Erie. Branches of that canal reached Lake Ontario and the so-called Finger Lakes region. The Champlain Canal connected Canada with New York City via Lake Champlain.

While the mid-Atlantic states included the highly productive coal mining regions of eastern and western Pennsylvania, as well as West Virginia, the products of New York State and New England were farm and dairy produce and the manufactured goods of thousands of factories and mills. Much of this manufacturing would be established and would expand as a result of the railroad age.

In addition, railroads had opened up the recreational and vacation areas of the Atlantic coastline, the White Mountains of New Hampshire, the Green Mountains of Vermont, and the Adirondack and Catskill Mountains of New York State. Late in the nineteenth century, an extensive and densely built commuter rail network had begun to develop around New York City and Boston. By 1946, however, especially in New England, abandonments resulting from automobile and truck competition had already begun to reduce active railroad mileage. Soon, what had once been a significant part of railroad operations, less than carload rail shipments, would be gradually eliminated completely.

Seven states are included in Volume 2: Connecticut, Maine, Massachusetts, New Hampshire, New York, Rhode Island, and Vermont. Of the total Class I railroad mileage in the United States in 1946, 14,591 miles—or 15.6 percent—were within these seven states. Of the 137 Class I railroads operating in the United States in 1946, twenty-five operated within these seven states. (In 1946, a Class I railroad was one having an annual operating revenue of more than $1 million.)

As in Volume 1, this atlas includes parts of states adjacent to its focal states. In this volume, parts of Pennsylvania, New Jersey, and the Canadian provinces of New Brunswick, Ontario, and Quebec are included.

Following is a brief summary of the major geographic and railroad characteristics of each of the seven states in the New York & New England volume.

Connecticut

Connecticut is the third smallest state in the union after Delaware and Rhode Island. It ranks forty-sixth in area, with a total of 4,965 square miles. It has generally hilly terrain, and its entire southern border is the northern shoreline of Long Island Sound. This characteristic gave rise to the popular name of the New Haven Railroad, the "Shore Line Route," and it was also the origin of the name of the famous, elegantly streamlined steam locomotives, the ten "Shore Liners," built by the Baldwin Locomotive Works and numbered 1400 through 1409. Later, the self-propelled Budd rail diesel cars were also called Shore Liners.

The principal railroad city was New Haven, home to the main offices of the New Haven Railroad and the sprawling, five-mile-long Cedar Hill freight classification yards (Map 154 NW). Built in 1920 with two separate "humps," Cedar Hill was then the largest such yard east of the Pennsylvania Railroad's Enola Yard in Harrisburg, Pennsylvania.

In 1946, only two Class I railroads operated in Connecticut, for a total of 1,767 miles: the Central Vermont (60 miles), and the New York, New Haven & Hartford (1,707 miles). Major main lines included the New Haven's Shore Line route from New York City to Boston, which had nine draw bridges in 102 miles. It had two main tracks east of New Haven and four main tracks, all electrified, west of New Haven. At Devon, twelve miles west of New Haven, the double-track Maybrook freight line ran north, then northwest through Danbury and into New York State over one hundred miles to Maybrook yard, where connections were made to five Class I railroads: the Erie; Lehigh & Hudson River; Lehigh & New England; New York Central; and New York, Ontario & Western railroads. In 1946, the Maybrook Line was the primary, all-rail freight route to and from New England.

One of the oldest railroad tunnels in the United States was located on the New Haven, built in 1835 northeast of Norwich, Connecticut, at Tafts (Map 145 SE). The Central Vermont Railway, controlled by the Canadian National Railways, linked the port of New London, Connecticut, with Canada to the north and provided considerable freight interchange with the New Haven and the Boston & Maine railroads.

Maine

Maine ranks thirty-sixth in area among the states, with 33,040 square miles. It is generally hilly, with many miles of delightful coastline and remote coastal islands. Five Class I railroads were operating in Maine in 1946: the Bangor & Aroostook (602 miles); the Boston & Maine (114 miles); the Canadian National (90 miles); the Canadian Pacific (247 miles); and the Maine Central (796 miles), for a total of 1,849 miles.

The major main lines of railroad in Maine were the double-track Boston & Maine "Western Route," from Boston to Portland, and the partly double track Maine Central, from Portland eastward to Vanceboro on the Canadian border. Others included the Canadian National's former Grand Trunk main line from Portland northwestward to Montreal, Canada, and the Bangor & Aroostook's main line from Searsport (on upper Penobscot Bay) north to Fort Kent on the Canadian border.

A historic deep-draft seaport, Portland (Map 71) attracted three Class I railroads: the Boston & Maine (three routes—two from Boston and one from Worcester, Massachusetts); the Canadian National, and the Maine Central. Both the Boston & Maine and the Maine Central shared terminal facilities by jointly controlling the Portland Terminal Railroad. The Canadian National entered Portland separately, to its own terminal on the northeast waterfront.

The other significant railroad "place" was Northern Maine Jct. Here, the east-west Maine Central interchanged with the north-south Bangor & Aroostook. In 1946, much of the annual Maine potato crop moved through this interchange to national markets. The famous Bar Harbor Express, also known as the "State of Maine" express, which ran from Washington, D.C., Philadelphia, and New York City to Maine during the summer months, still ran in 1946, but to Ellsworth station on the Calais Branch rather than to Mount Desert Ferry.

The Maine Central Railroad prior to 1946 had controlled an extensive system of narrow gauge lines, including the Sandy River & Rangeley Lakes and the Rangeley Lakes & Megantic, which served the forest and lake resorts of Oxford and Franklin counties.

Massachusetts

Massachusetts ranks forty-fourth among the states in size, with a total area of 8,266 square miles. It has an extensive coastline, with low hills along a coastal plain and higher hills in the center and the west. The principal railroad center was Boston, a deep-water port and an old and historic city center with many buildings and institutions dating back to the early days of American history.

In 1946, an extensive rail commuter network, dating from the late nineteenth century, radiated outward from the two large "stub-end" passenger terminals of North Station and South Station. Regretably, these two stations remain unconnected despite recent multi-billion dollar highway construction ("Big-Dig") in the same corridor. In 1946, however, the street industrial trackage of the 4.5-mile Union Freight Railroad did provide a slender, freight-only, local switching interconnection between the New Haven and Boston & Maine railroads.

Four Class I railroads operated in Massachusetts in 1946, with a total of 1,397 route miles: the Boston & Albany (294 miles); the Boston & Maine (655 miles); the Central Vermont (55 miles) and the New Haven (393 miles). Major main lines included the two-, three- and four-tracked New Haven main lines from New York to Boston and the double-tracked routes to Portland, Maine, and Concord, New Hampshire. Also, there were two east-west main lines: the Boston & Albany through Worcester, Springfield, and Pittsfield, to Albany, New York; and the Boston & Maine through Lowell, Greenfield, and the Hoosac Tunnel, to both Troy and Rotterdam Jct. (west of Schenectady), New York.

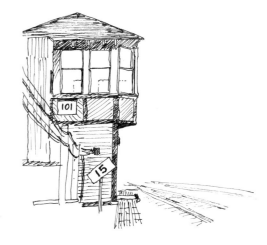

There were no large hump classification yards in Massachusetts. Major locomotive shops were located at Billerica (near Lowell) on the Boston & Maine and at Readville (southwest of Boston) on the New Haven.

New Hampshire

New Hampshire ranks forty-third in area among the states, with 9,341 square miles. Beyond the coastal plain along the Atlantic shoreline, the state is hilly; in the north—where Mount Washington, at 6,288 feet above sea level, is the highest point in New England—it is mountainous.

In 1946, four Class I railroads served the state with a total of 979 route miles: the Boston & Maine (795 miles); the Maine Central (100 miles); the Canadian National (51 miles); and the Central Vermont (33 miles). The principal main lines were all Boston & Maine: the double-tracked "Western Route" from Boston to Portland, through Dover, New Hampshire; the New Hampshire Division from Boston to White River Junction, Vermont, through Manchester and Concord; and finally that portion of the double-tracked "Connecticut River Line," which ran in New Hampshire from North Brattleboro to just southeast of Windsor, Vermont.

There were no major freight classification yards in New Hampshire, but there was a historic Boston & Maine locomotive and car shop in Concord. Unique "ball signals" were located at railroad crossings at both Whitefield and Waumbek Jct., as well as at Coos Jct.

New York

New York ranks twenty-ninth among the states in area, with 47,654 square miles. It extends from Montauk Point on the Atlantic Ocean (at longitude 71° 50' west) to Ripley on the Pennsylvania border (at longitude 79° 45' west). Except for the coastal plains of Long Island, it is generally hilly. Two mountainous areas exist—the Adirondacks northwest of Albany and the Catskills southwest of Albany.

The Erie Canal opened in 1825 and boldly connected the Hudson River at

Troy with Lake Erie at Buffalo. Dramatically lowering the cost of transporting goods to eastern markets at New York City, the Canal helped open the midwest territory of the United States to settlement.

By 1851, an all-rail route had opened from Piermont, on the Hudson River, to Dunkirk, on Lake Erie, forty miles southwest of Buffalo. This was the Erie Railroad, built to a wide six-foot gauge—a characteristic that some 150 years later, due to higher and wider clearances, would enable the first "double-stack" container trains to reach New York City via this route.

Twenty Class I railroads operated in New York State in 1946, with a total of 7,573 miles: the Boston & Albany (58 miles); Boston & Maine (97 miles); Baltimore & Ohio (182 miles); Canadian National (23 miles); Delaware & Hudson (645 miles); Delaware, Lackawanna & Western (493 miles); Erie (948 miles); Lehigh & New England (31 miles); Lehigh & Hudson River (25 miles); Lehigh Valley (589 miles); Long Island (376 miles); New Haven (108 miles); New York Central (2,669 miles); New York, Chicago & St. Louis (70 miles); New York, Ontario & Western (449 miles); New York, Susquehanna & Western (99 miles); Pennsylvania (424 miles); Pittsburgh, Shawmut & Northern (89 miles); Rutland (169 miles); and the Staten Island Rapid Transit Railway (29 miles). In 1946, four of these Class I railroads directly competed for traffic between New York City and Buffalo: the Delaware, Lackawanna & Western (395 miles between the two cities); Erie (423 miles); Lehigh Valley (447 miles); and New York Central (439 miles). Two additional railroads, the Baltimore & Ohio and the Pennsylvania, both indirectly served both markets but by circuitous routes through New Jersey and central Pennsylvania.

In 1946, all of the large hump classification yards in New York State were located on the New York Central: Selkirk (Albany), DeWitt (Syracuse), and Gardenville (Buffalo). The Delaware, Lackawanna & Western, the Erie, and the Pennsylvania all had smaller classification yards at East Binghamton and East Buffalo (DL&W), Hornell and East Buffalo (Erie), and at Ebenezer, southeast of Buffalo (Pennsylvania). Major locomotive shops were located at Hornell on the Erie, at West Albany on the New York Central, and at Colonie (near Troy) on the Delaware & Hudson.

Notable examples of railroad engineering could be found at Portage (Portage Viaduct [over 190 feet high] on the Erie), Selkirk (A. H. Smith Bridge on the New York Central), the Poughkeepsie Bridge (212 feet high) on the New Haven, and Suspension Bridge on the Michigan Central and New York Central at Niagara Falls.

Rhode Island

Rhode Island is the smallest state in the union, at 1,248 square miles. It has a beautiful, scenic shoreline and low, gentle hills. One Class I railroad served Rhode Island in 1946—the New Haven (240 miles). One of the first railroads to be built in the United States, the New York, Providence & Boston Railroad—popularly known as the "Stonington Railroad"—was built during the period from 1832 to 1837, from Fields Point in Providence southwestward to Westerly, then to Stonington, Connecticut, from which passengers and freight traveled by steamboat to New York City. This route avoided the longer water passage around Point Judith, where the weather was often stormy and dangerous, especially in winter. The alignment of Stonington Railroad has essentially remained unchanged. Today, the railroad line between milepost 170, east of Davisville (map 147), and milepost 162, near Slocums (map 146), is where the highest speed (150 miles per hour) is permitted for Acela passenger trains on Amtrak's Northeast Corridor high-speed rail line.

In 1946, there was a medium-sized hump classification yard at Northrup Avenue, just north of Providence. Despite its small size, there were four short line railroads in the state: the Moshassuck Valley (2 miles), the Narragansett Pier (8 miles), the United Electric Railways (later part of the Warwick Railroad, 2 miles), and the Wood River Branch (5.6 miles).

Vermont

Vermont ranks forty-second among the states in size, at 9,564 square miles. It is all hills and mountains—the Green Mountains plus the beautiful islands of Lake Champlain.

In 1946, eight Class I railroads served Vermont: the Boston & Maine (104 miles); Canadian National (32 miles); Canadian Pacific (86 miles); Central Vermont (248 miles); Delaware & Hudson (38 miles); Maine Central (34 miles); Quebec Central (6 miles); and the Rutland (238 miles), for a total of 786 route miles.

All the rail lines through Vermont were single-track, except for the double-track Boston & Maine "Connecticut River Line" between Brattleboro and Bellows Falls. The north-south Rutland Railroad extended from Chatham, New York, through Bennington to Rutland, where it was joined by a branch from Bellows Falls and continued northward to the Canadian border at Alburgh. The Central Vermont also ran north and south, alternately sharing its rails with or using the rails of the Boston & Maine between the Massachusetts border and White River Junction. Finally, the Canadian National's former Grand Trunk main line between

Portland, Maine, and Montreal, Quebec, crossed the "Northeast Kingdom" part of Vermont, with a remote crew change point at Island Pond, Vermont.

Interesting railroad engineering features included the single-track causeways of the Rutland between the north and south end of Grand Isle in Lake Champlain and the elaborate switchbacks of the granite quarries around Barre. No classification yards or shops of significant size were located in Vermont. Unique "ball signals" were located at Bellows Falls, White River Junction, Wells River, and St. Albans, Vermont.

How to Use This Atlas

FINDING GEOGRAPHICAL AREAS

Railroads within specific geographical areas may be found by consulting the Key Map, which shows state lines, the boundaries of all 162 standard atlas maps, and the locations of detail maps. The boundaries of each of these standard atlas maps represent 30-minute quadrangles between each full degree of latitude and longitude.

Each map has a geographical name and a consecutive number. The name reflects a prominent railroad station within the quadrangle. The number represents the consecutive arrangement of maps from west to east and from north to south. These numbers are also the page numbers for the atlas. The maps in this New York & New England states volume begin with number 1, Kent, Maine, and end with number 162, East Hampton, New York. Areas of complexity are shown on detail maps, which also have a geographical name and a number as well as a letter. The number is the standard atlas map number of the map within which the detail map or maps are located. This number is followed by a letter distinguishing this map as a detail map.

FINDING STATIONS

Particular railroad stations may be found by consulting the alphabetical index. Each index entry gives a station name, followed by the abbreviation for the state in which it is located, followed by the consecutive number of the standard atlas map and detail map on which it is to be found. This number is followed by one of four map quadrant designations, as follows: upper-left quadrant, NW; upper-right quadrant, NE; lower-left quadrant, SW; and lower-right quadrant, SE.

Also included in this index are the names of interlocking stations, both existing and former, coaling stations, track pans, tunnels, and major viaducts and bridges. On every atlas map, all of these names are given in quotation marks to distinguish them from ordinary stations.

RAILROAD COLORS

Each railroad company is identified by one of ten colors: red, yellow, orange, brown, green, light blue, dark blue, purple, gray, and black. Every effort has been made to choose a color for each railroad company that has a historic association with that railroad. For example, red has been used for the Pennsylvania Railroad and the Canadian Pacific Railway; green represents the Erie Railroad and the Central Vermont Railway; blue has been used for the Boston & Maine Railroad and the Delaware, Lackawanna and Western Railroad; and orange has been used for the Lehigh Valley Railroad and the New York, New Haven & Hartford Railroad. However, in order to preserve a graphic color contrast between adjacent railroads, the choice of an unrelated color has sometimes been necessary. Railroads controlled or leased by a major railroad are represented by the same color.

CROSS REFERENCE WITH OTHER MAPS

This railroad atlas can easily be cross-referenced and compared with United States Geological Survey topographic maps, the DeLorme Mapping Company state atlases, and any atlas map that includes latitude and longitude lines, as those lines serve as the basis for this atlas.

The Key Map identifies the names of the USGS United States 1:250,000 scale series, which were used as the base map for the 162 standard atlas maps. These maps, with a scale of one inch equals four miles, were originally produced by the U.S. Army Map Service during the 1940s and 1950s. This map series was the first USGS map product to completely cover the United States, and it has provided a consistent mapping base for this 1946 railroad atlas.

Each standard atlas map measures 30 minutes of latitude and longitude and includes tick marks defining the four possible USGS 15-minute topographic quadrangles that could be contained within. In turn, each of these 15-minute quadrangles may contain as many as four 7.5-minute USGS topographic maps. The United States 15-minute series was the USGS standard from 1910 to 1950, with many early maps dating from 1890 onward. In about 1950, the United States 7.5-minute quadrangle maps, having first been widely produced in the 1940s, became the major product of the USGS. The United States was finally completely covered by these 7.5-minute maps in the 1990s.

Nearly all of the DeLorme Mapping Company state map pages are defined by 30 minutes of north latitude and 26.25 minutes of west longitude.

Finally, nearly all Hammond and Rand McNally state atlas maps include latitude and longitude lines normally drawn to a 30-minute grid.

LIST OF RAILROADS

Each of the railroad companies shown in this atlas are identified by abbreviations or "reporting marks." These are inscribed along each discrete line or branch segment on each map. An alphabetical list of these reporting marks, followed by the railroad name as it existed in 1946 or at the date of abandonment prior to 1946, is provided in the appendix.

MAPPING STYLE

This atlas shows state and county lines as well as major rivers in the interest of geographic orientation. Roads are not shown in the interest of simplicity. Cross-referencing to USGS maps or the DeLorme Mapping Company state atlases is recommended for finding contemporary road network details.

MILEPOSTS

All railroad lines have railroad milepost tick marks every five miles. Nearly all of these mileposts have remained unchanged since long before 1946. The place names along the outside of the map neat line indicate the zero milepost location and/or the furthest point of that particular series of mileposts. On the New Haven's New York-to-Boston line (which was only partly mileposted in 1946), signal number mileage has been added.

SOURCES

Sources used in the compilation of this atlas include a wide range of USGS topographic maps, Rand McNally and Hammond atlases, official railroad company maps, employee timetables, and track charts. Particularly valuable resources included *Moody's Steam Railroads*—1946; *Poor's Manual of Railroads*—1923; William D. Edson's *Railroad Names*; and many individual railroad histories. A complete list may be found in the "References" section.

ACCURACY

While I have taken every reasonable step to ensure accuracy, there may be some errors or omissions which are my own responsibility. Corrections and additional information are welcomed .

Acknowledgments

In producing this re-creation of the very complex rail network of the New York & New England states, I have benefited from the research and generosity of many individuals—fellow railroad historians, railroad officials and workers, and others.

Particular thanks should be given to the late Henry T. Wilhelm and the late Bruce Coughlin for generosity in sharing their extensive track and signal drawings of U.S. railroads from the 1920s through the 1970s. Also, Ben F. Anthony has always been ready to provide information regarding rail operations and signaling. My thanks to Charles Smith for assistance regarding New York Central history and to Vincent H. Bernard for Boston & Maine historical information. Finally, particular thanks are given to Robert Gambling for allowing me to use his collection of first and second editions of United States 1:250,000-scale USGS maps.

Among the sources of inspiration for this railroad atlas were the Railway Clearing House's *Railway Junction Diagrams* of pre-grouping British railroads and Jowett's *Railway Atlas of Great Britain and Ireland*. Alan Jowett generously shared with the author advice concerning his methods and graphic style.

Acknowledgment is also due the helpful staff at the following map libraries: Detroit, Michigan, Public Library; New York Public Library—Map Division; Map and Geographic Information Center, University of Connecticut, Storrs; and Wesleyan University Science Library.

Special thanks go to Richard B. Hasselman, retired Senior Vice President, Operations–Conrail, for his advice and encouragement and for New York Central historical information.

Finally, I wish to express special appreciation to George F. Thompson, president of the Center for American Places, for his encouragement and guidance; to Annette Evarts, for her indispensible word processing and dedication; to my son, John, for his editorial advice; to my daughter, Ellin, for her design advice; and to my wife, Mary Jane, for her support and counsel.

THE ATLAS

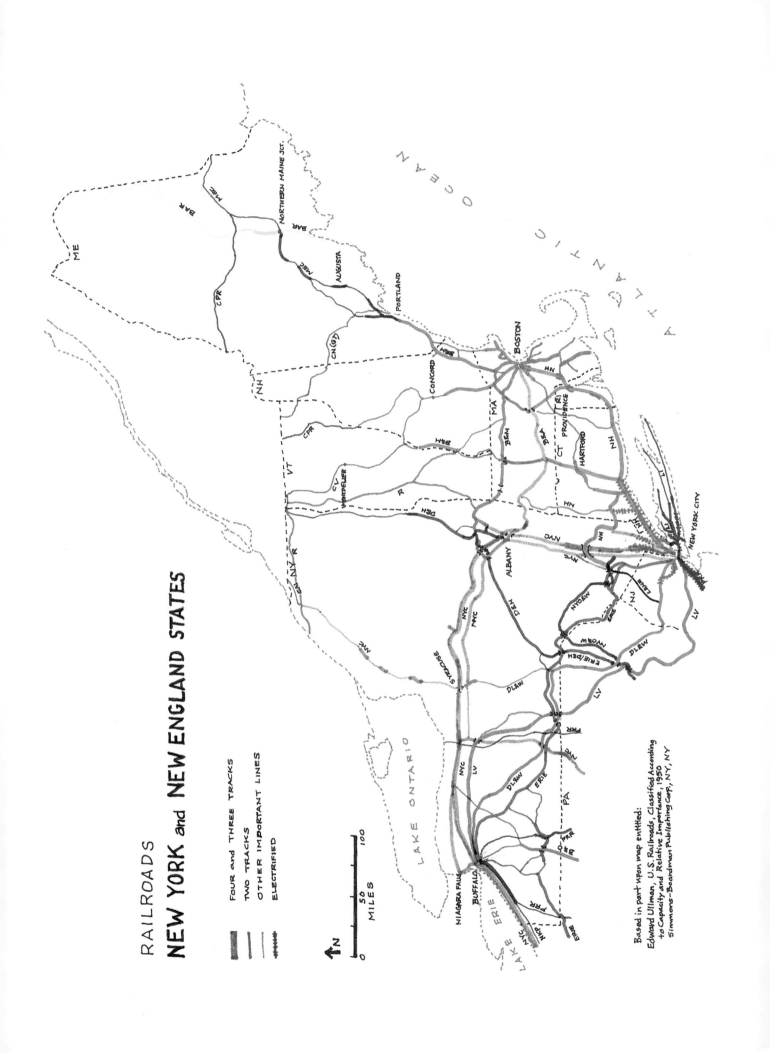

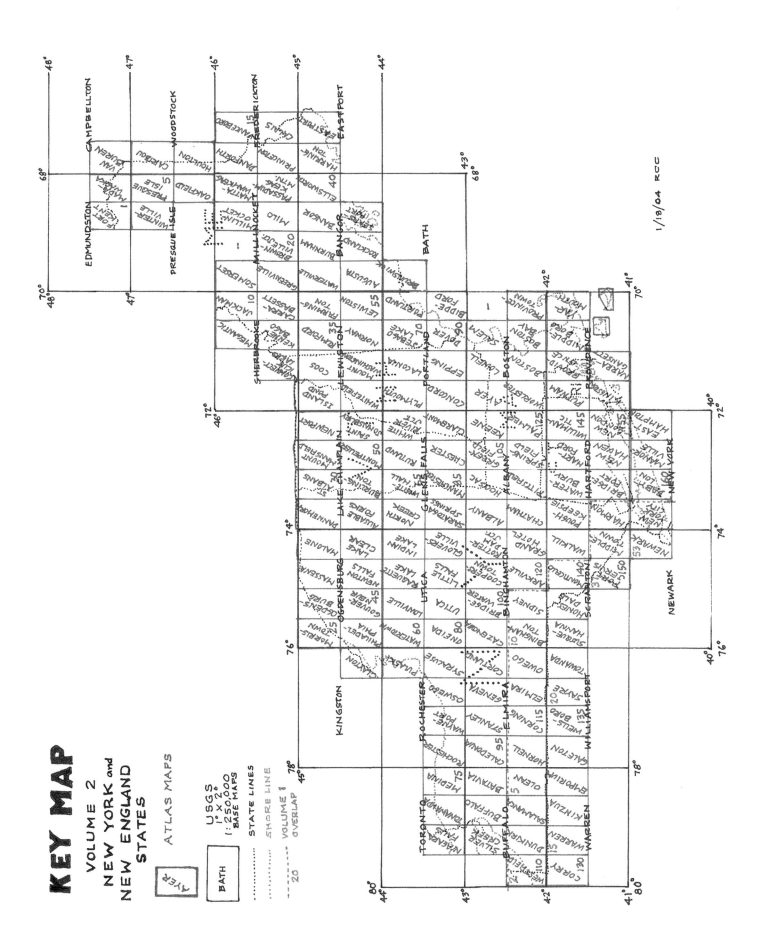

SYMBOLS AND ABBREVIATIONS

———————	RAILROAD LINE (In Service)
- - - - - - -	RAILROAD LINE (Abandoned)
———————	RAILROAD (Outline of Yard/Connecting tracks – In Service)
——○———	PASSENGER STATION
———┼———	NON-PASSENGER STATION
"Poughkeepsie"	VIADUCT/MAJOR BRIDGE (with name)
"Otisville"	TUNNEL (with name)
"Yosts"	TRACK PAN (with name)
"Midway"	COALING STATION/MAIN TRACK (with name)
———————	MILEPOST (every 5 miles standard maps; every mile on detail maps.)
120 Oswego	START/END MILEPOST MILEAGE
DL&W	1946 OWNER (See Appendix for reporting marks)
MEC (SRRL)	1946 OWNER (Previous owner or Associated line – " " " ")
ERIE/L&NE	1946 OWNER / (Trackage Rights – " " " ")
NKP / PRR	1946 JOINTLY OWNED & OPERATED (" " ")
"133" KINGSTON	INTERLOCKING TOWER (In service, with name)
□ CN CAMERON	INTERLOCKING TOWER (Abandoned, with name)
• AM	BLOCK STATION (In service, with telegraphic call letters)
EAST RIVER (R-92)	INTERLOCKING – REMOTE CONTROLLED (In service, with name and control point name)
•• PORT	BLOCK LIMIT STATION (In service, with name)
HORNELL	CREW CHANGE POINT
— — — —	NATIONAL BOUNDARY
— — — —	STATE/PROVINCIAL BOUNDARY
— — — —	COUNTY BOUNDARY
- - - - -	CITY/TOWN BOUNDARY
～→	RIVER (With direction of flow)
┬┬┬┬ ----	CANAL (In Service / Abandoned)

C ..... color of railroad company Roundhouse

LS / CS Existing or former locomotive and car shops with owning railroad

A Automatically controlled interlocking
D Draw, lift or swing bridge
G Railroad crossing at grade – Gate
S Railroad crossing at grade – Stop sign
T Railroad crossing at grade – Target

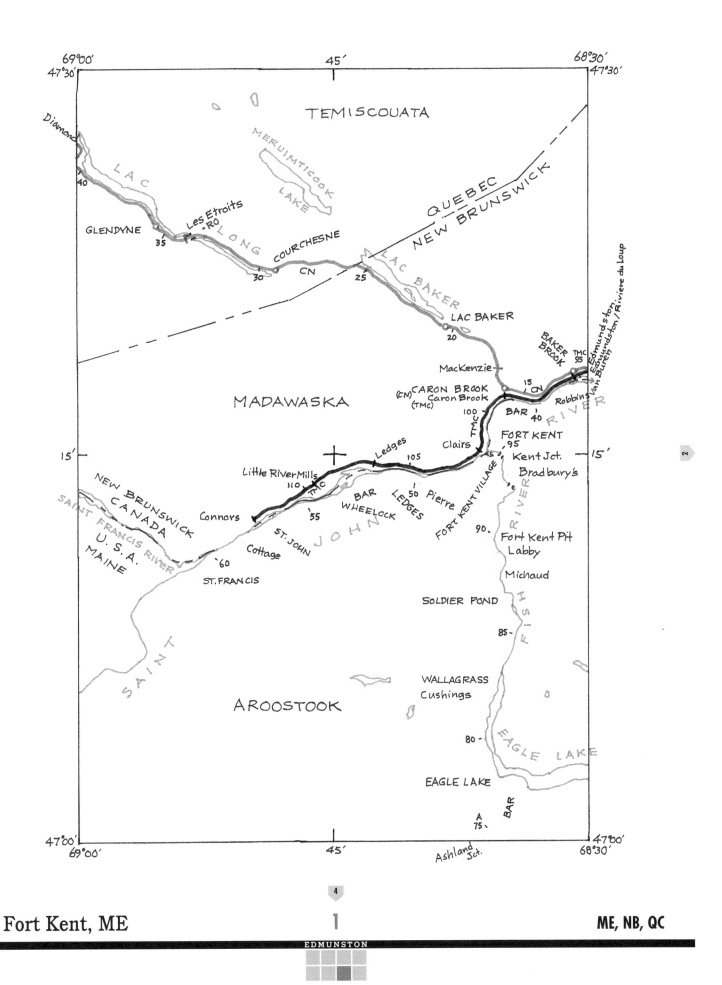

Fort Kent, ME — ME, NB, QC

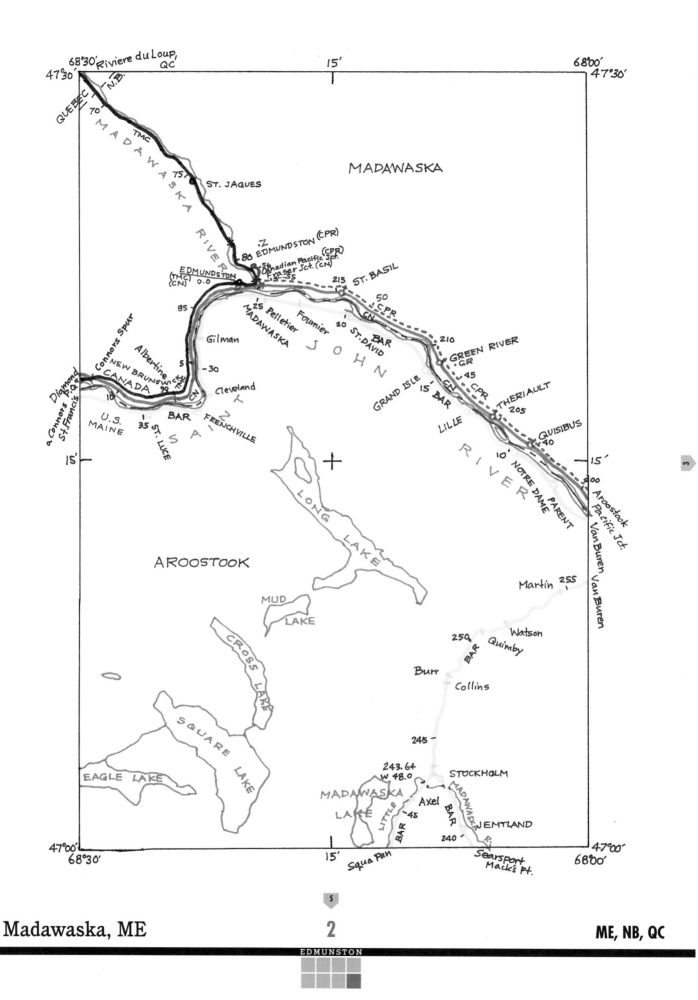

Madawaska, ME — 2 — ME, NB, QC

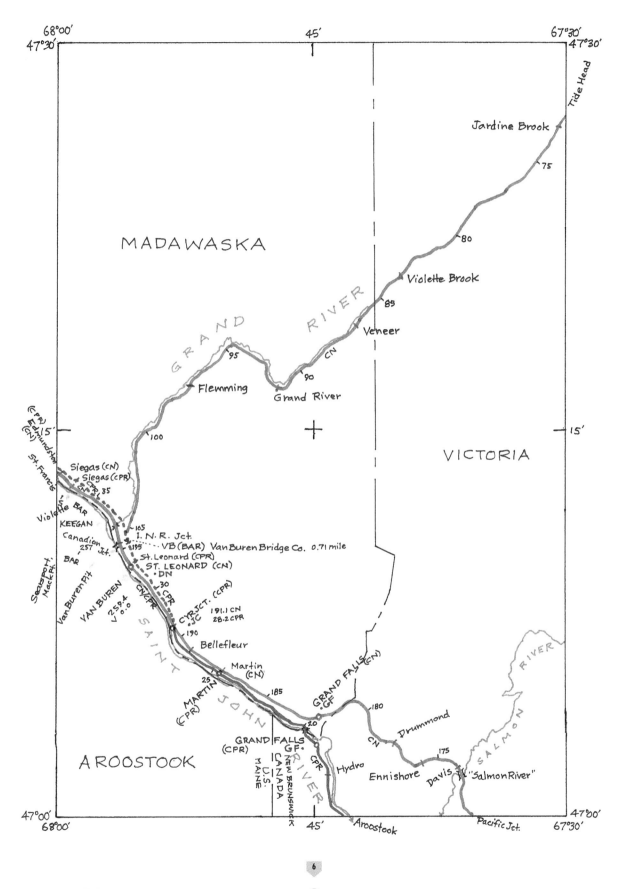

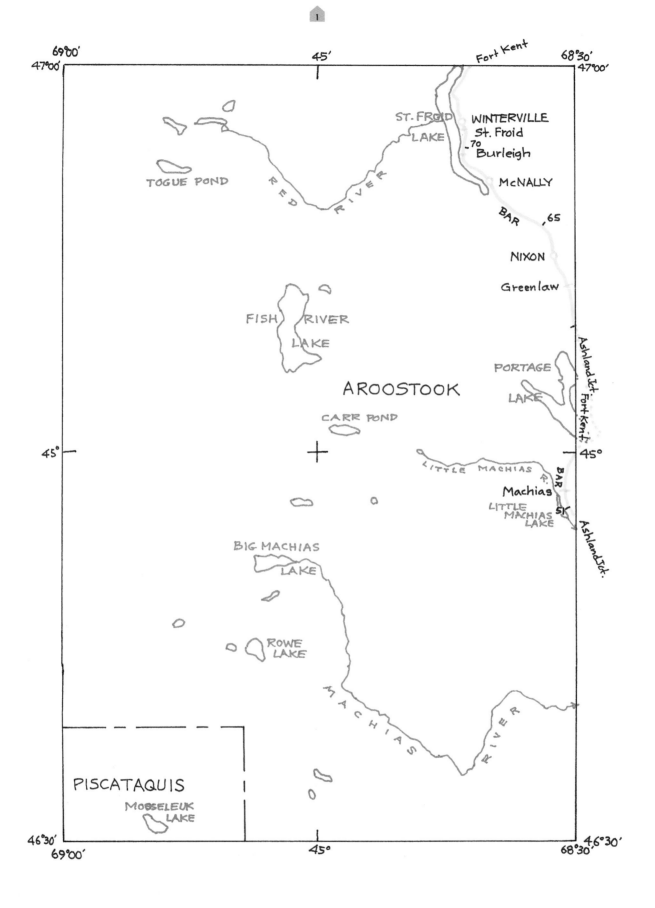

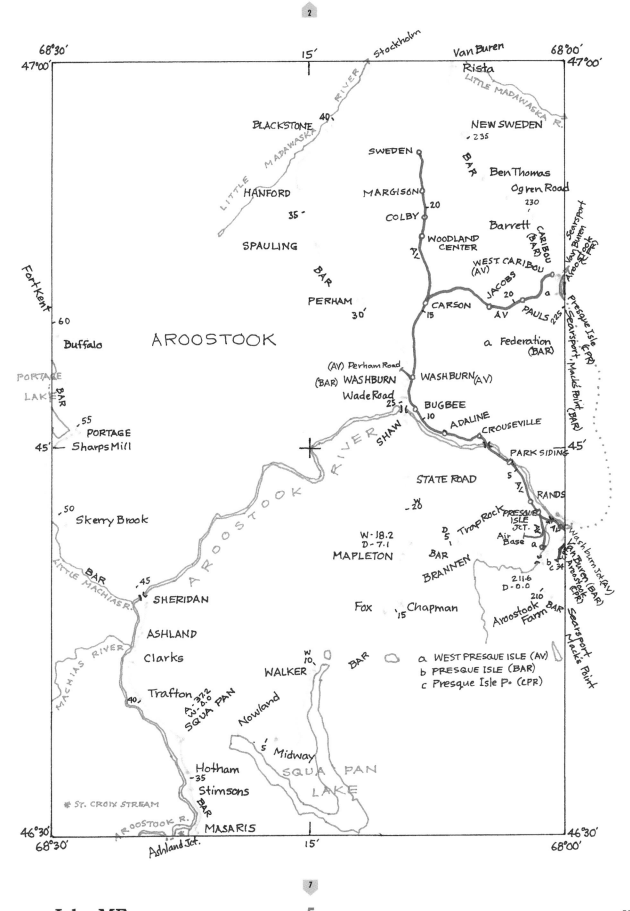

Presque Isle, ME

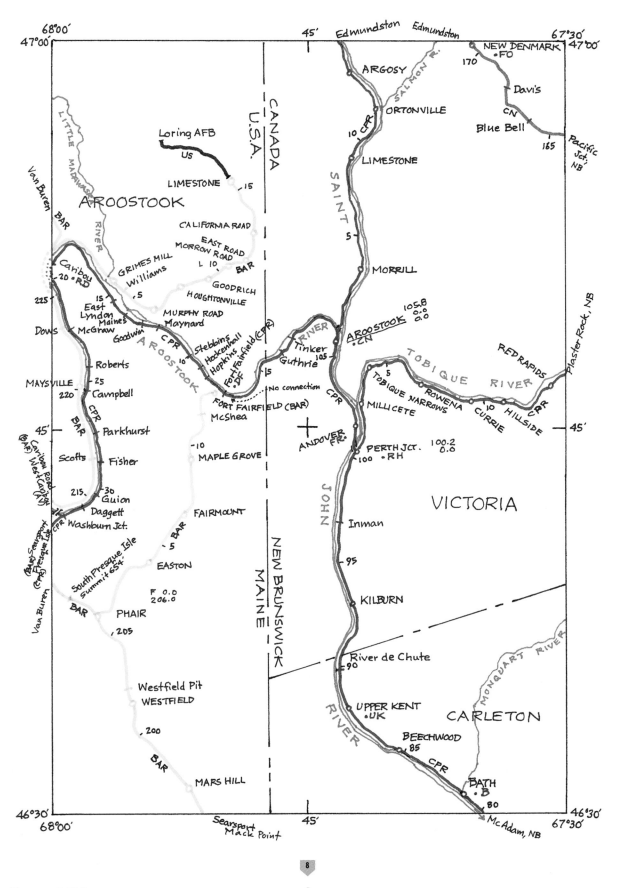

Caribou, ME

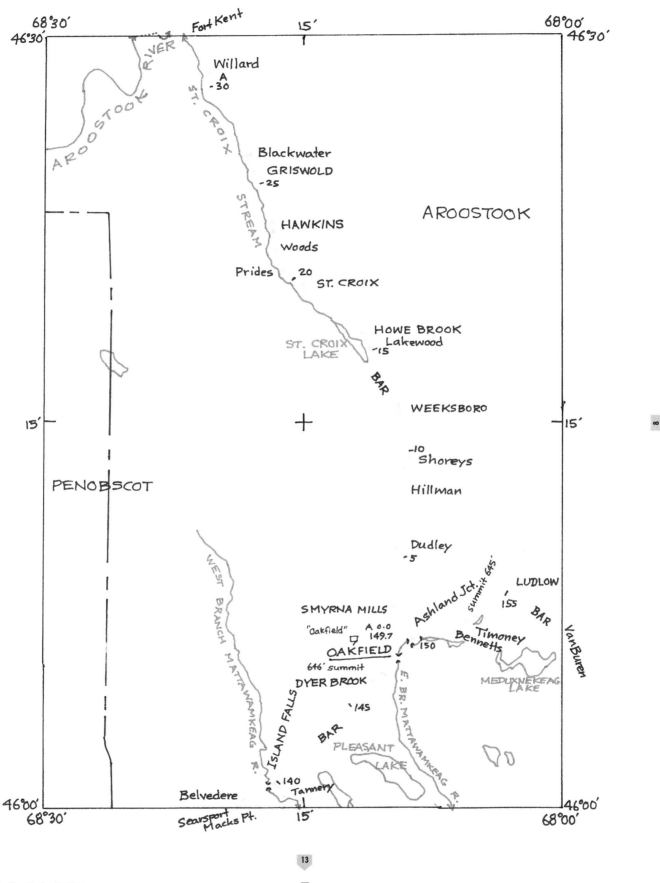

Oakfield, ME

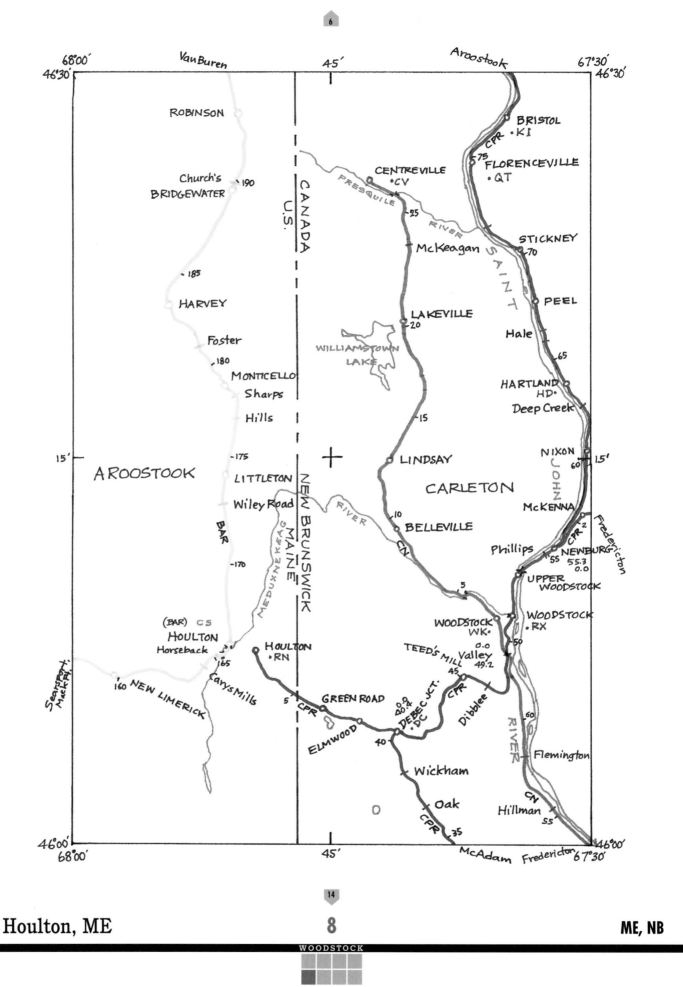

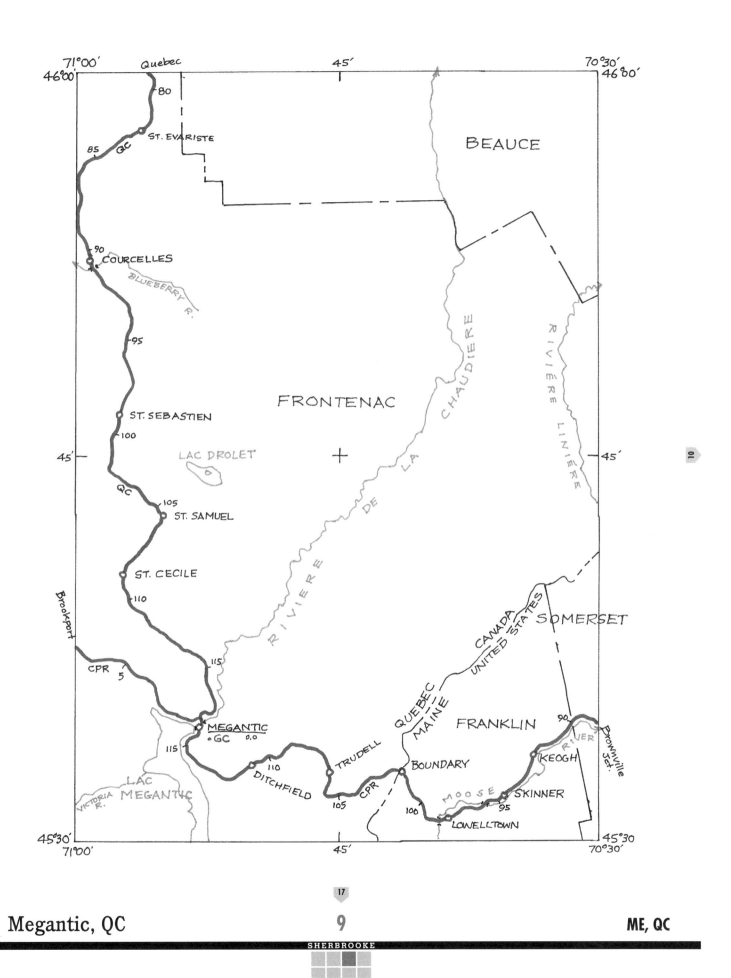

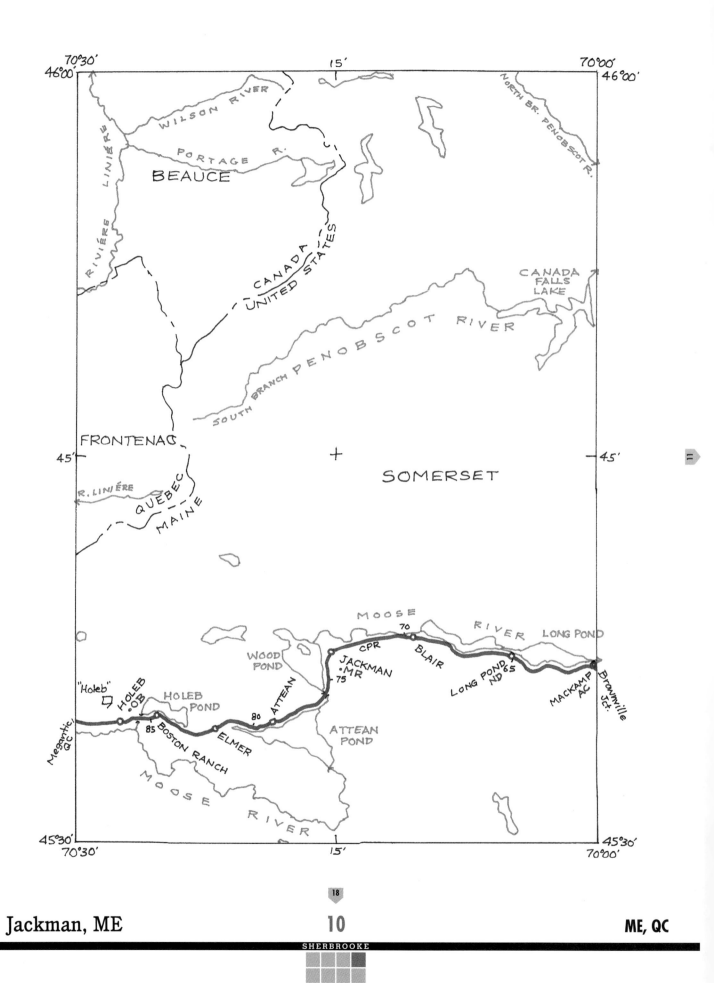

Jackman, ME — ME, QC

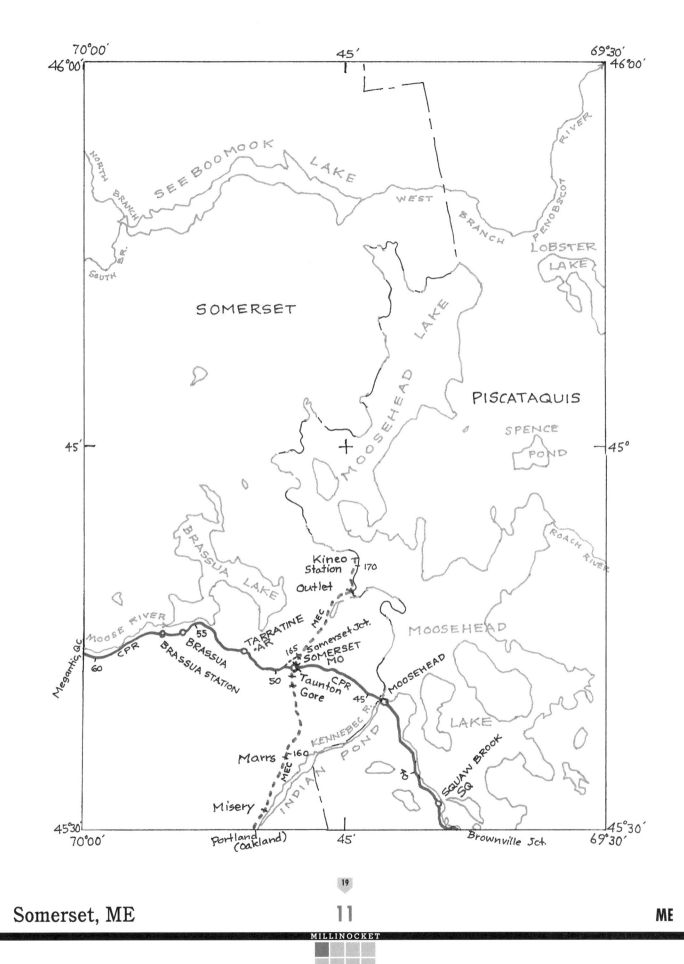

Somerset, ME

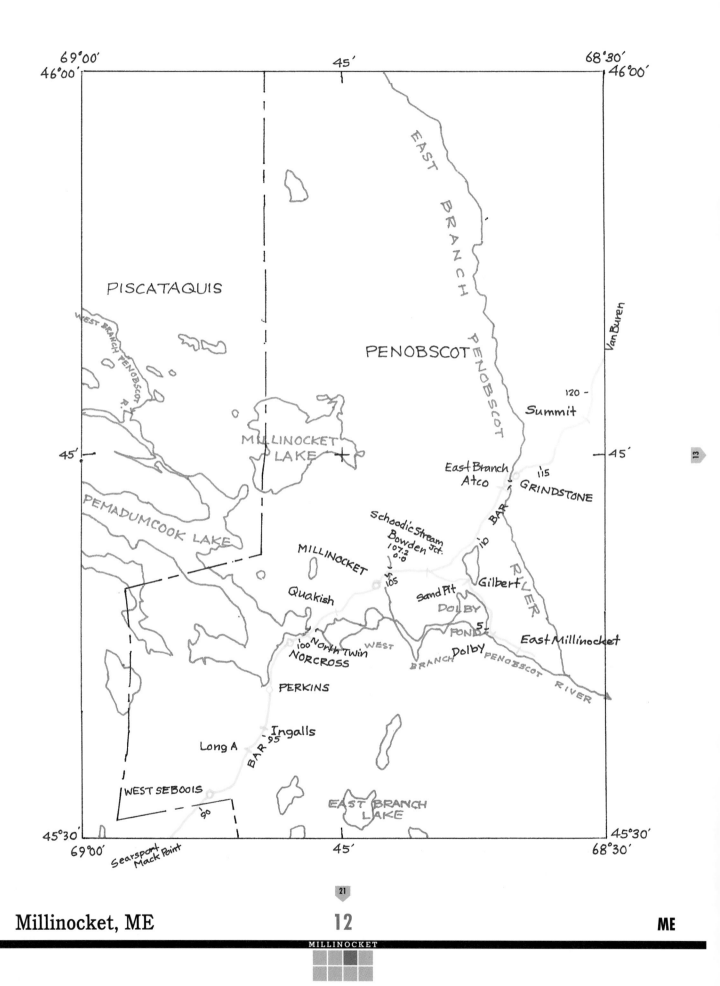

Millinocket, ME

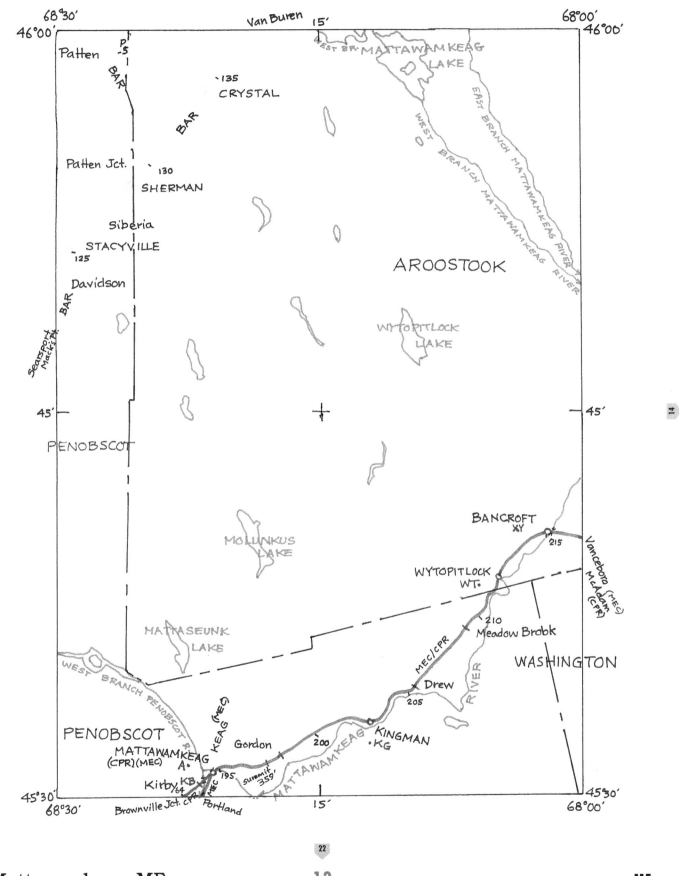

Mattawamkeag, ME

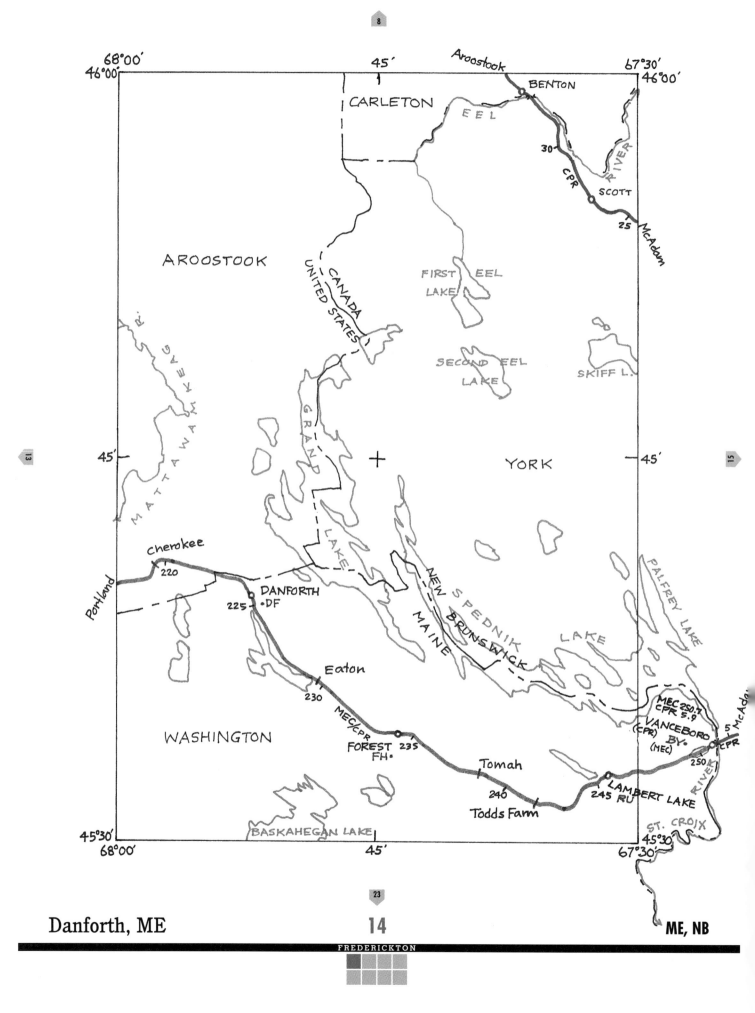

Danforth, ME — 14 — ME, NB
FREDERICKTON

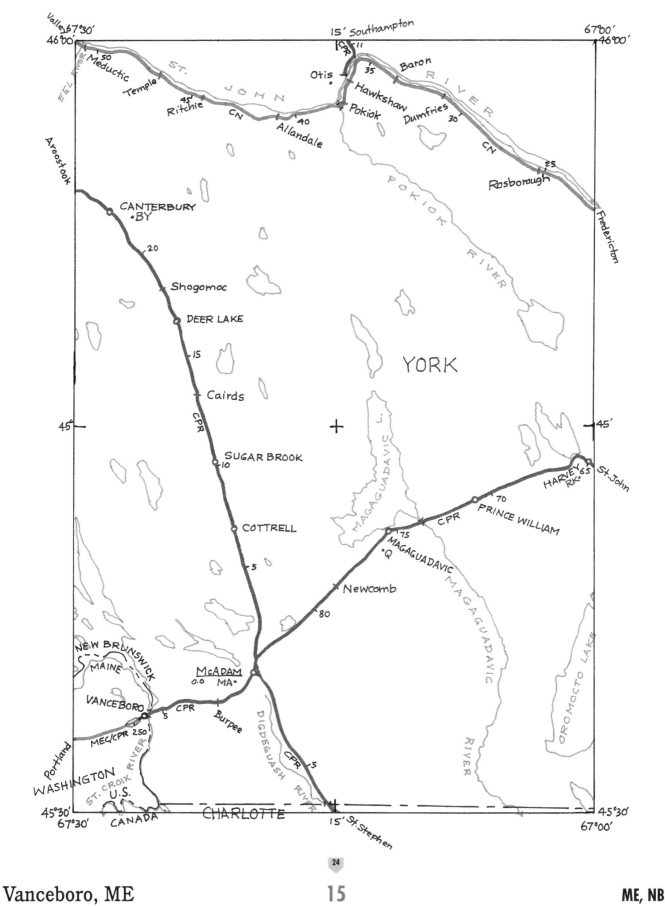

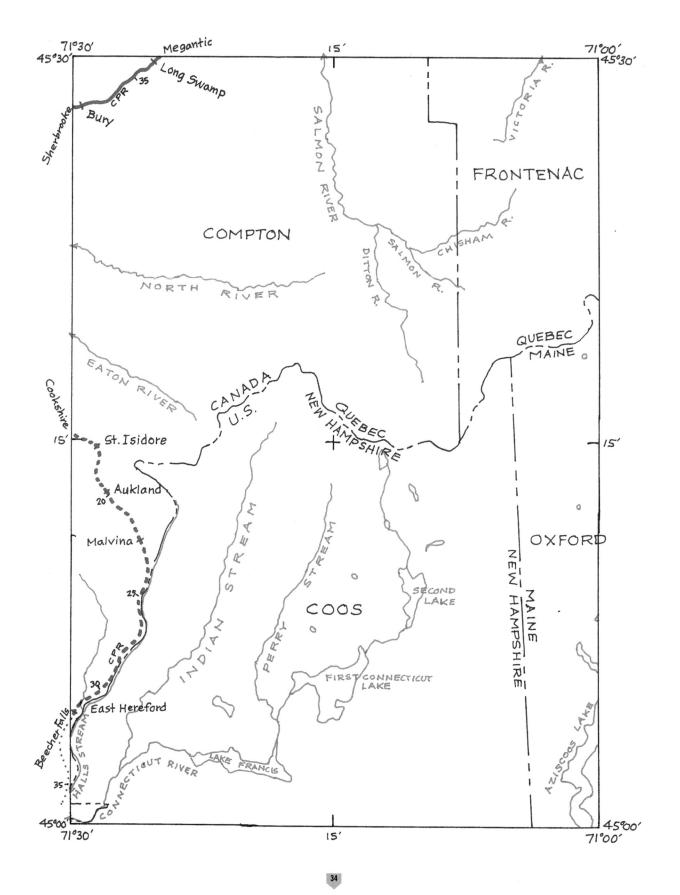

Connecticut Lakes, NH

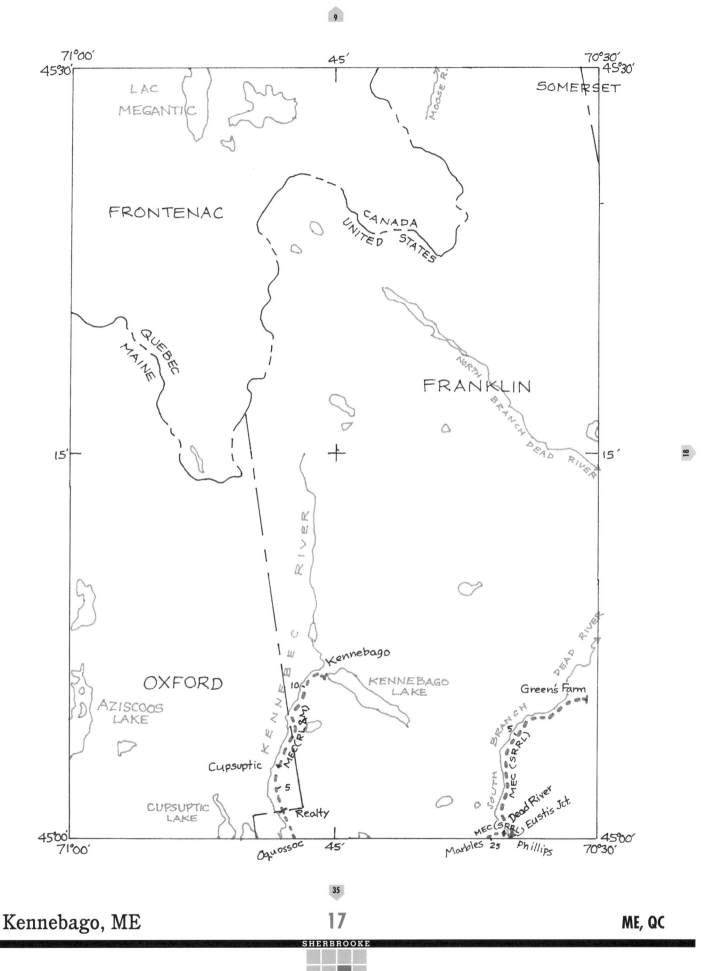

Kennebago, ME — 17 — ME, QC

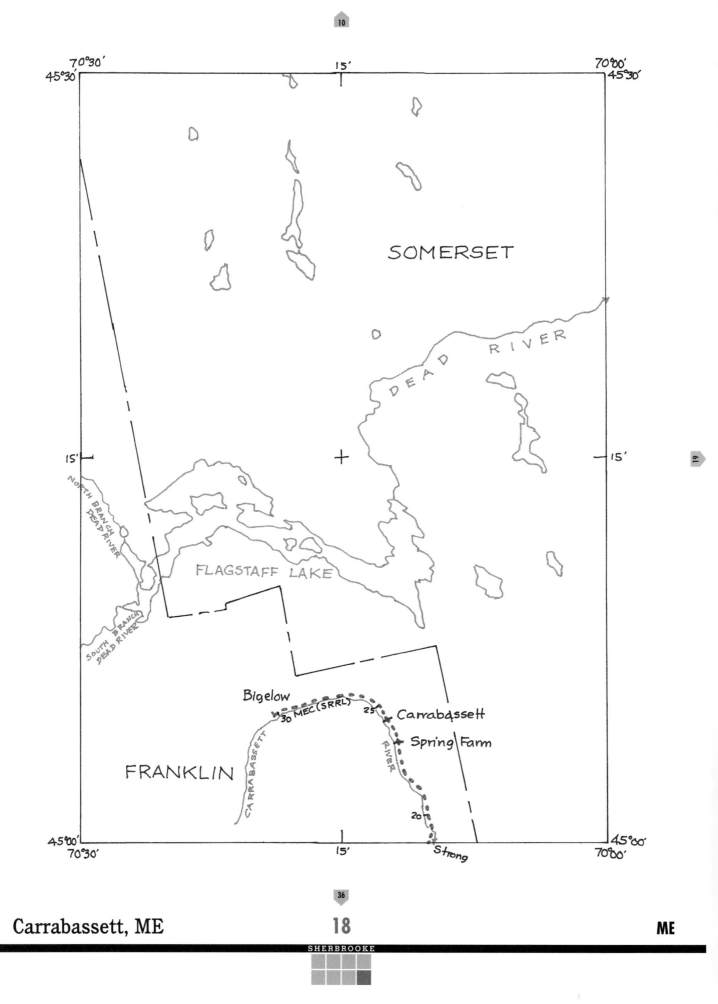

Carrabassett, ME

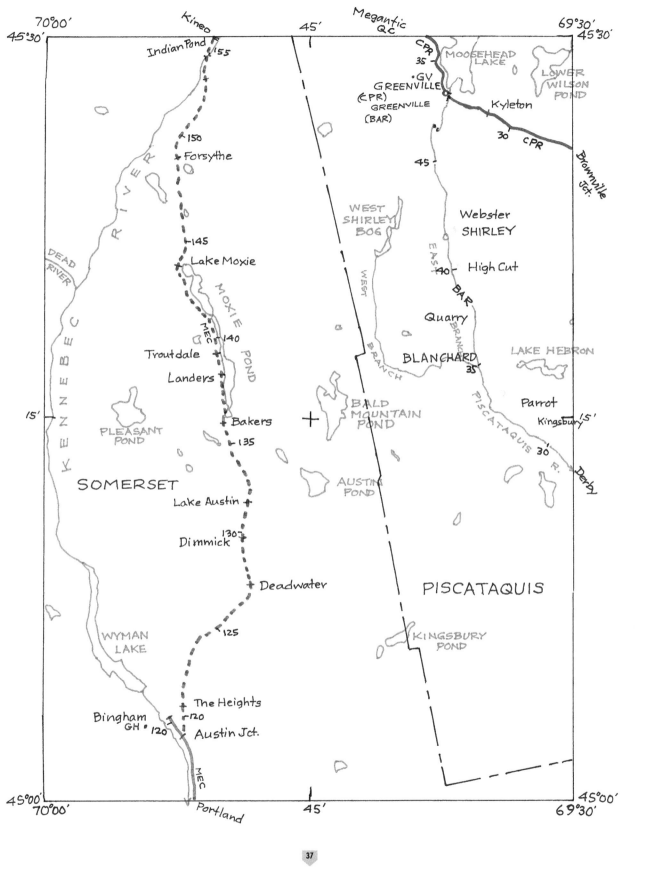

Greenville, ME

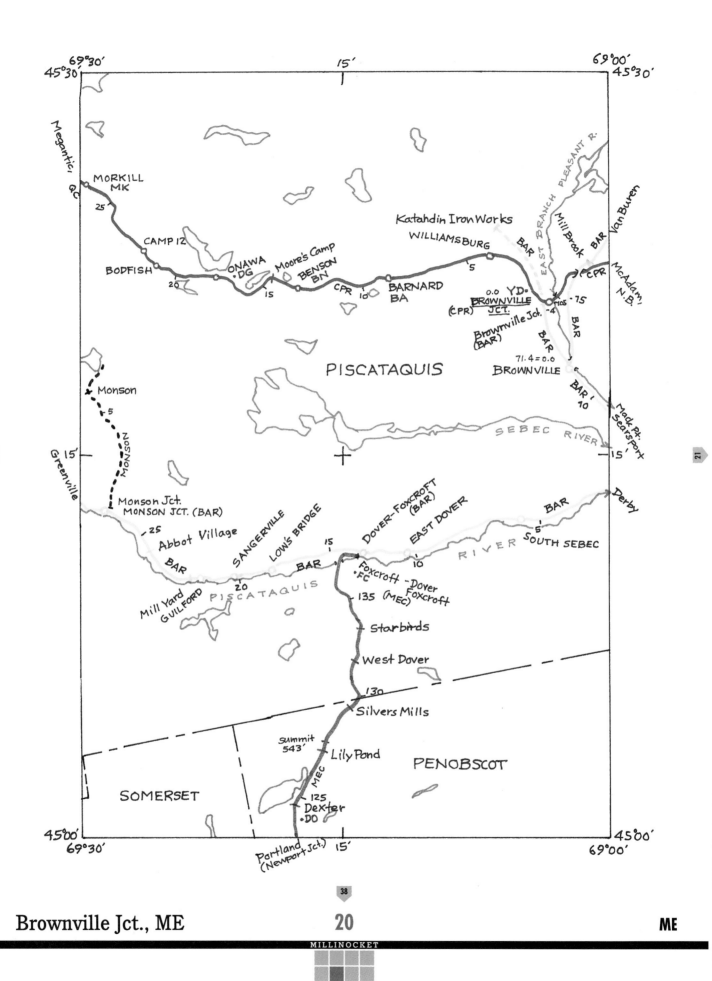

Brownville Jct., ME

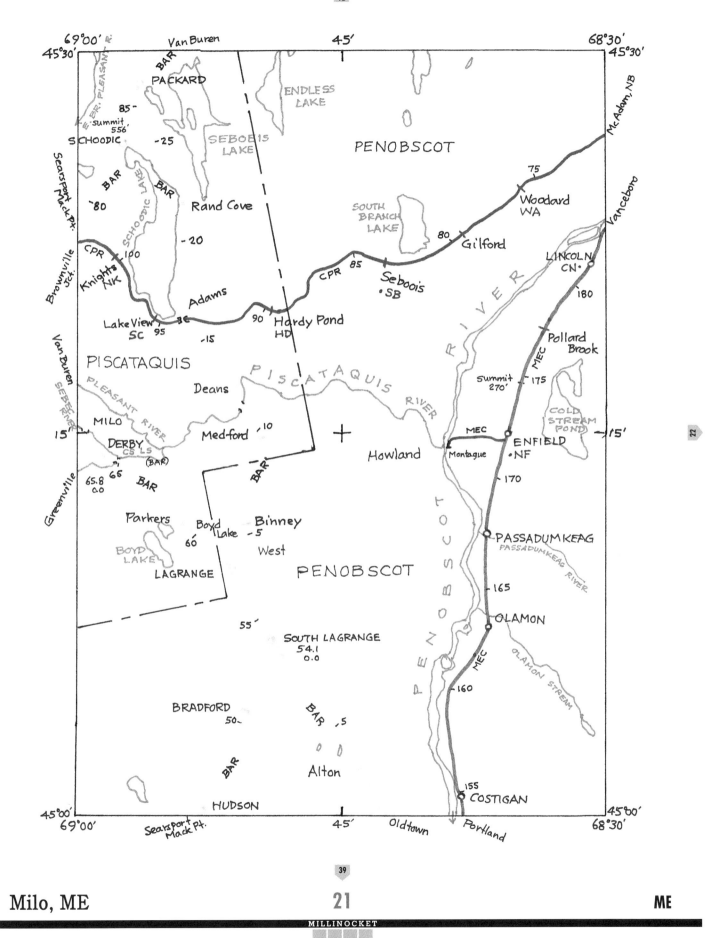

Milo, ME   21   ME

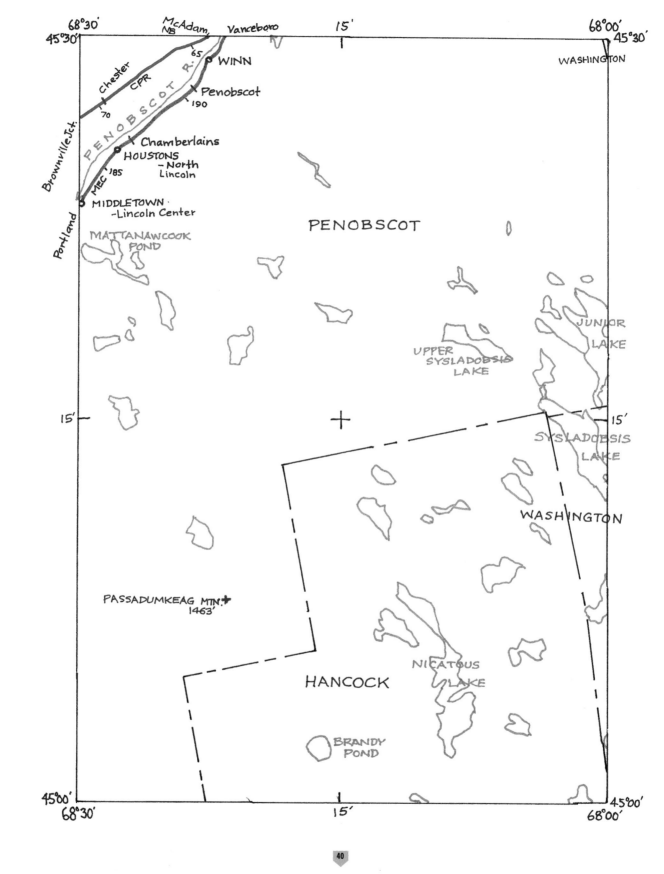

Passadumkeag Mtn., ME

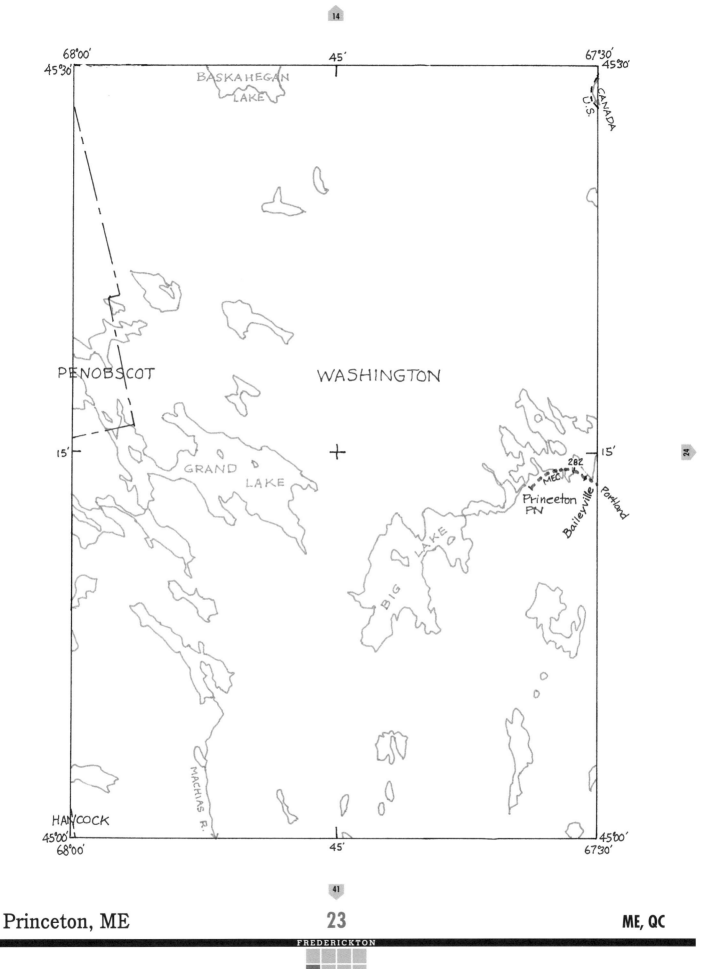

Princeton, ME   23   ME, QC

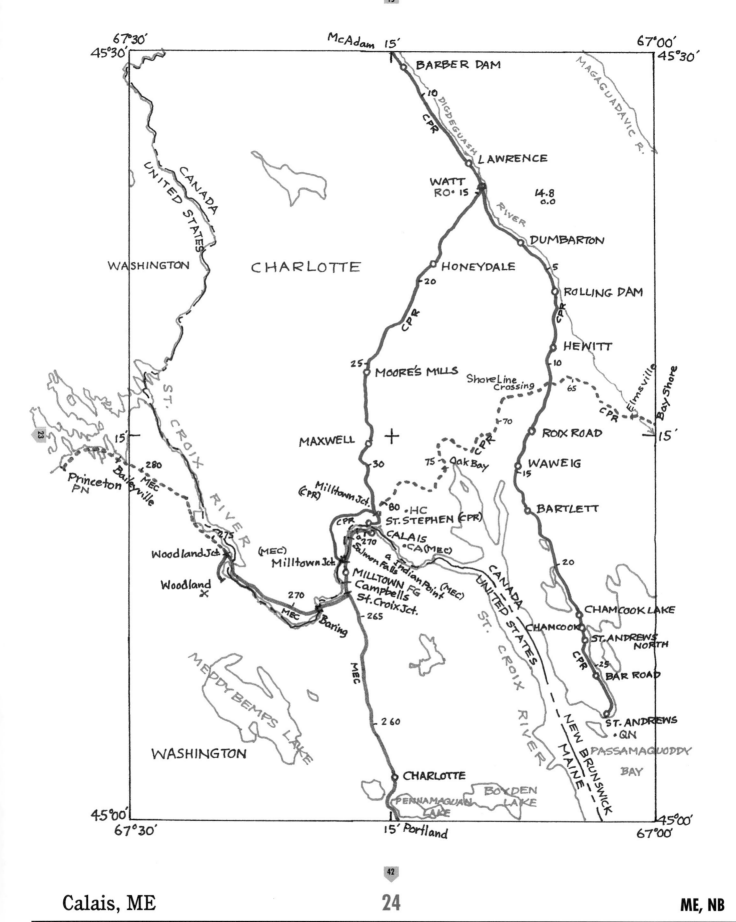

Calais, ME — 24 — ME, NB

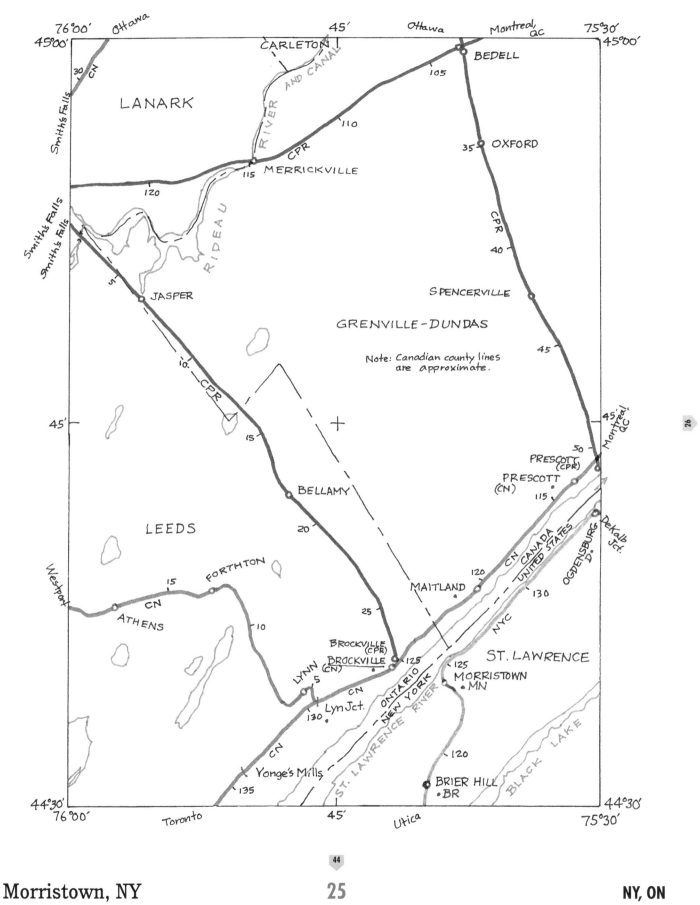

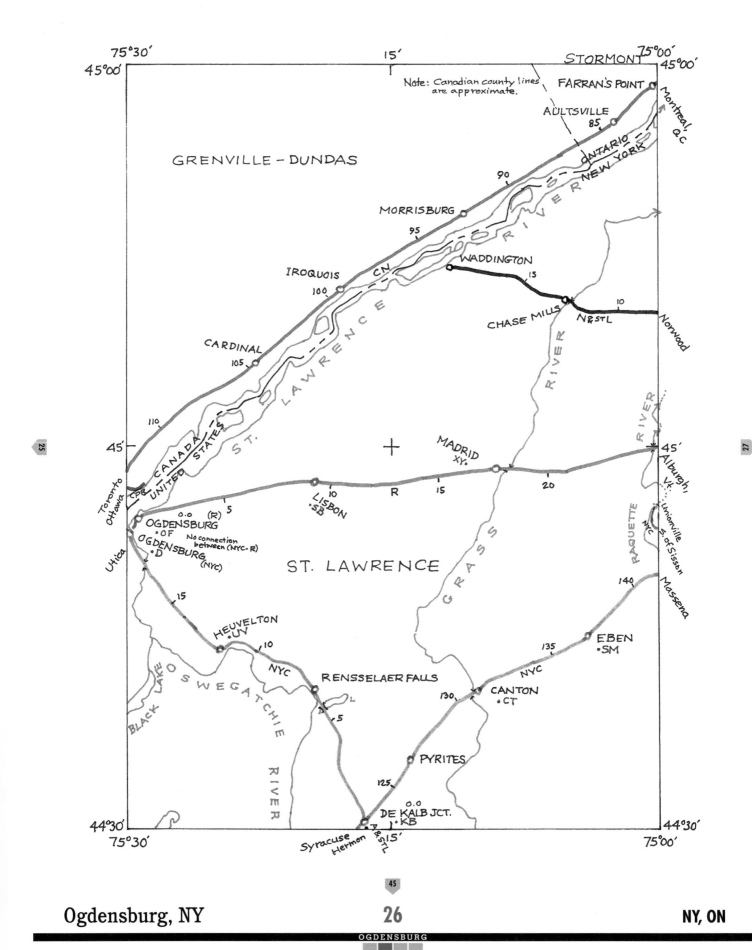

Ogdensburg, NY — NY, ON

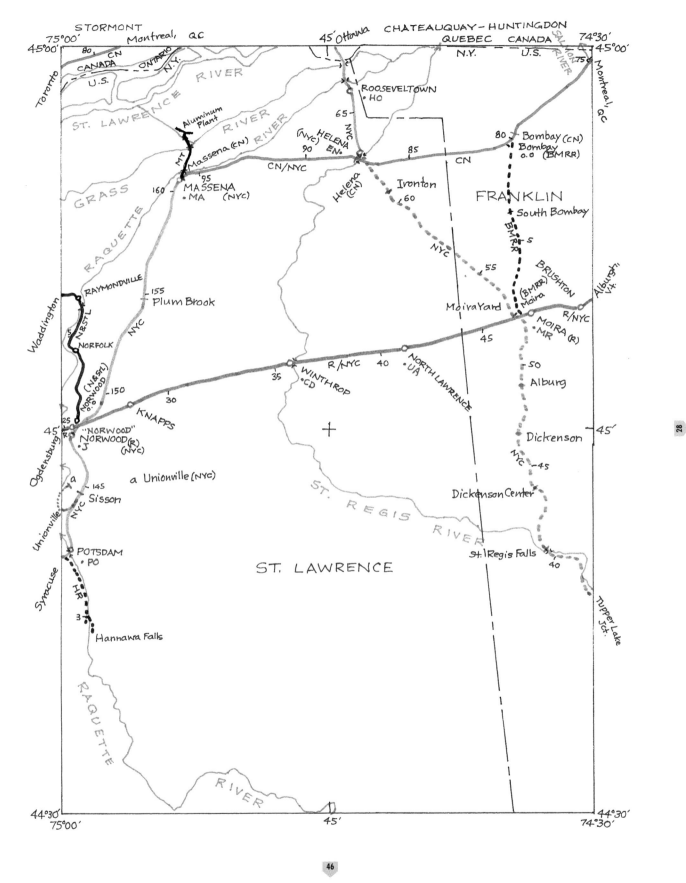

Massena, NY — 27 — NY, ON, QC

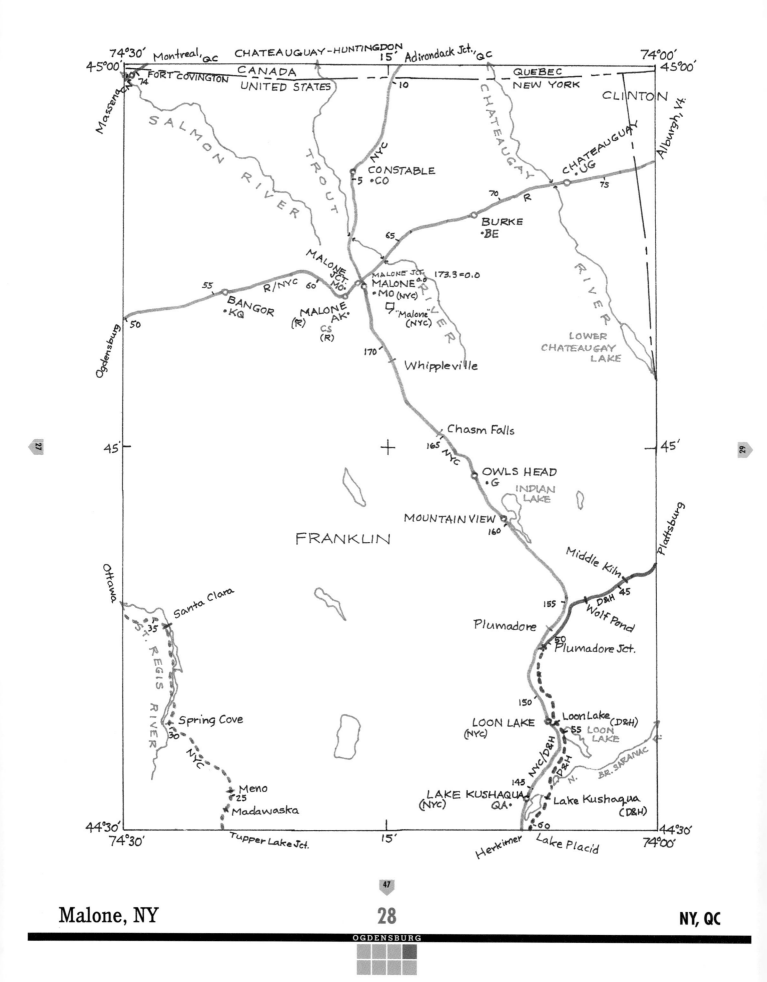

Malone, NY — 28 — NY, QC

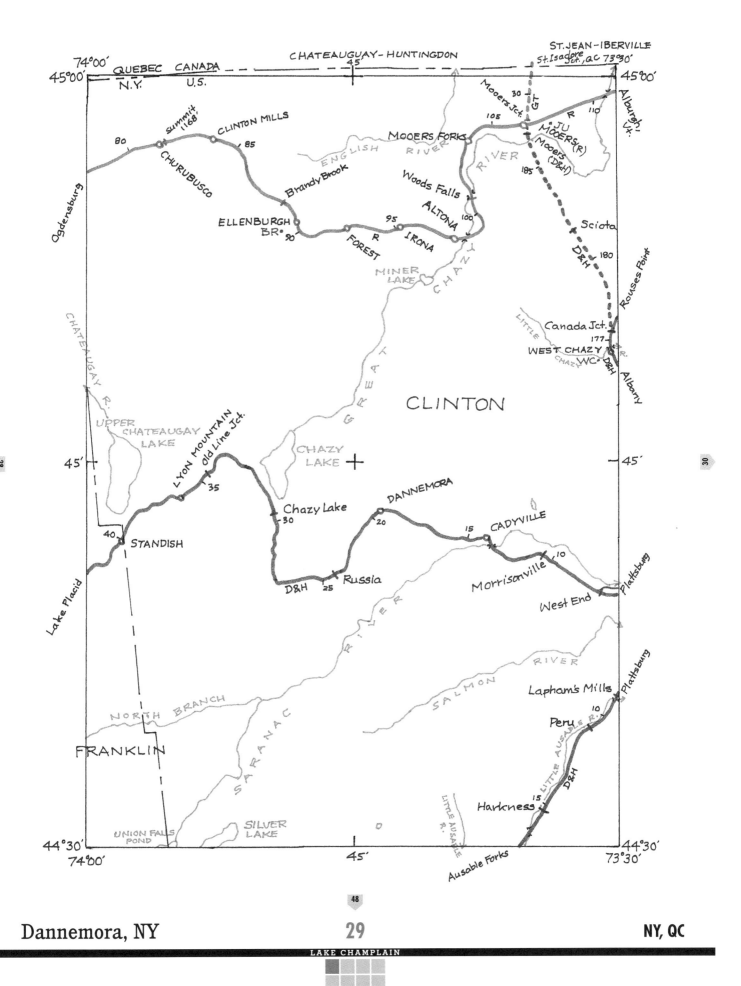

Dannemora, NY

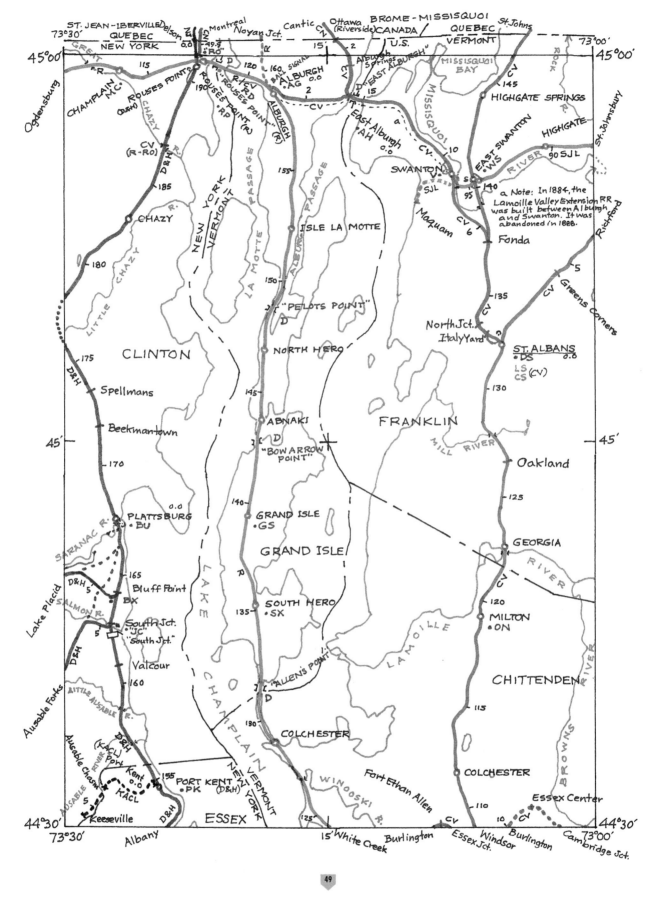

St. Albans, VT — 30 — NY, QC, VT

LAKE CHAMPLAIN

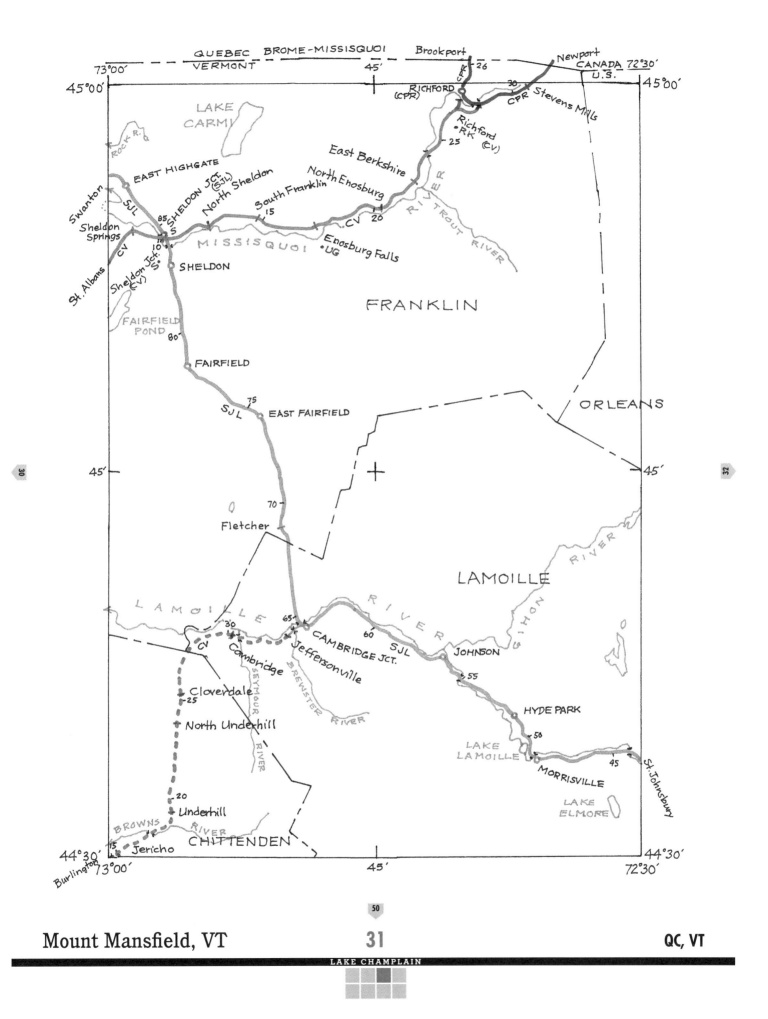

Mount Mansfield, VT

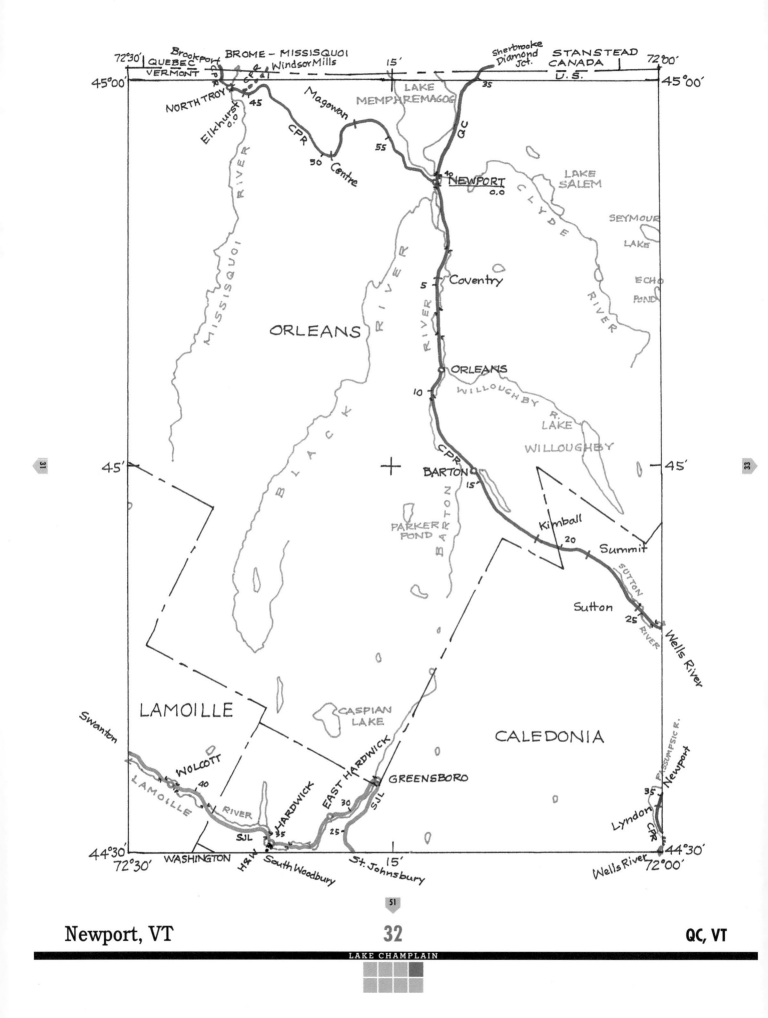
Newport, VT — 32 — QC, VT

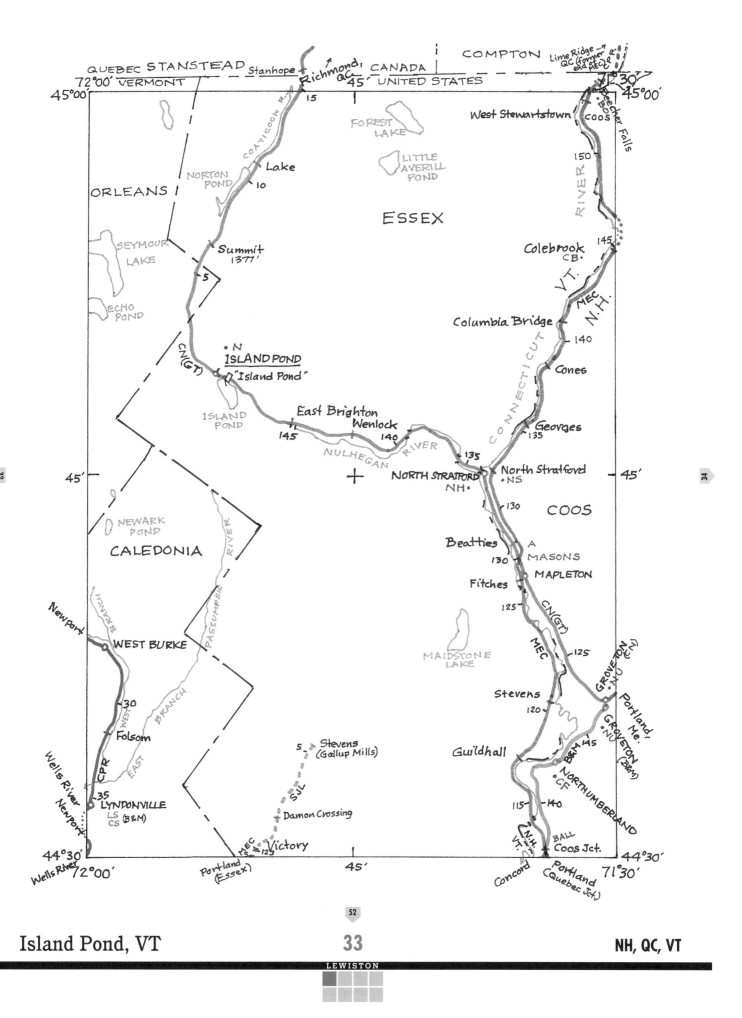

Island Pond, VT

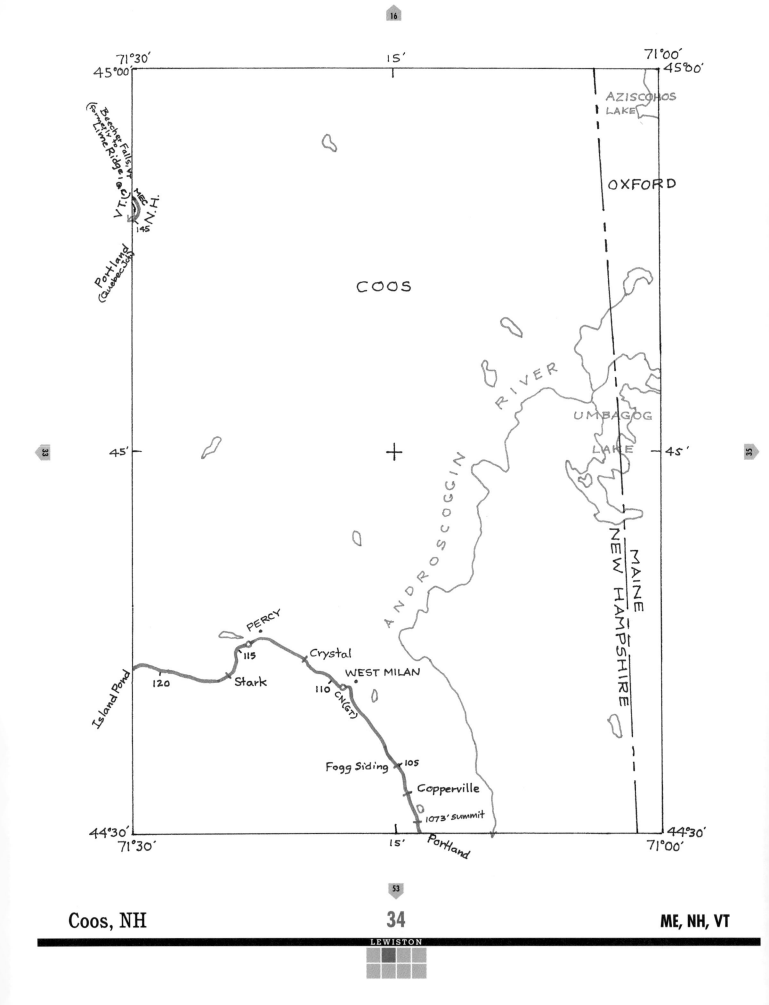

Coos, NH 34 ME, NH, VT

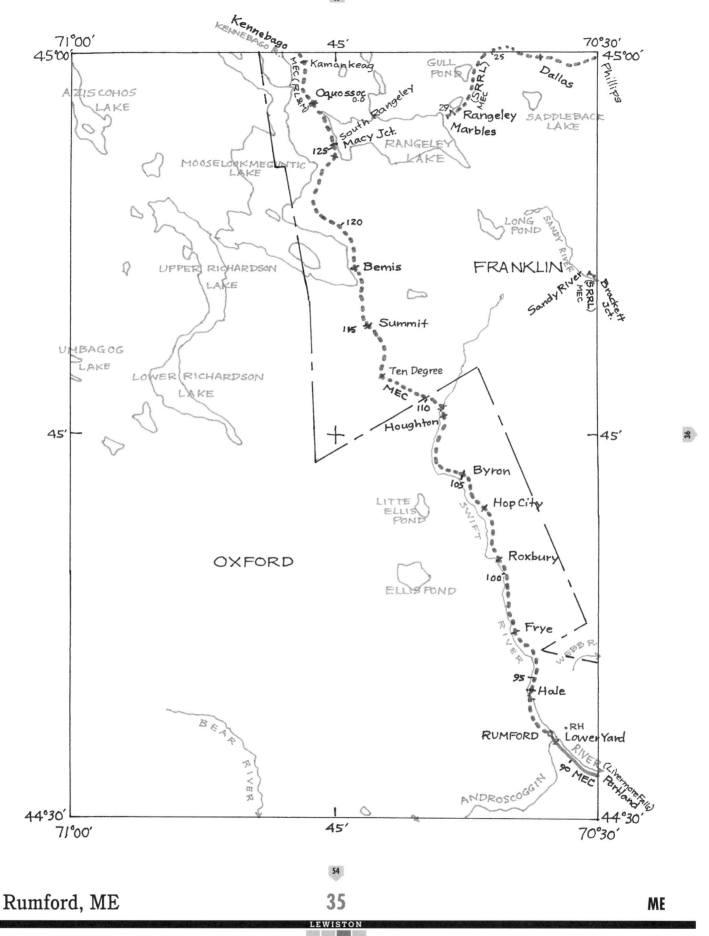

Rumford, ME

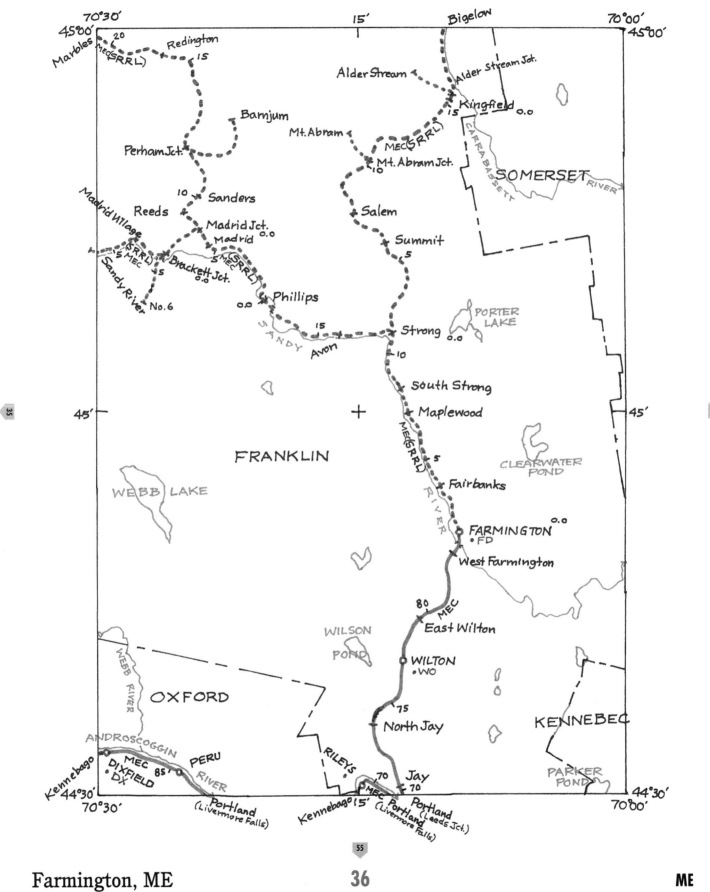

Farmington, ME

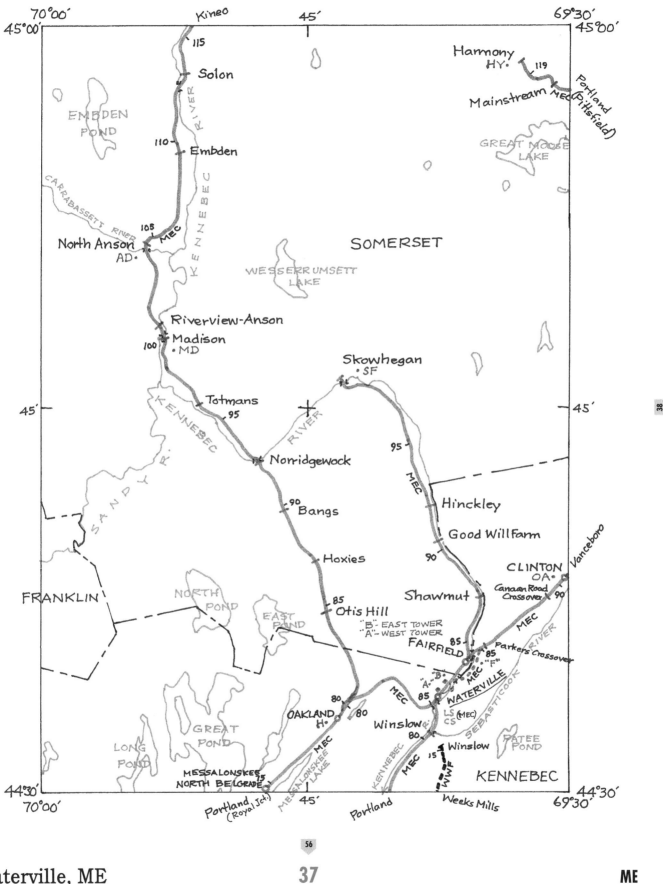

Waterville, ME

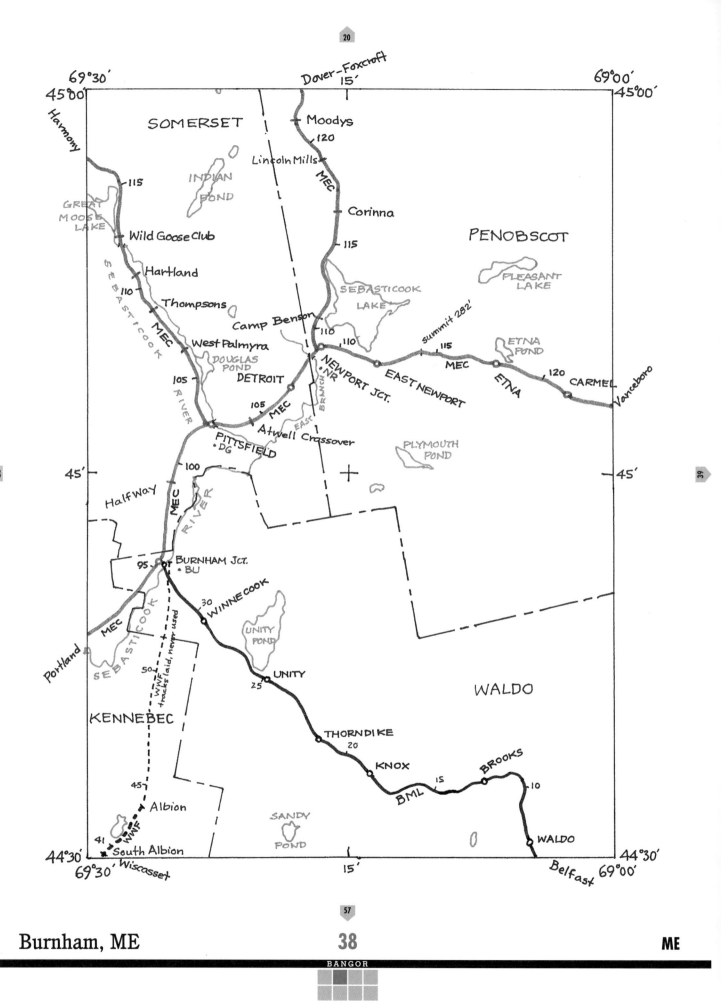

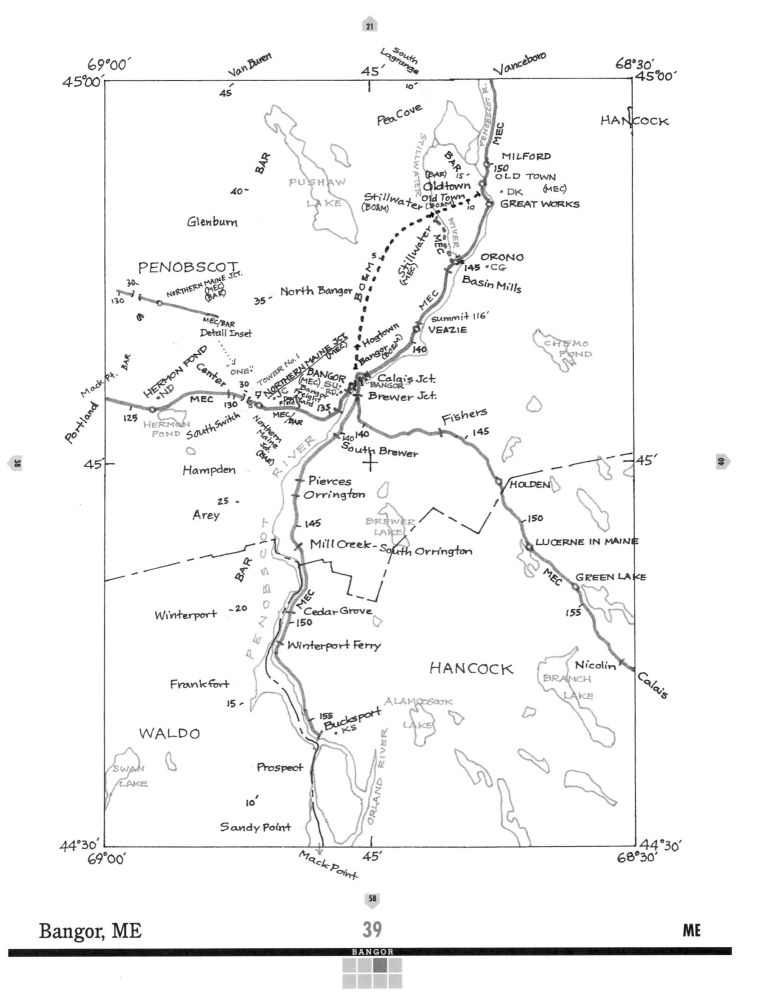

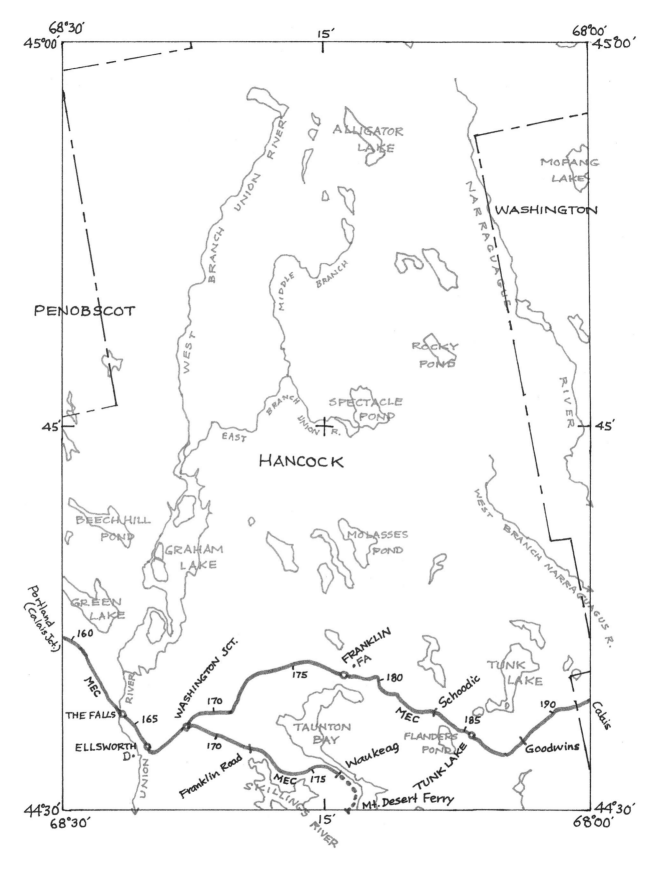

Ellsworth, ME

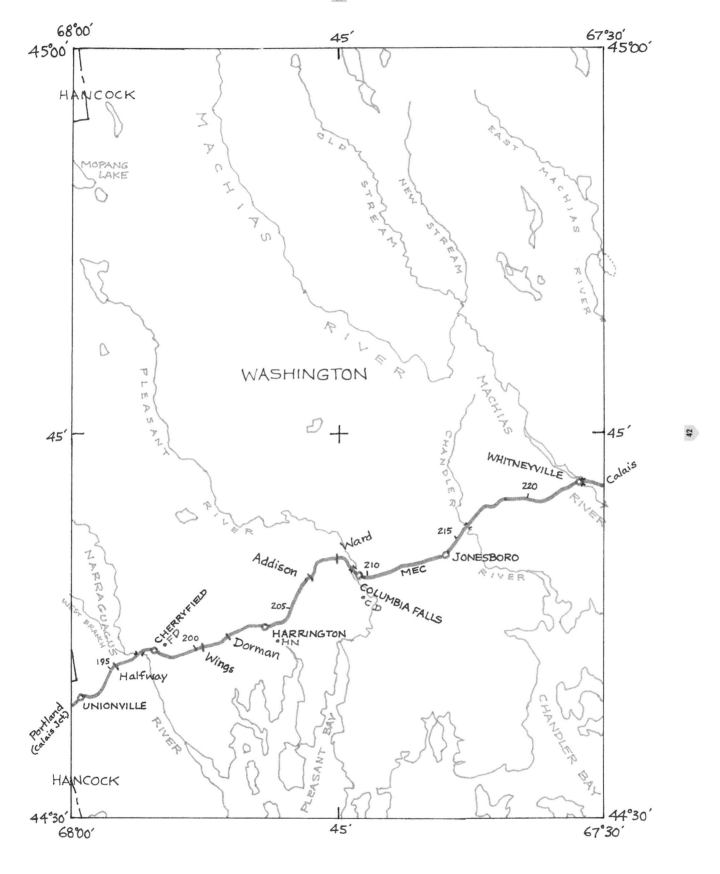

Harrington, ME

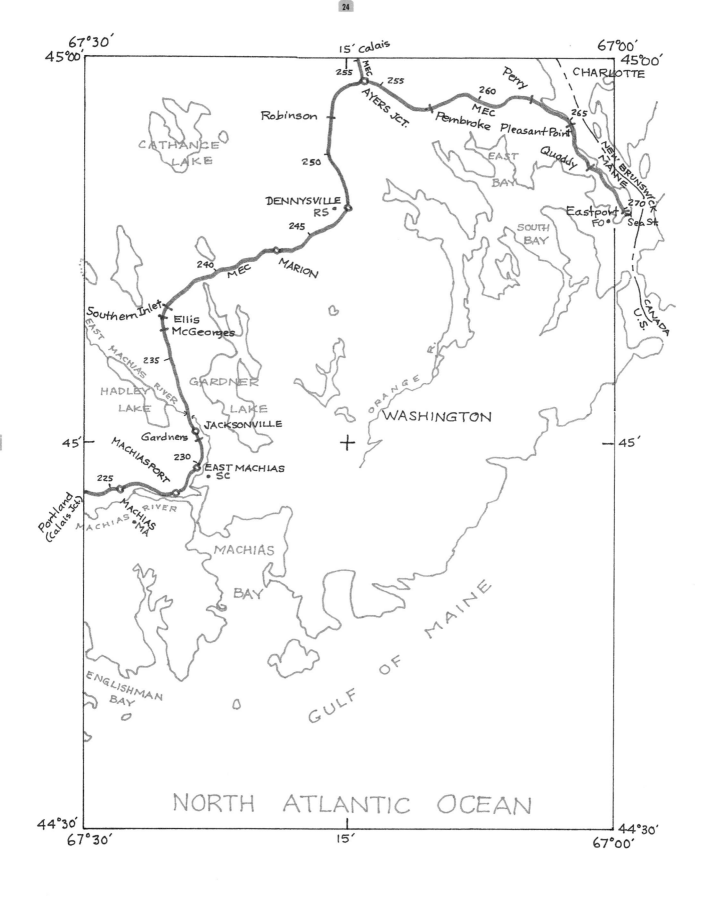

Eastport, ME   42   ME, NB

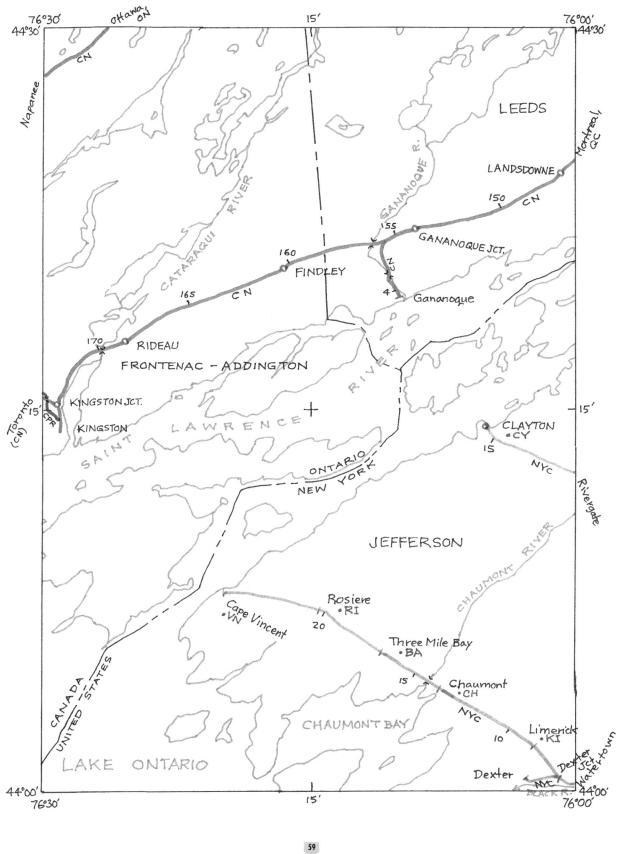

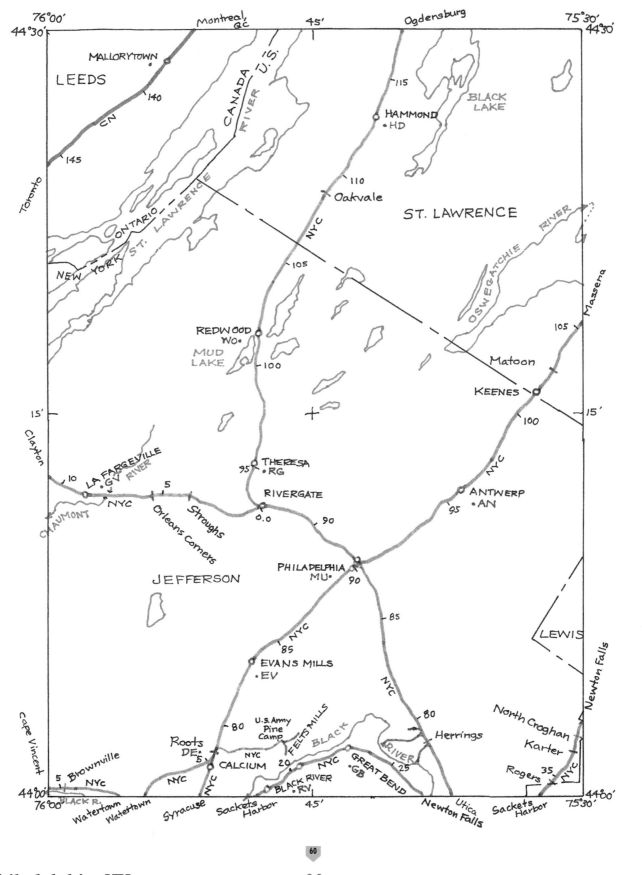

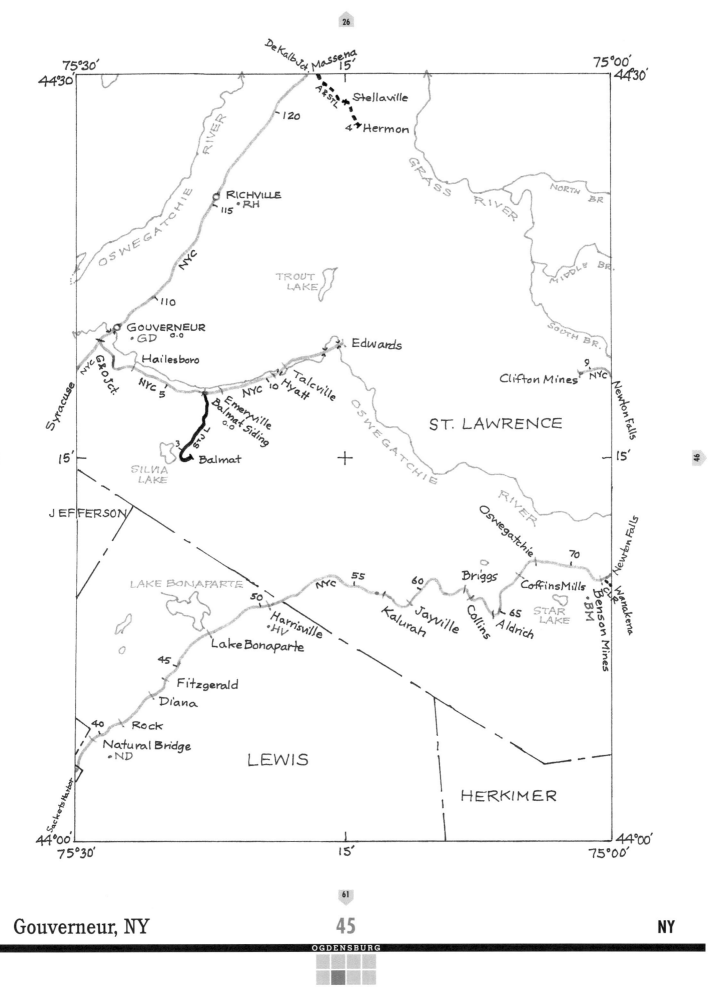

Gouverneur, NY 45 NY

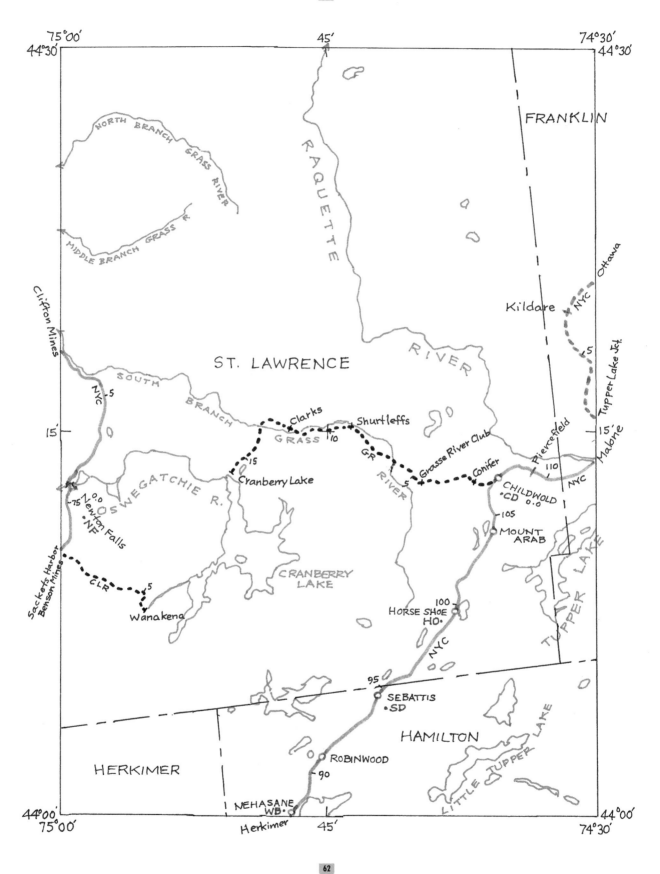

Newton Falls, NY   46   NY

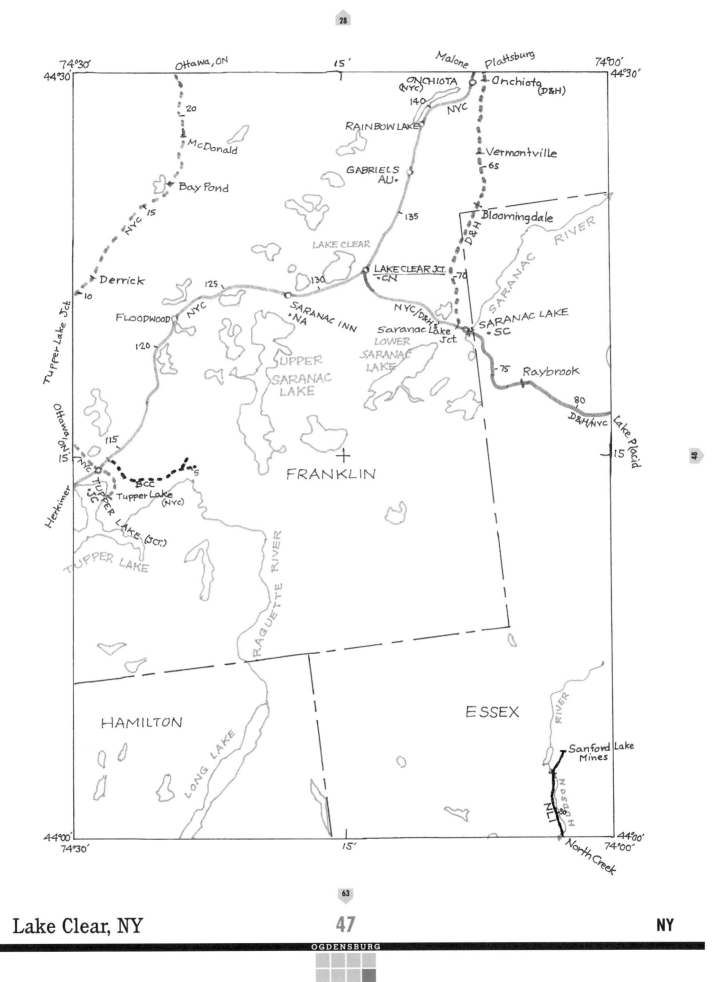

Lake Clear, NY

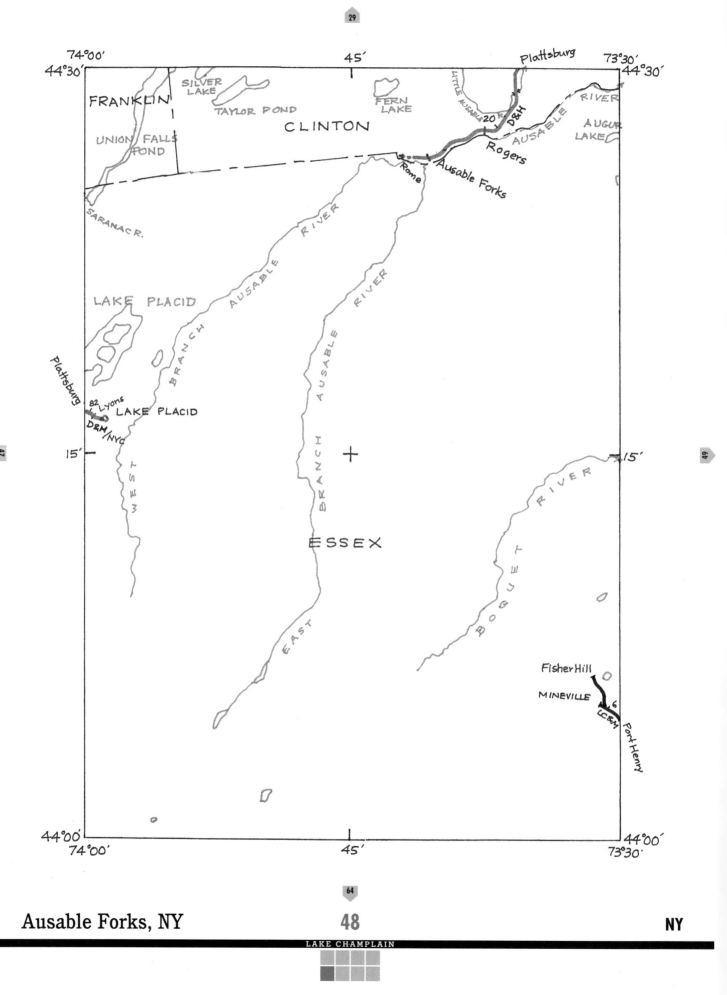

Ausable Forks, NY — 48 — NY

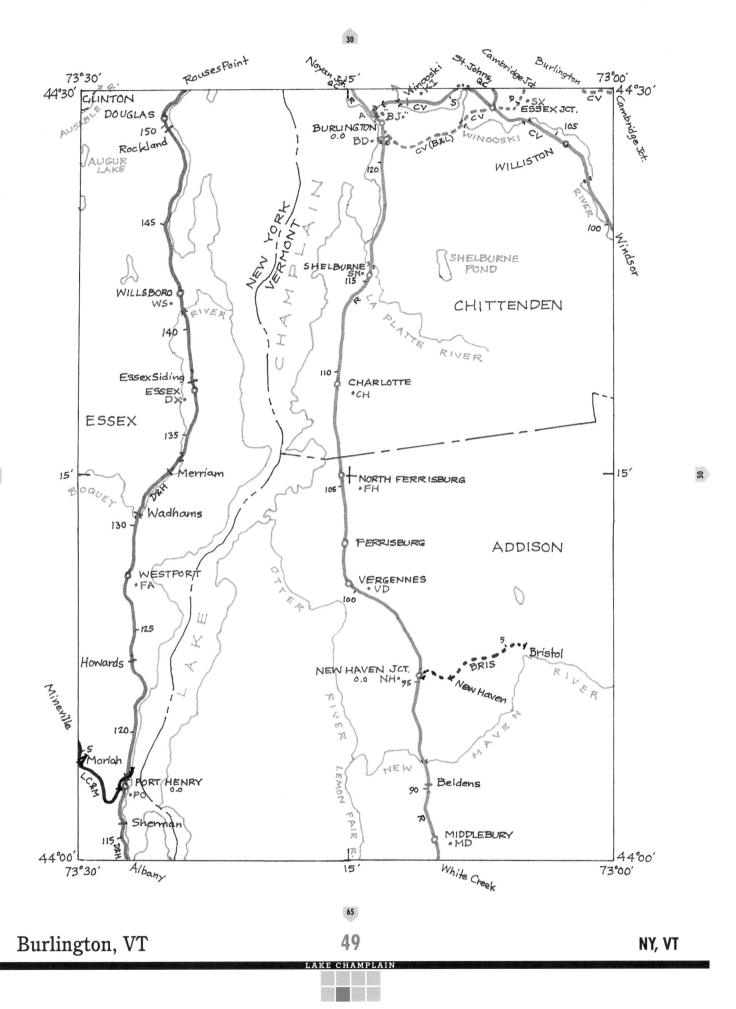

Burlington, VT

NY, VT

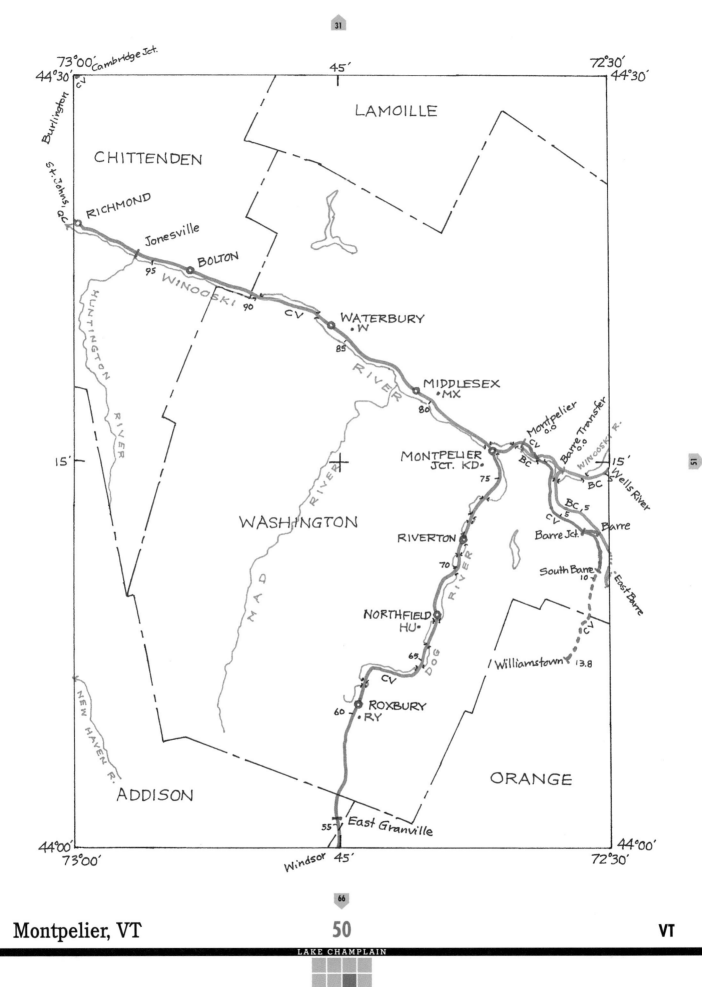

Montpelier, VT

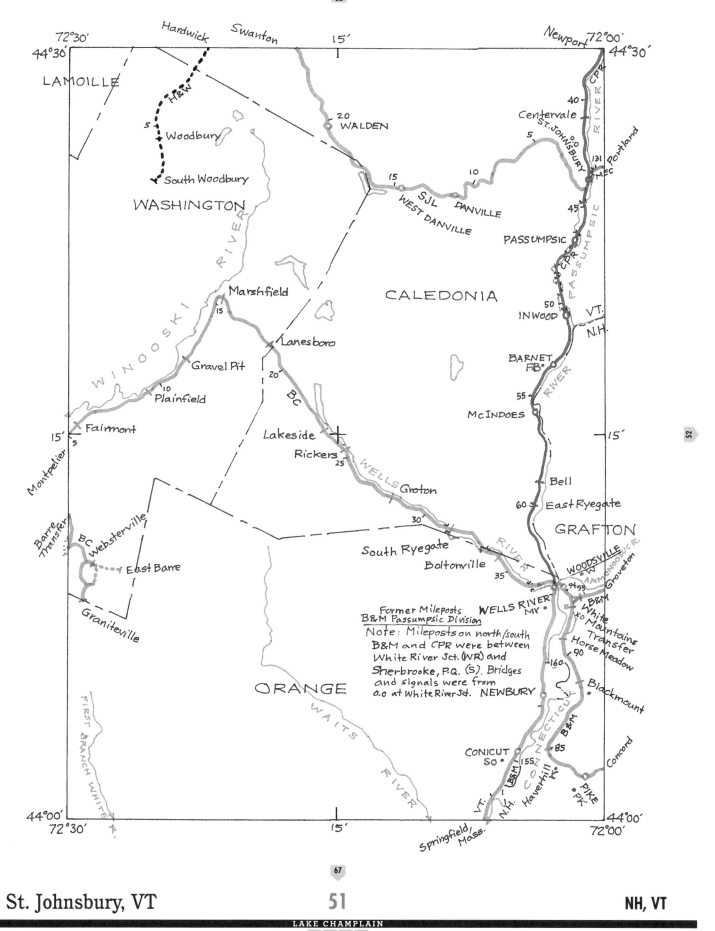

St. Johnsbury, VT

NH, VT

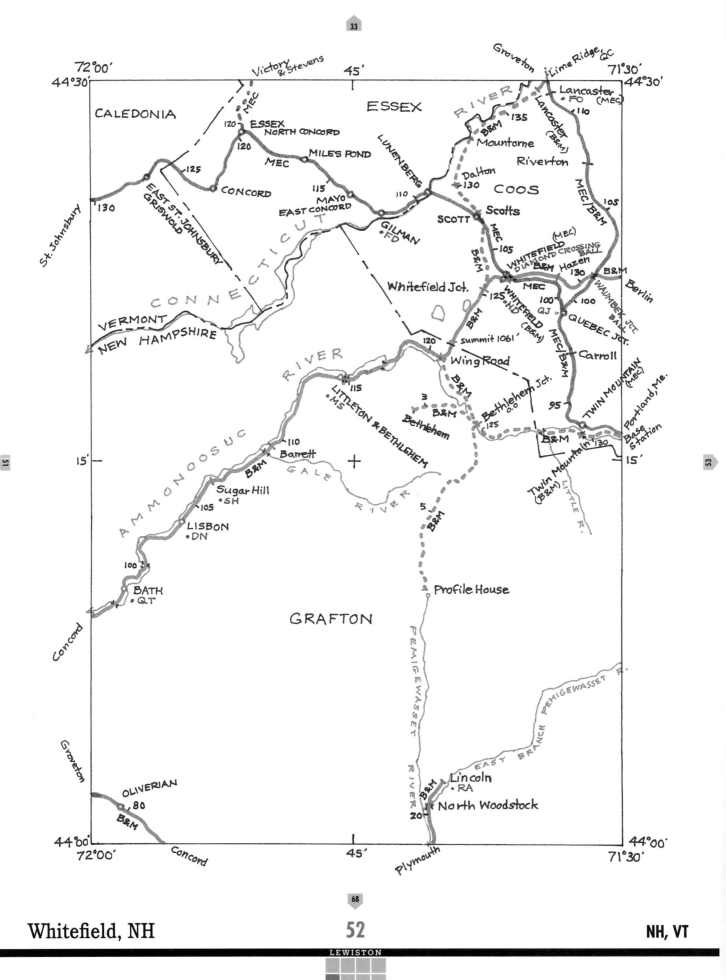

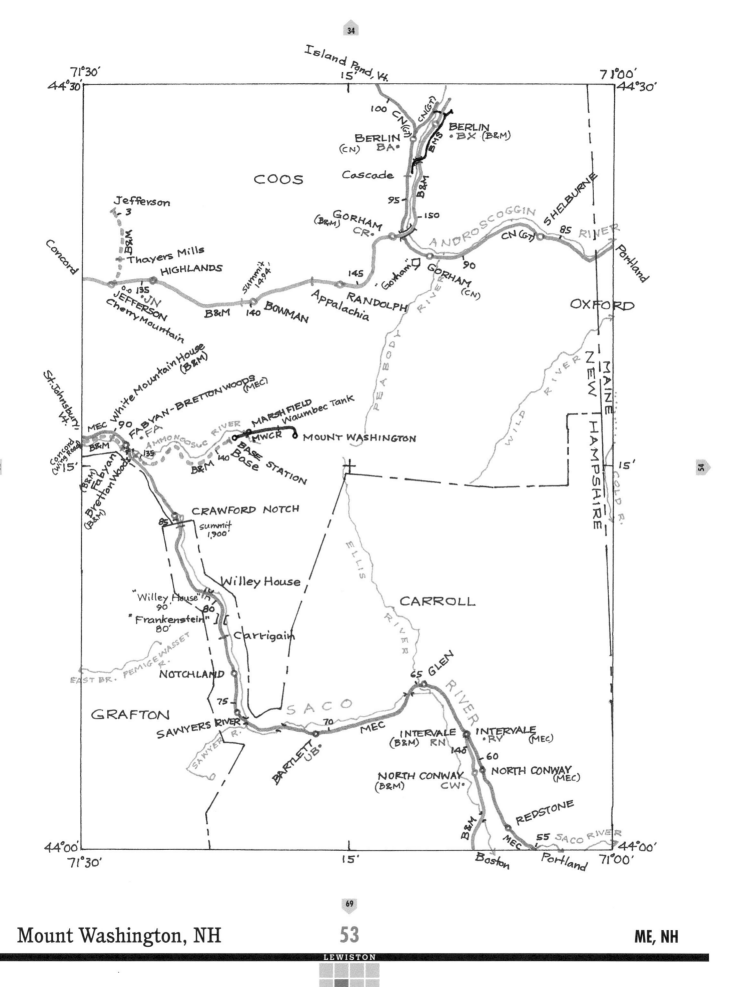

Mount Washington, NH

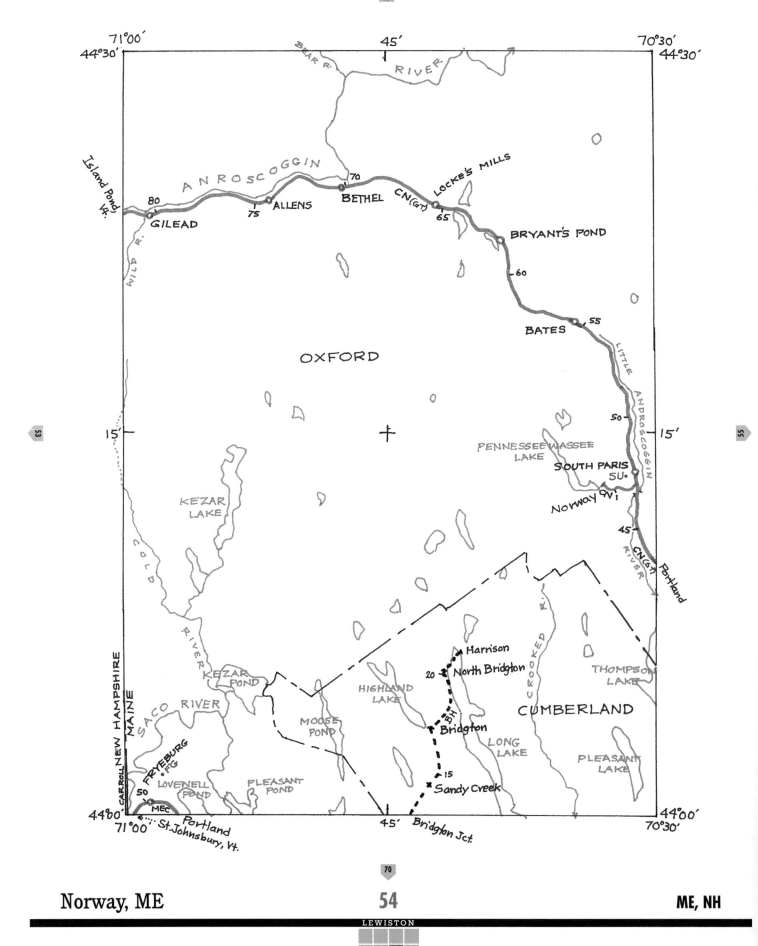

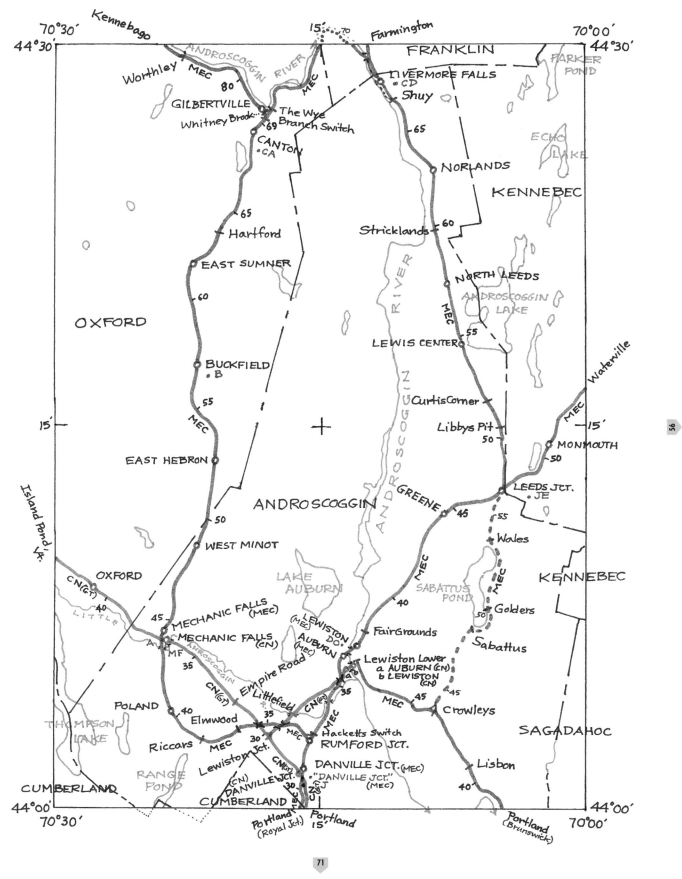

Lewiston, ME

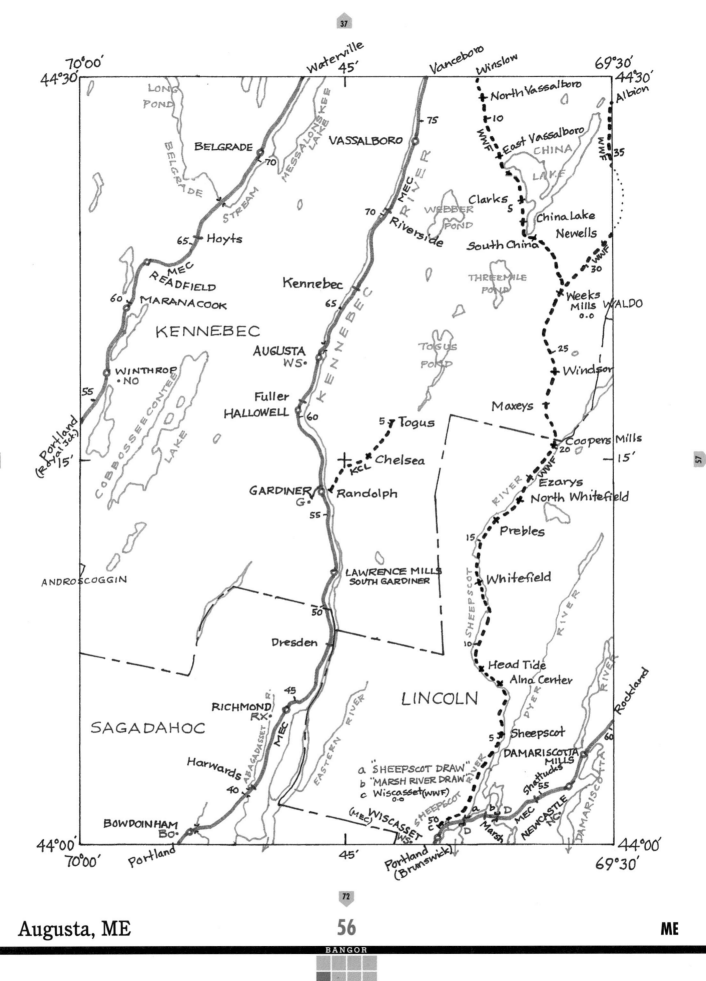

Augusta, ME 56 ME

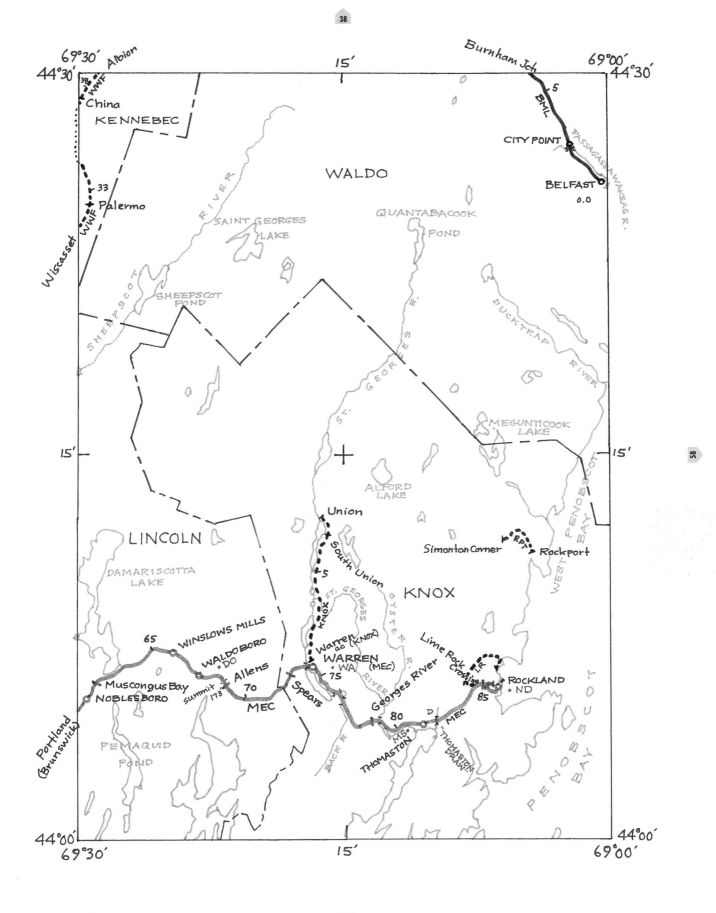

Rockland, ME

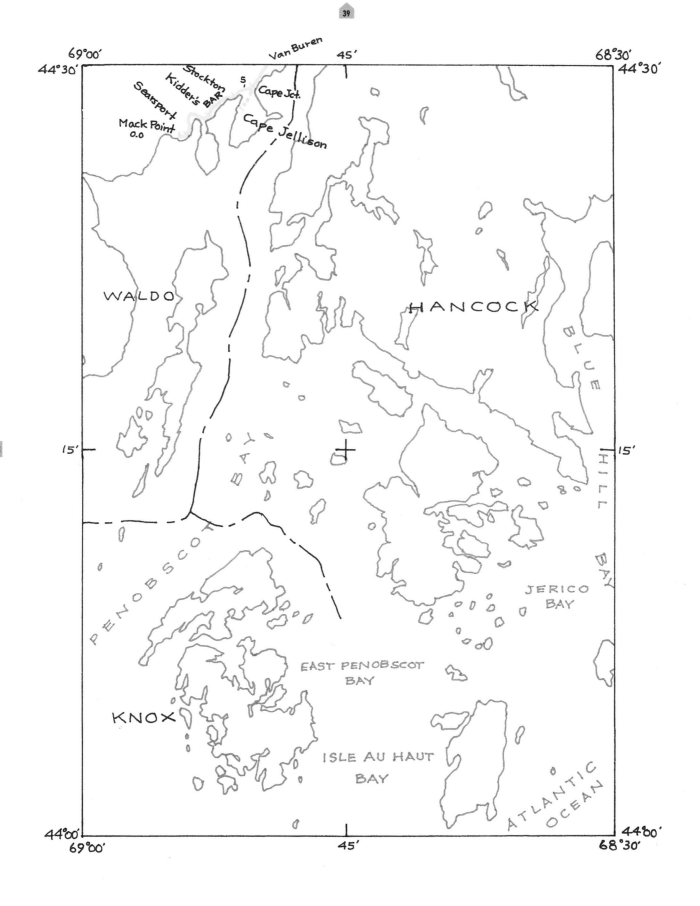

Searsport, ME

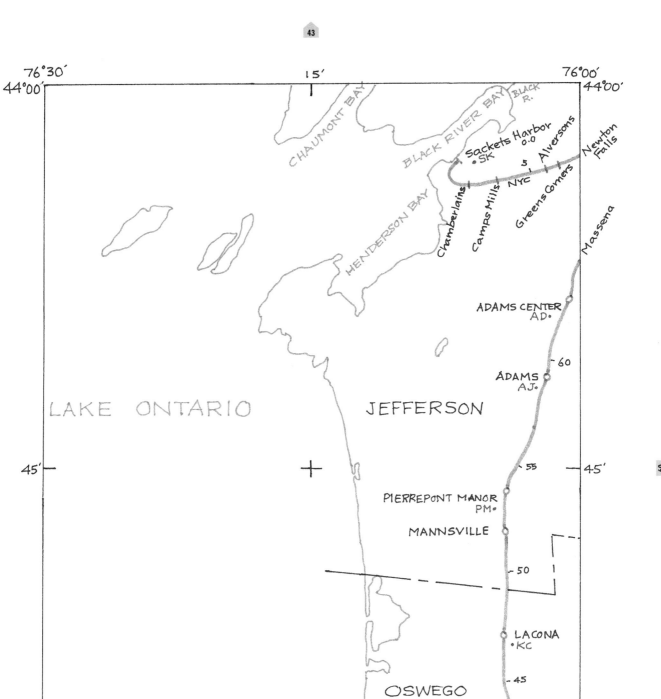

Pulaski, NY   59   NY

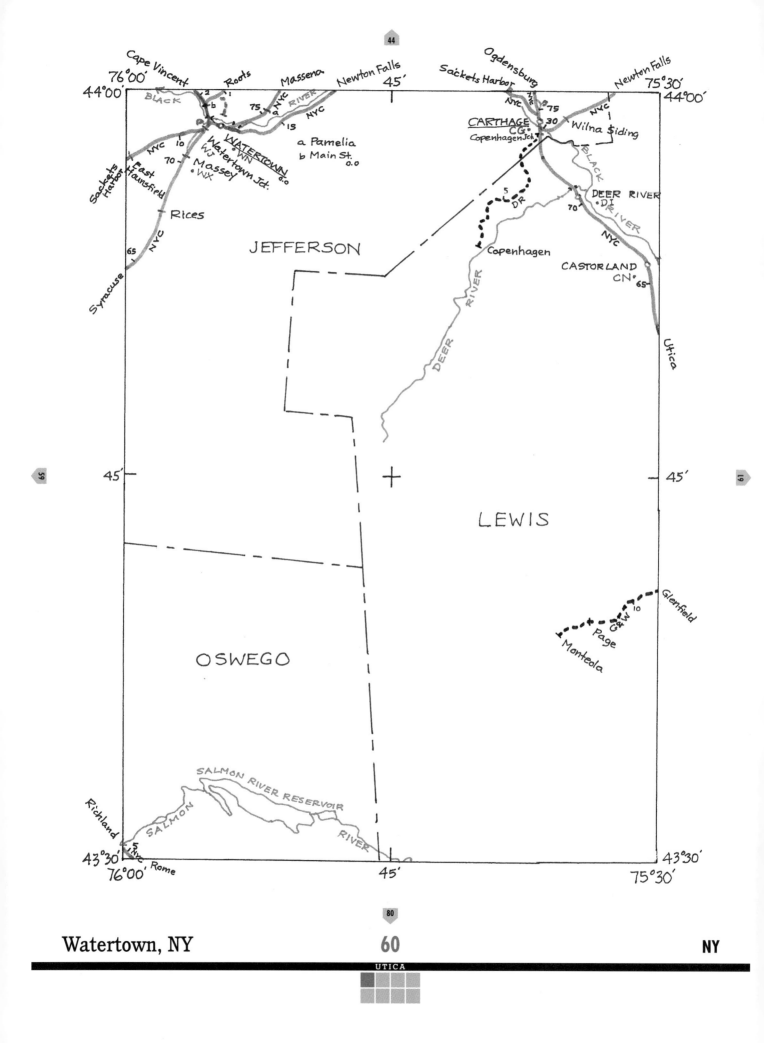

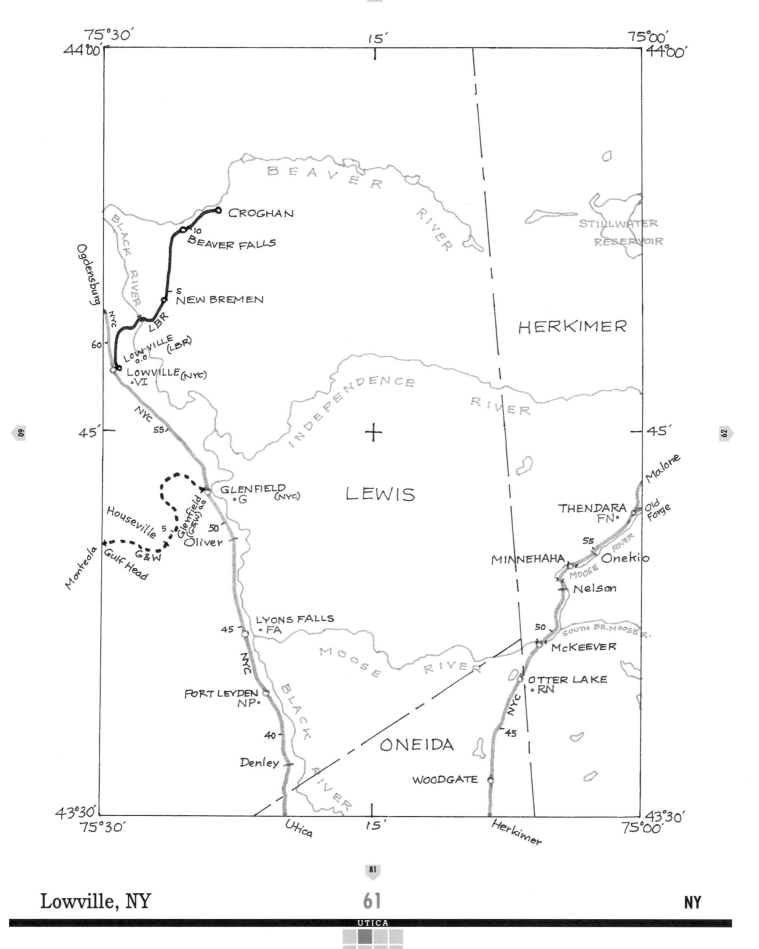

Lowville, NY

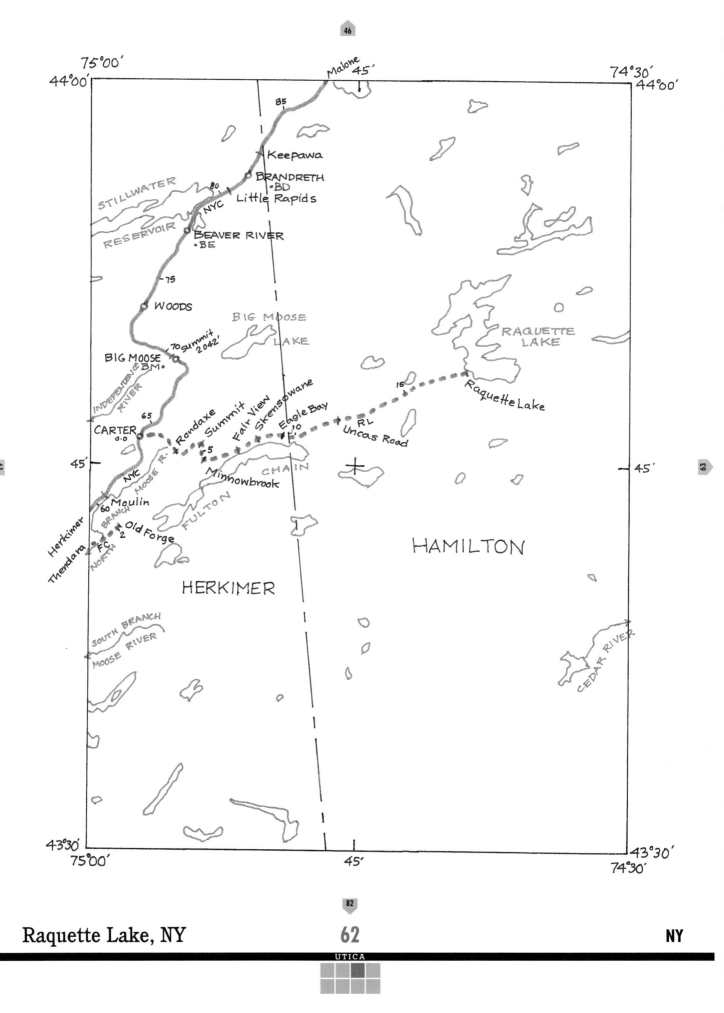

Raquette Lake, NY

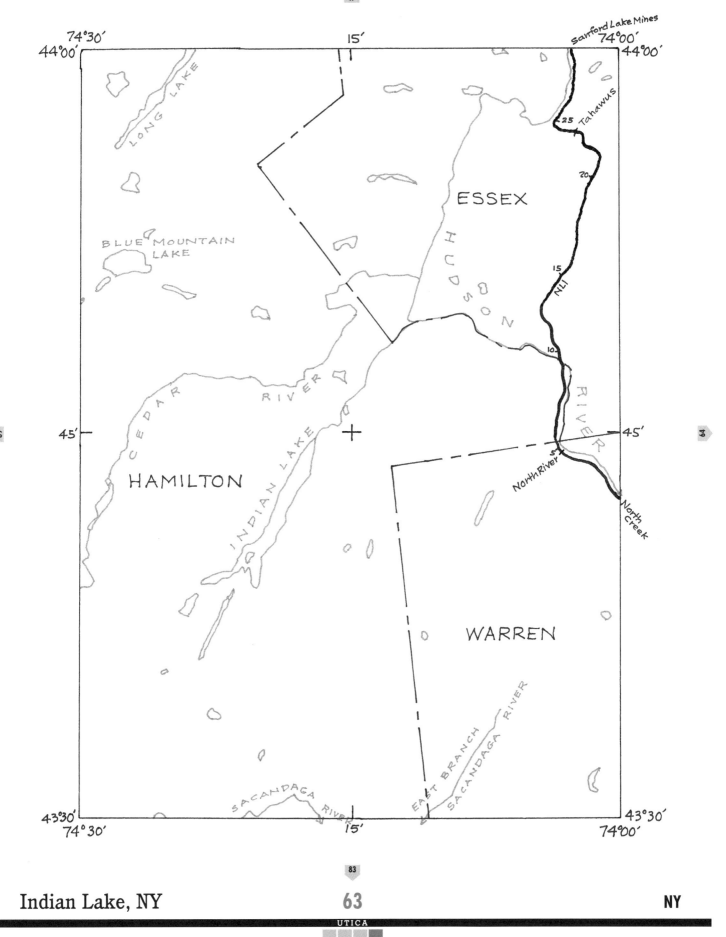

Indian Lake, NY

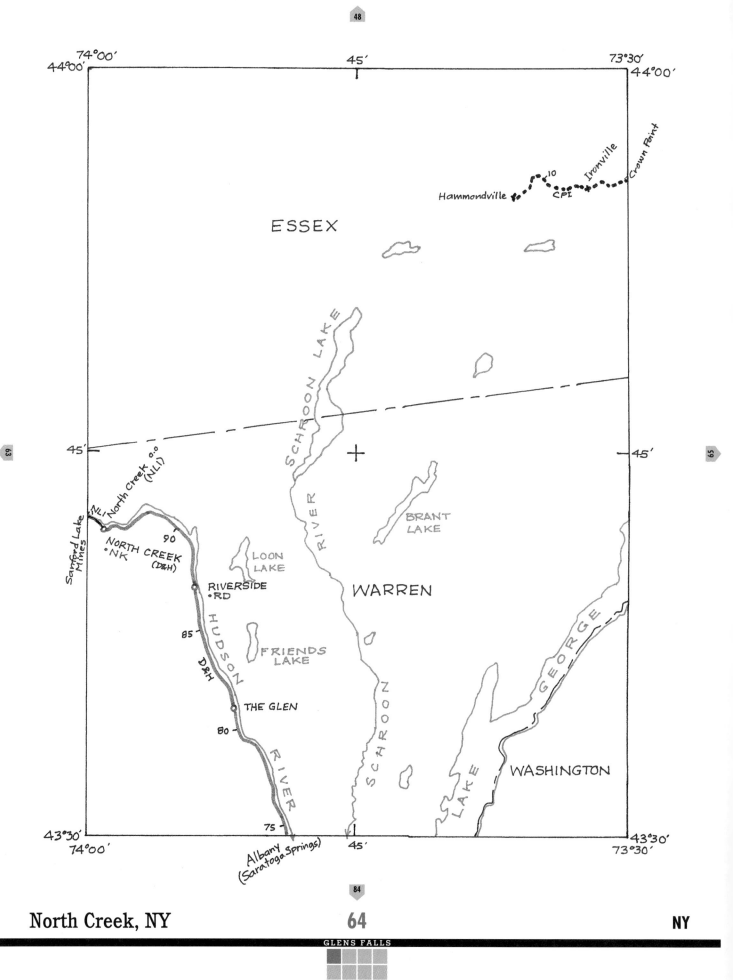

North Creek, NY 64 NY

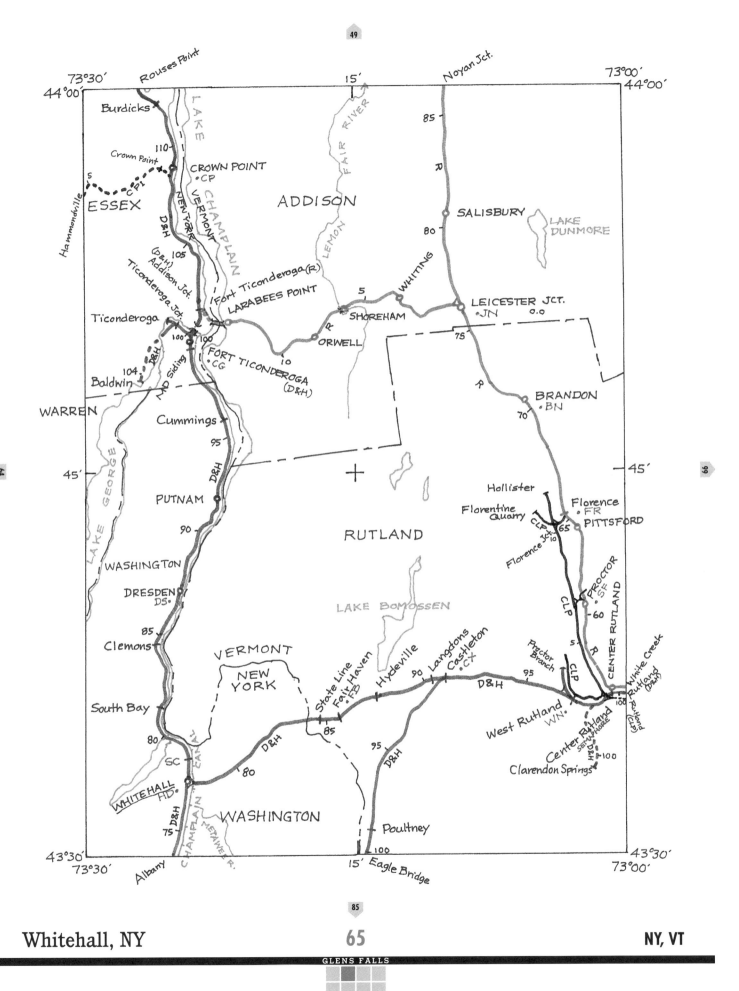

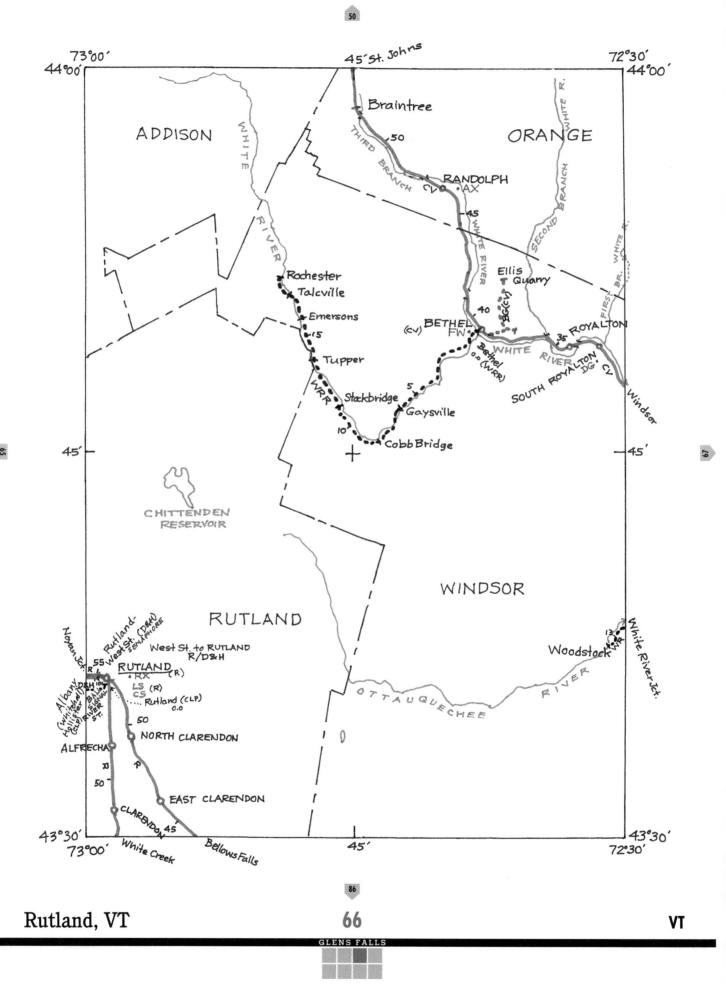

Rutland, VT   66   VT

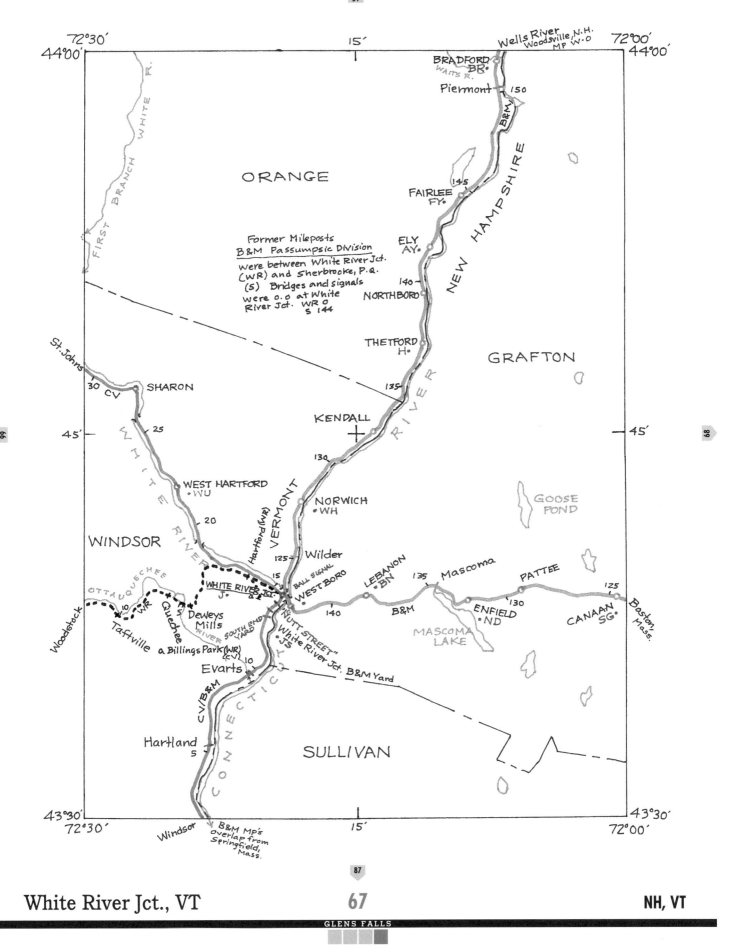

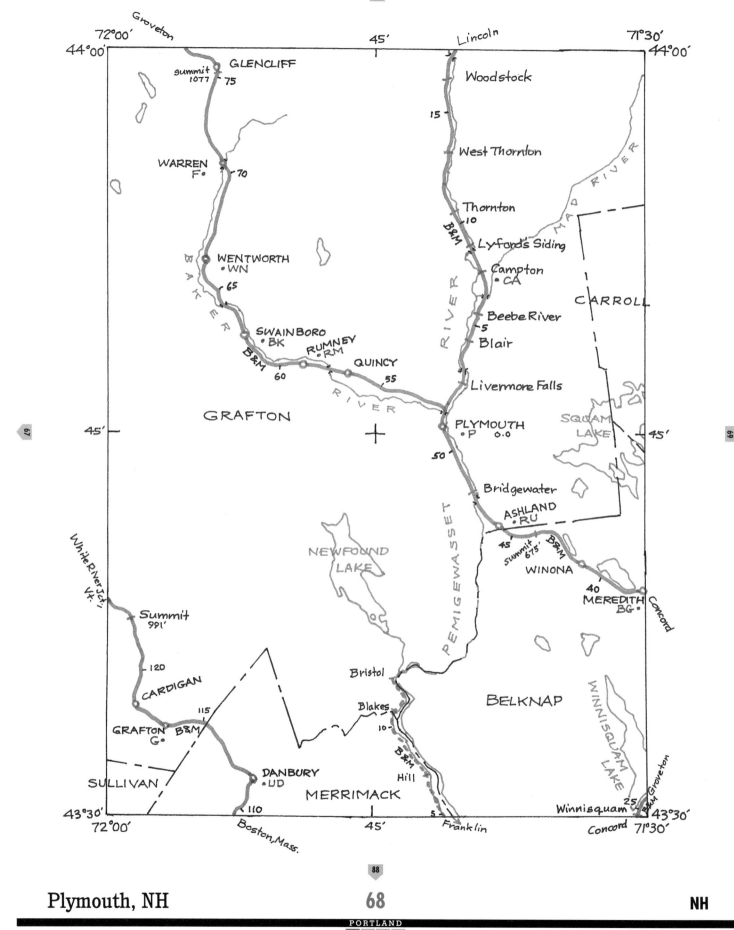

Plymouth, NH

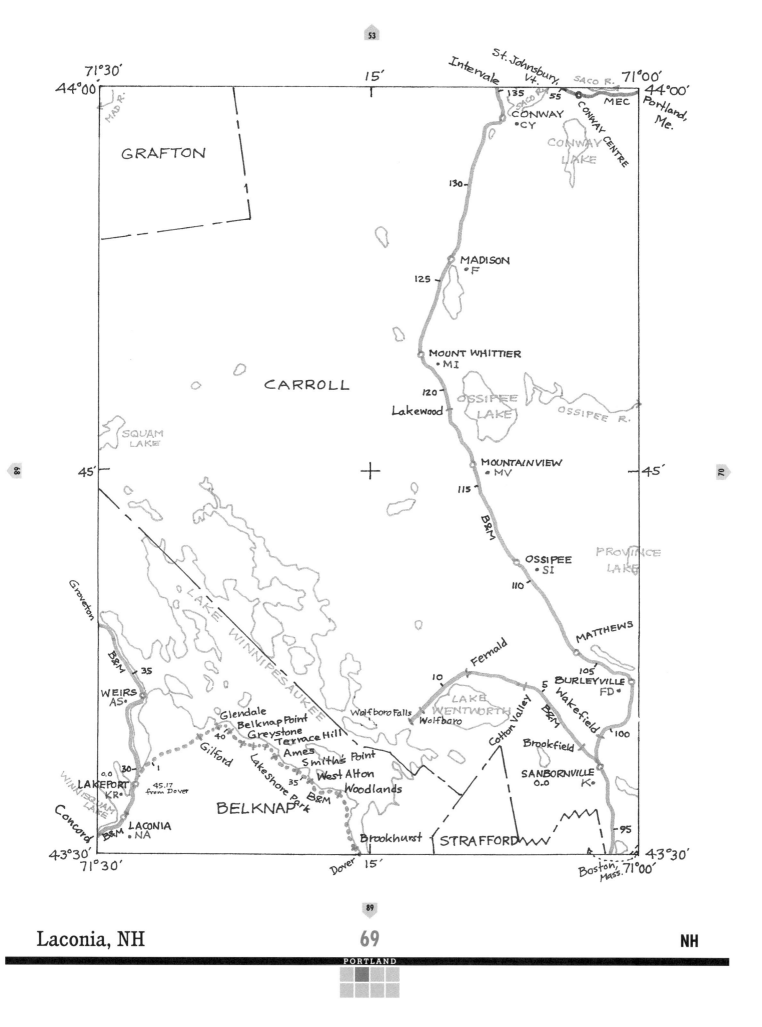

Laconia, NH

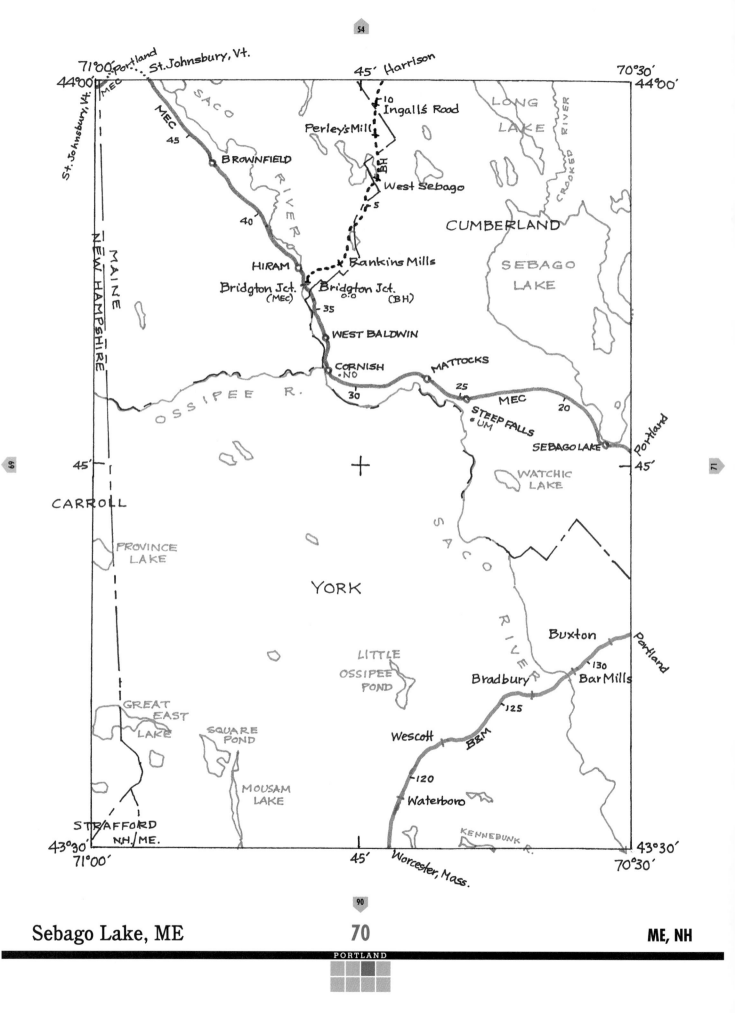

Sebago Lake, ME

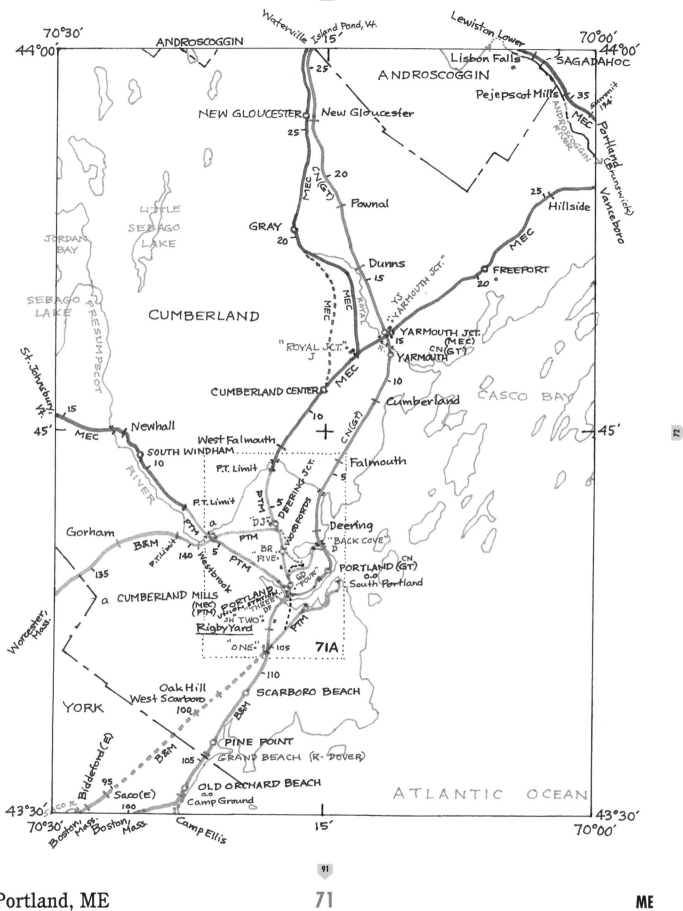

Portland, ME

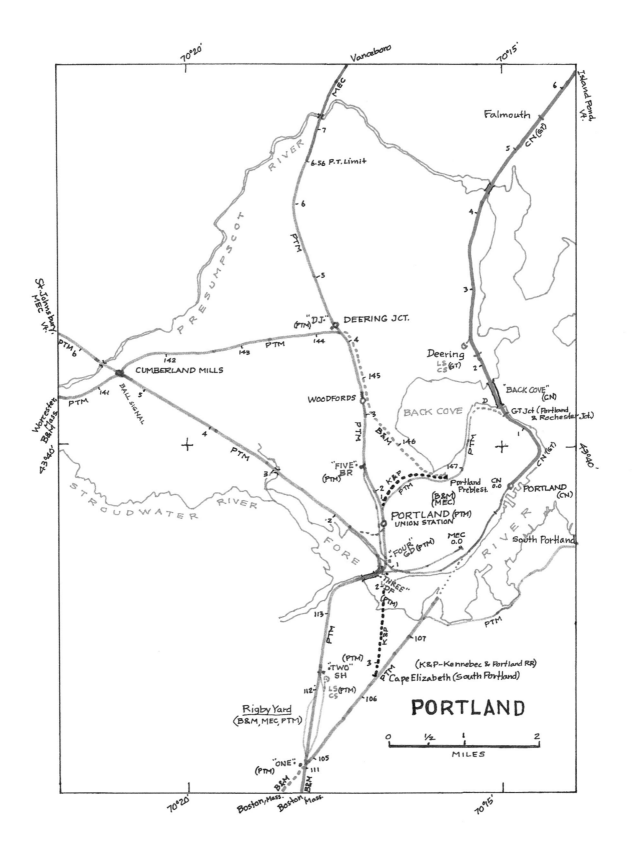

71A Portland, ME

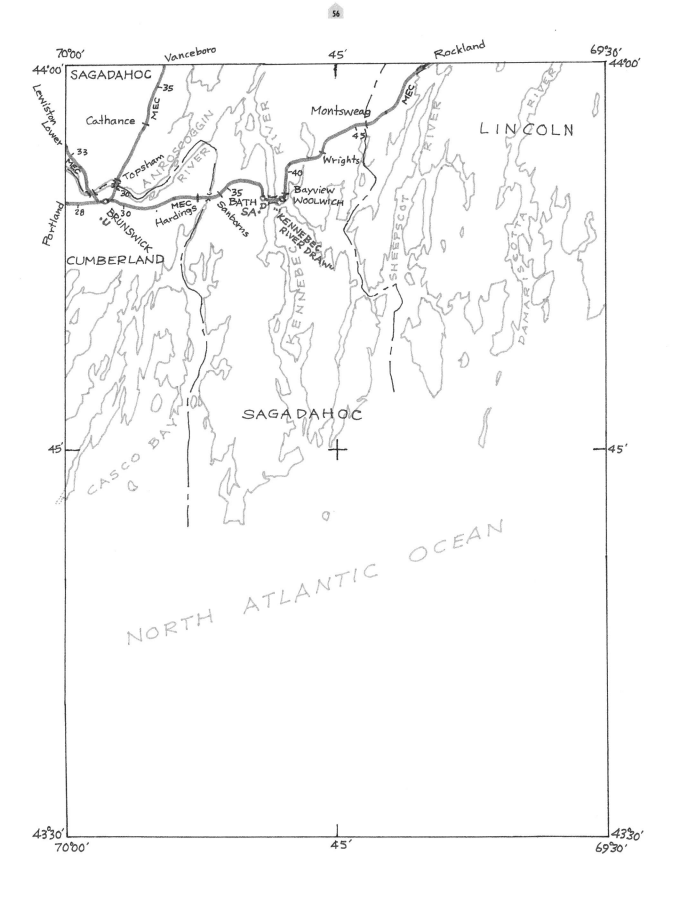

Brunswick, ME

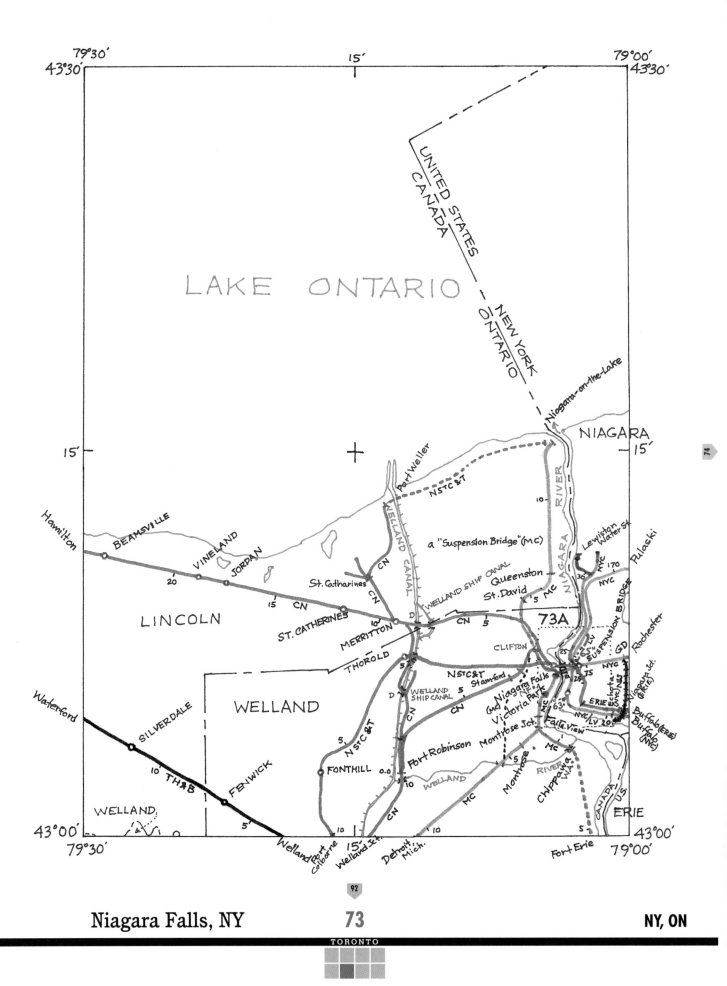

Niagara Falls, NY — 73 — NY, ON

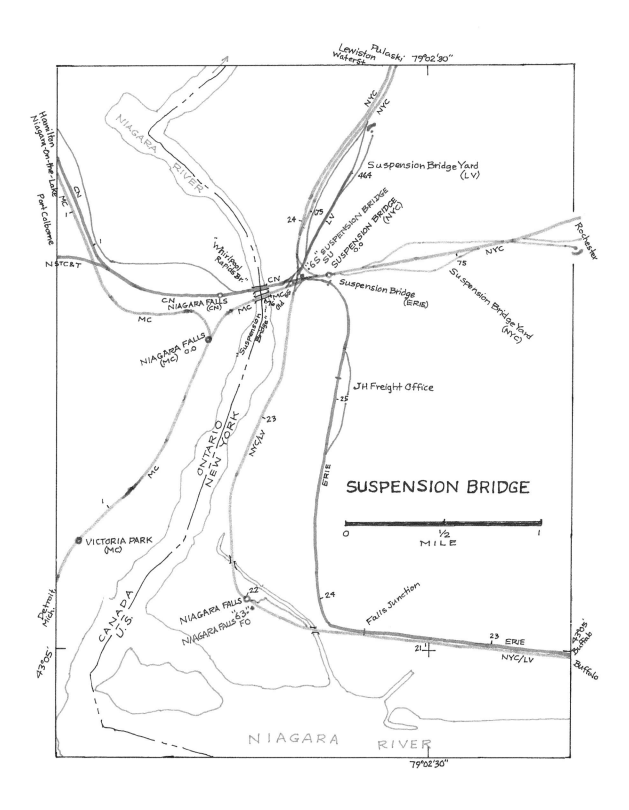

73A Suspension Bridge, NY

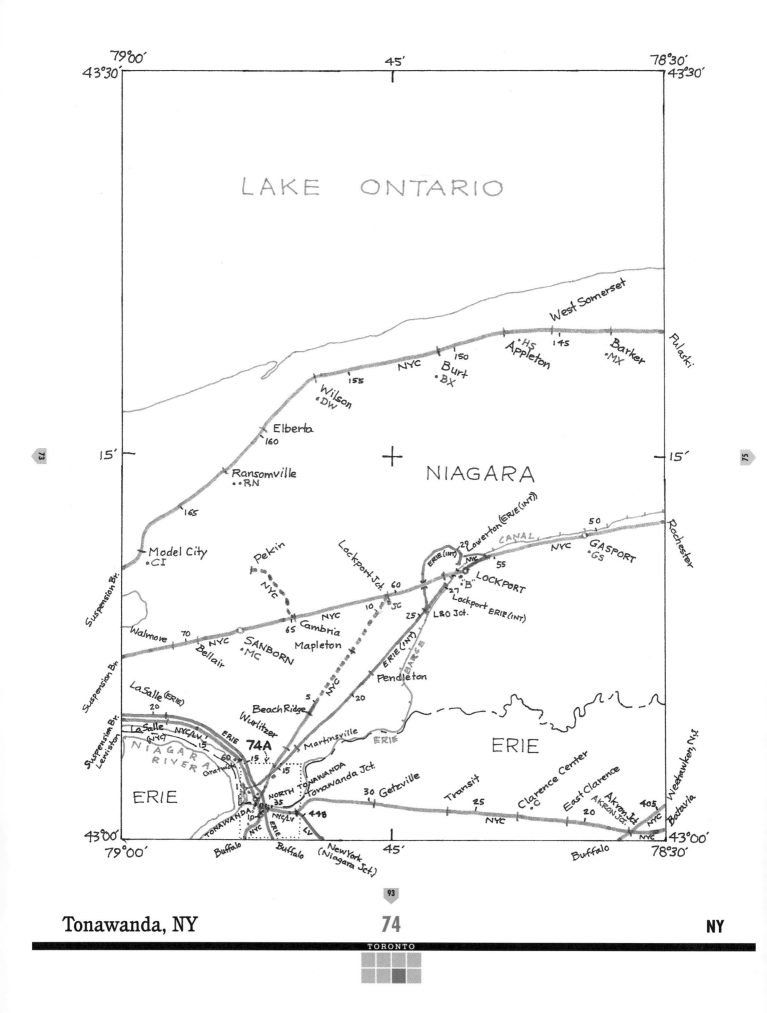

Tonawanda, NY

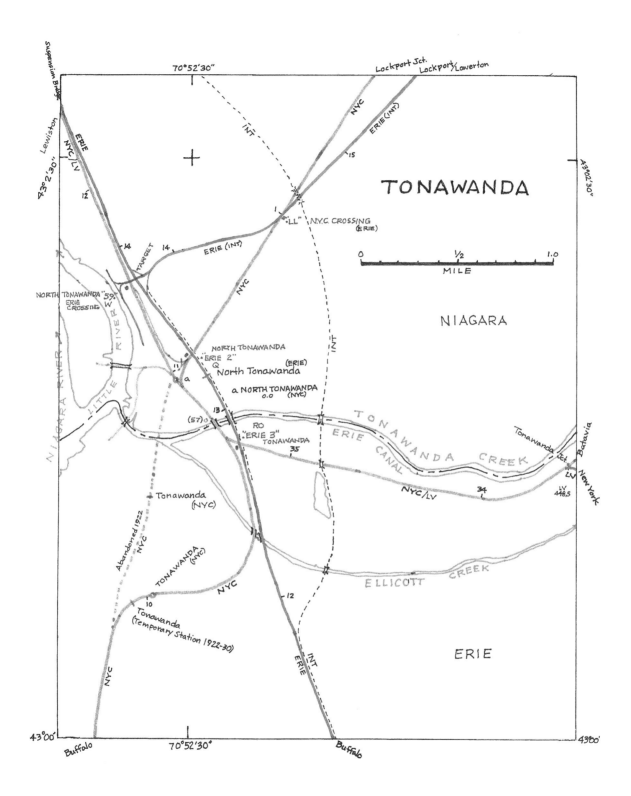

74A Tonawanda, NY

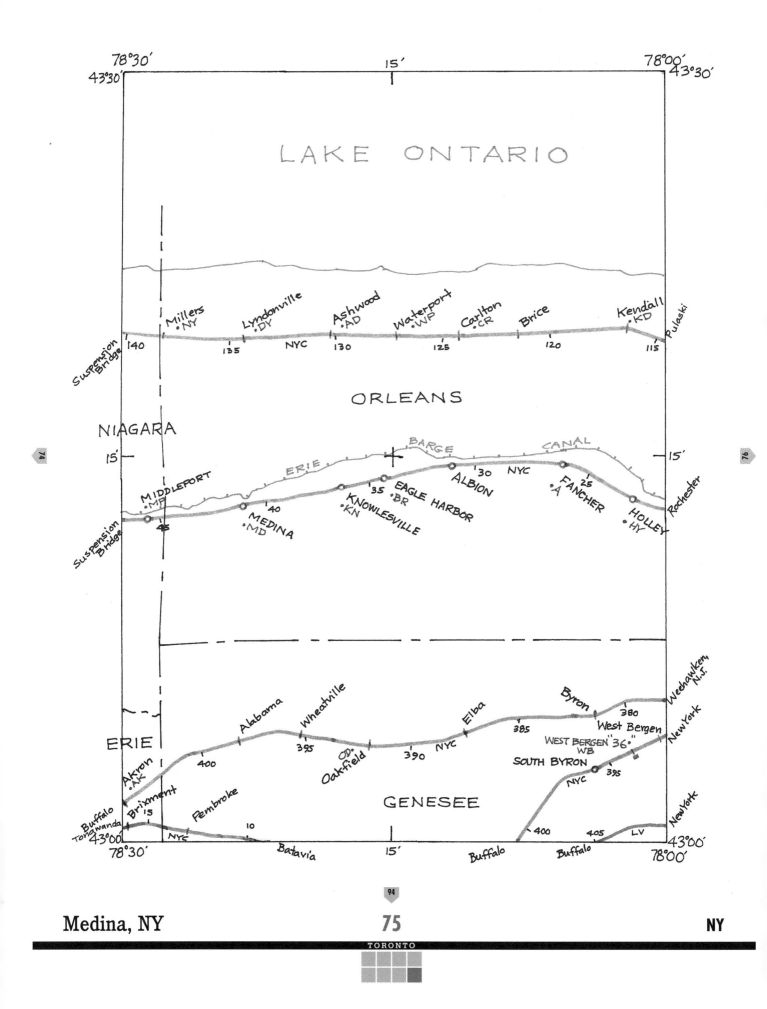

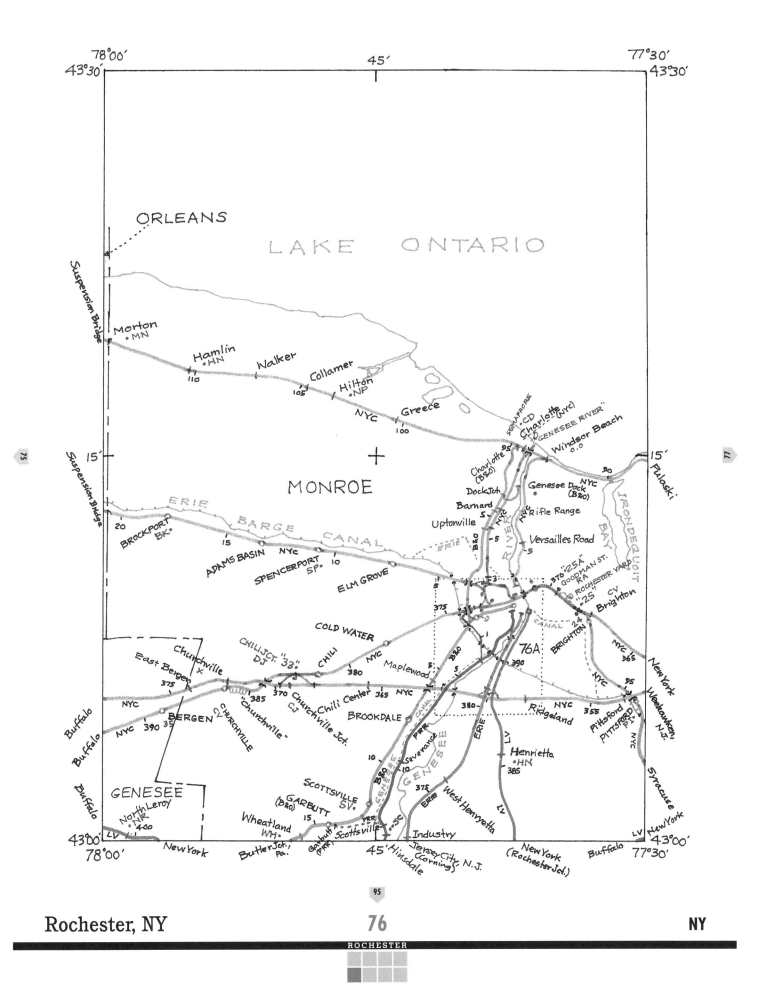

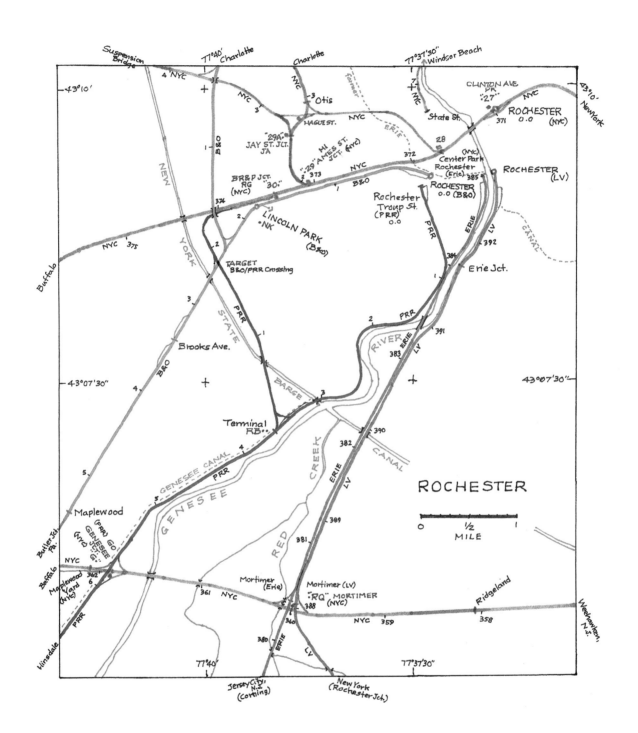

76A Rochester, NY

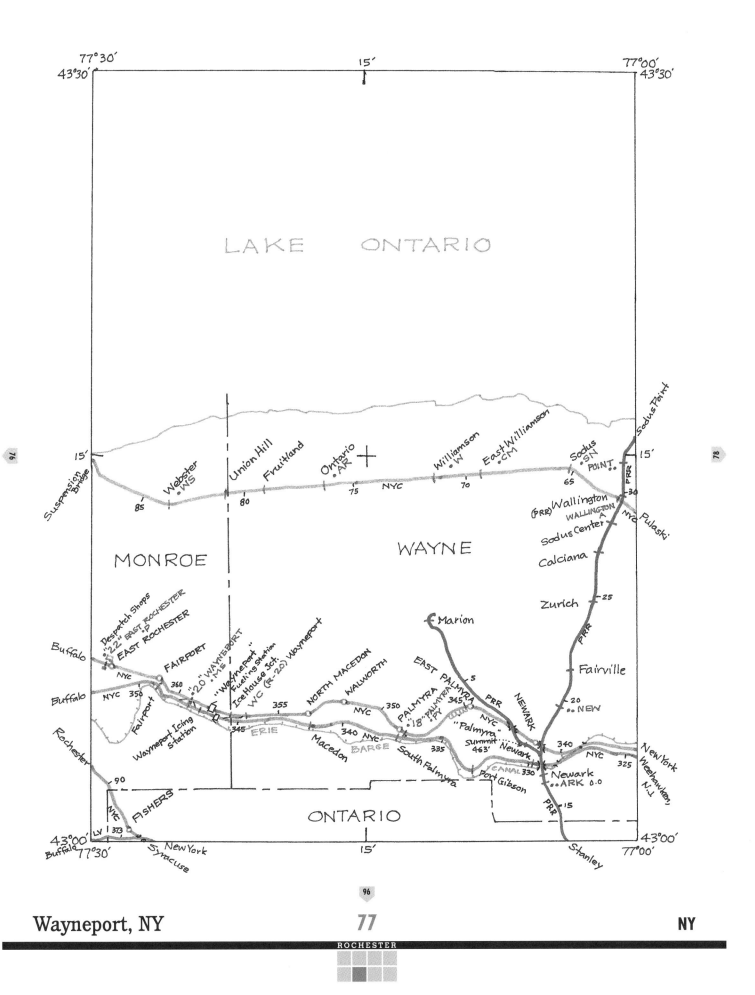

Wayneport, NY

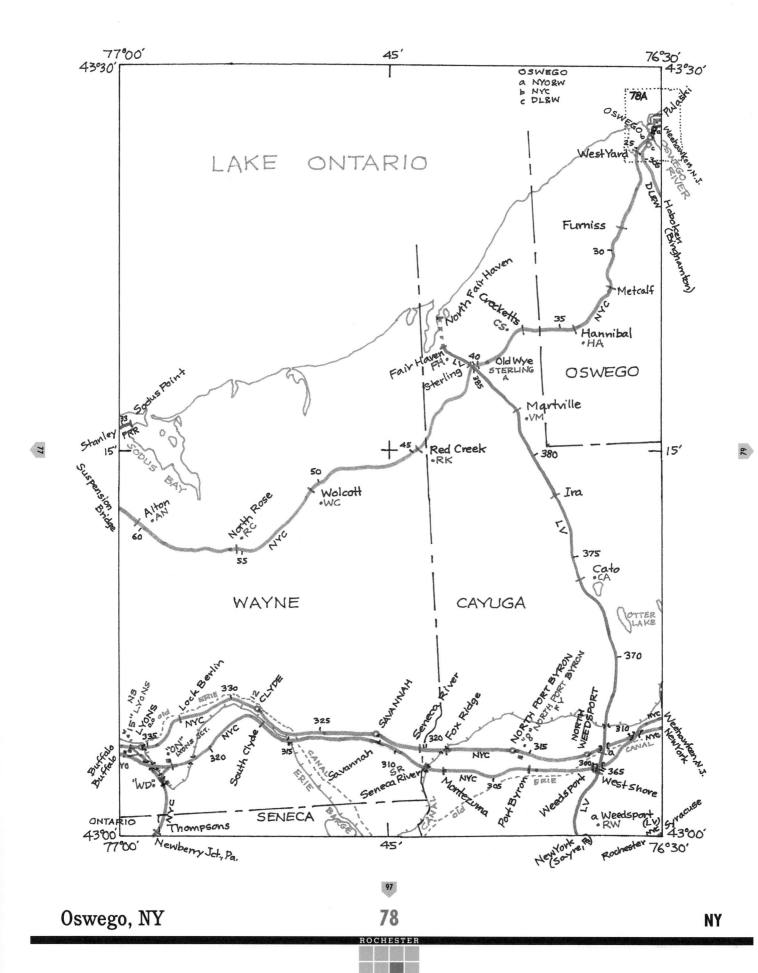

Oswego, NY   78   NY

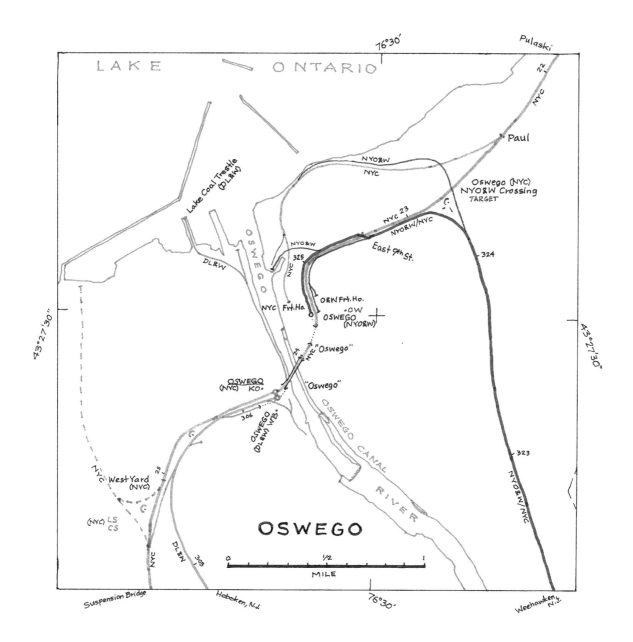

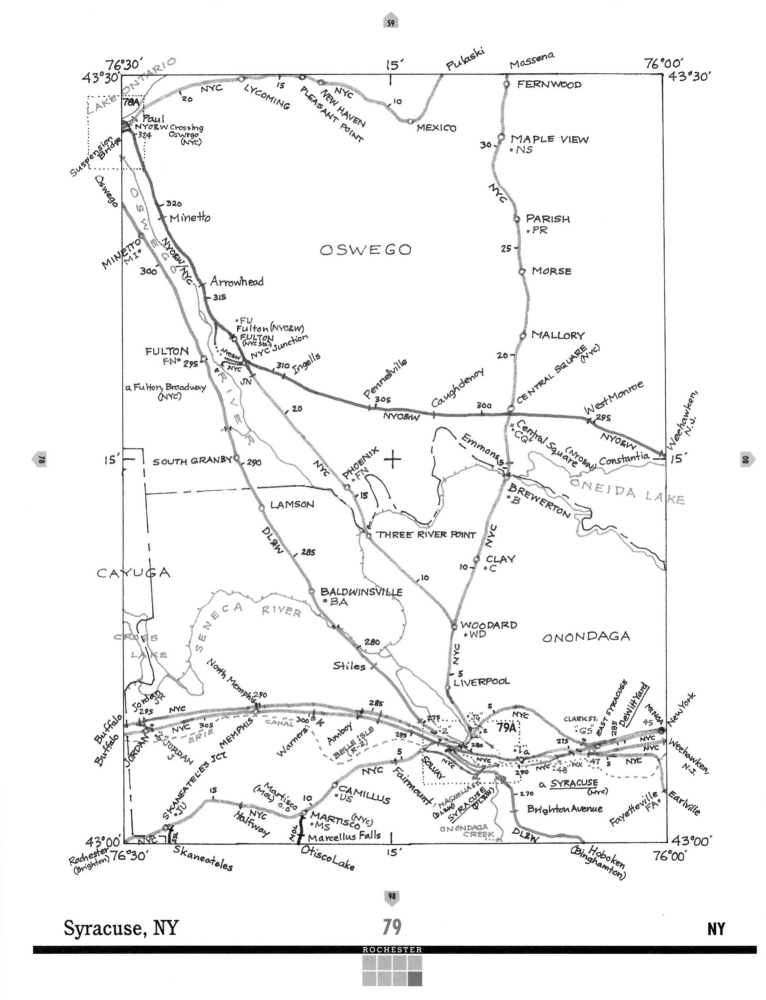

Syracuse, NY

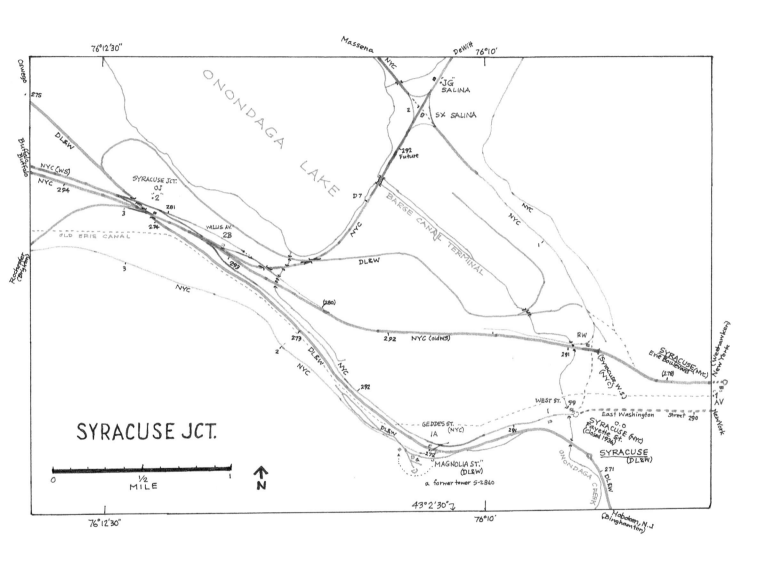

79A Syracuse Jct., NY

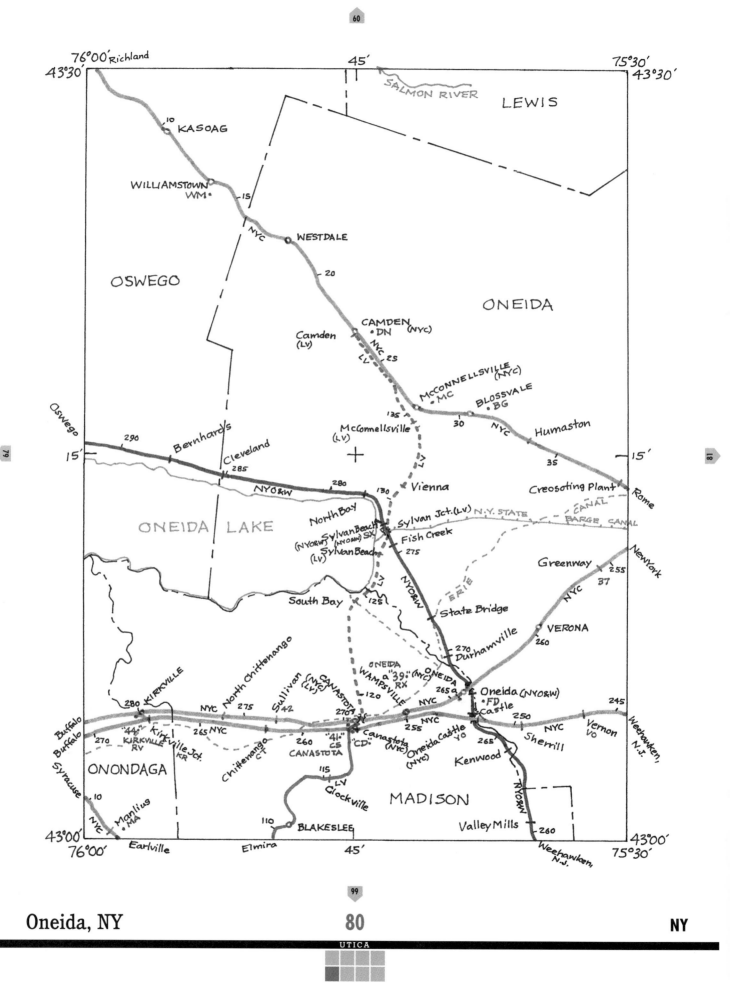

Oneida, NY

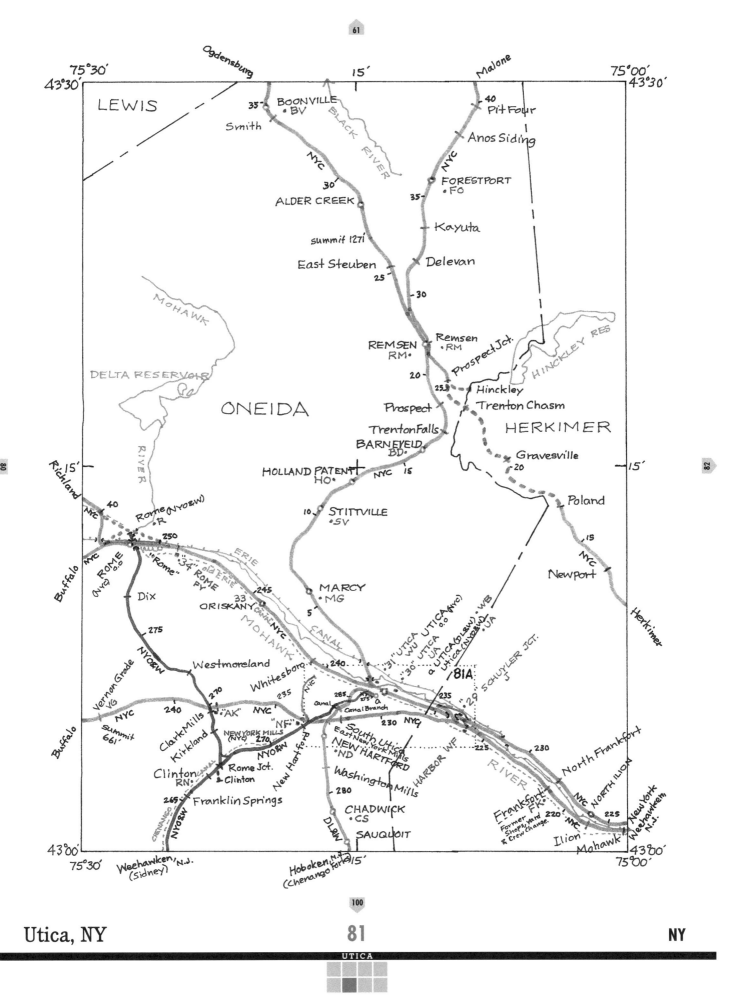

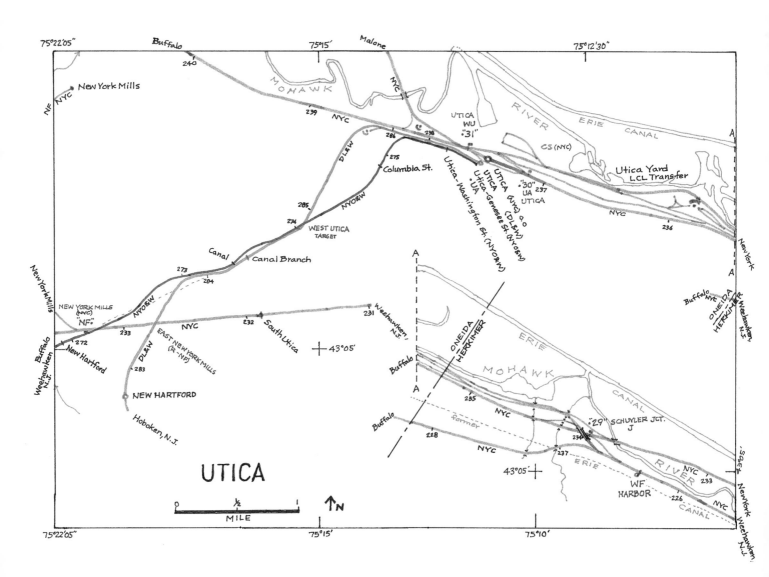

81A Utica, NY

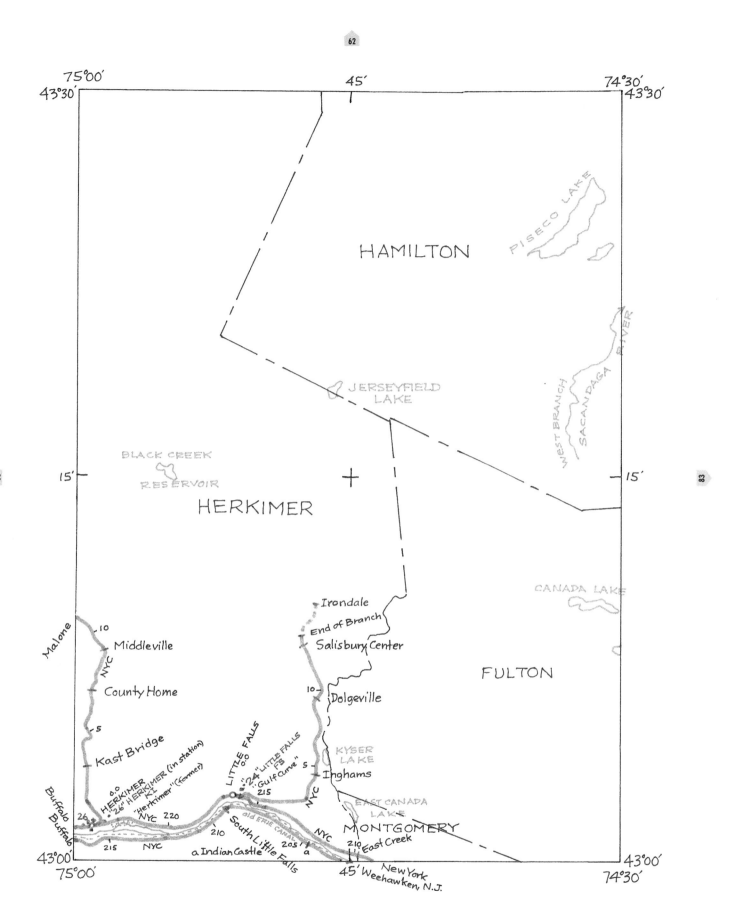

Little Falls, NY 82 NY

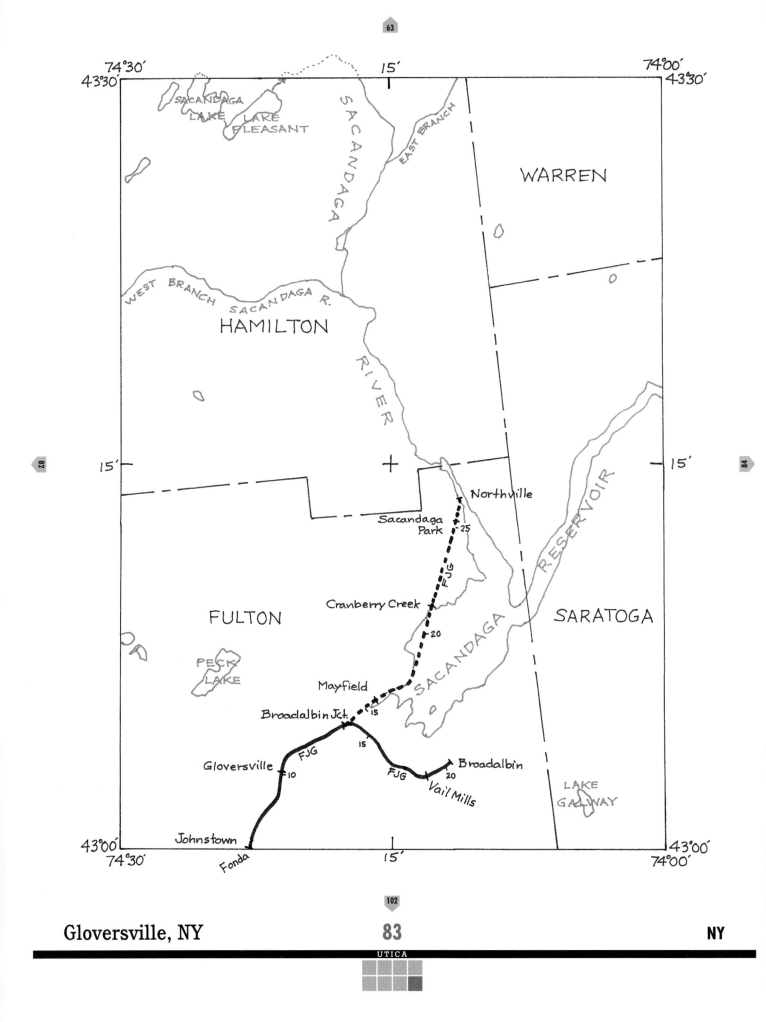

Gloversville, NY 83

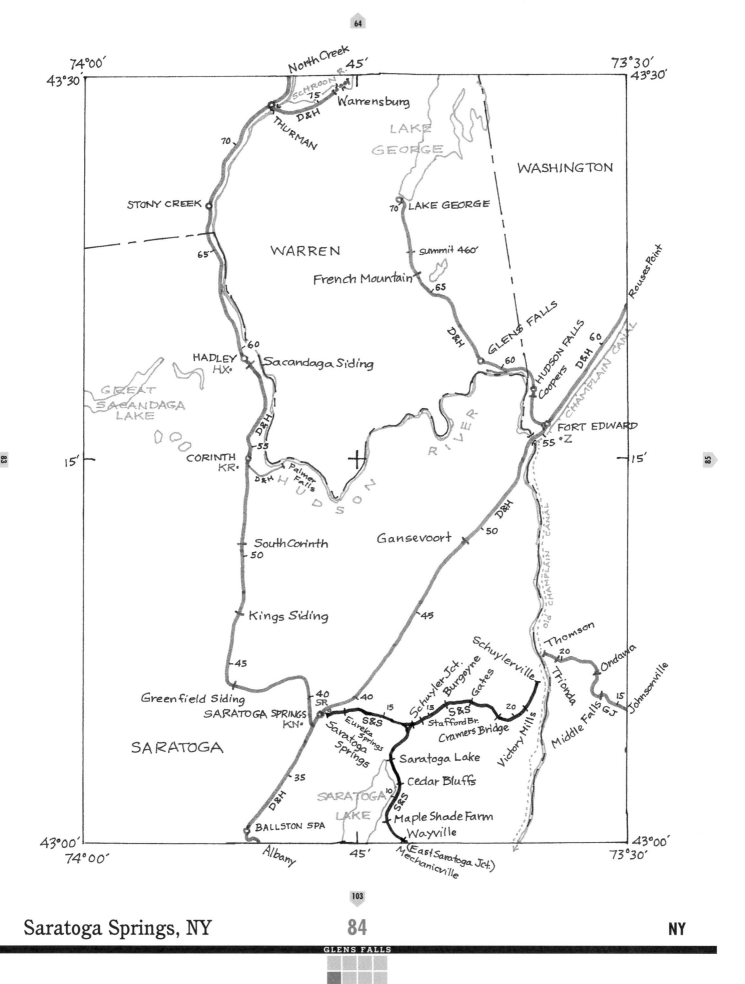

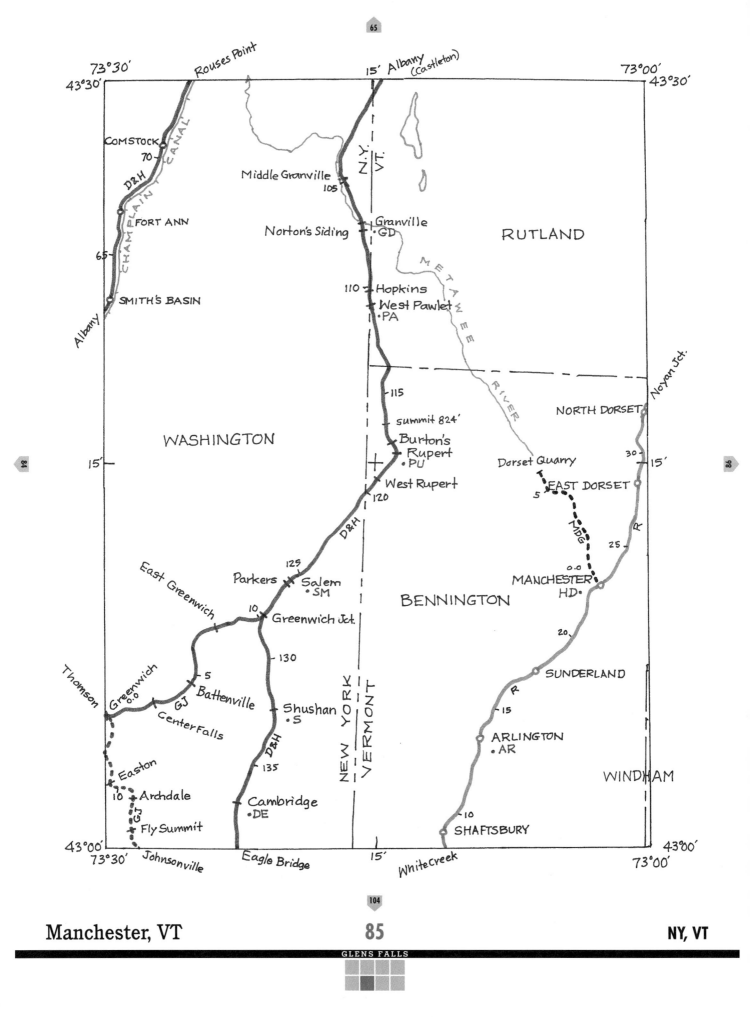

Manchester, VT — 85 — NY, VT

GLENS FALLS

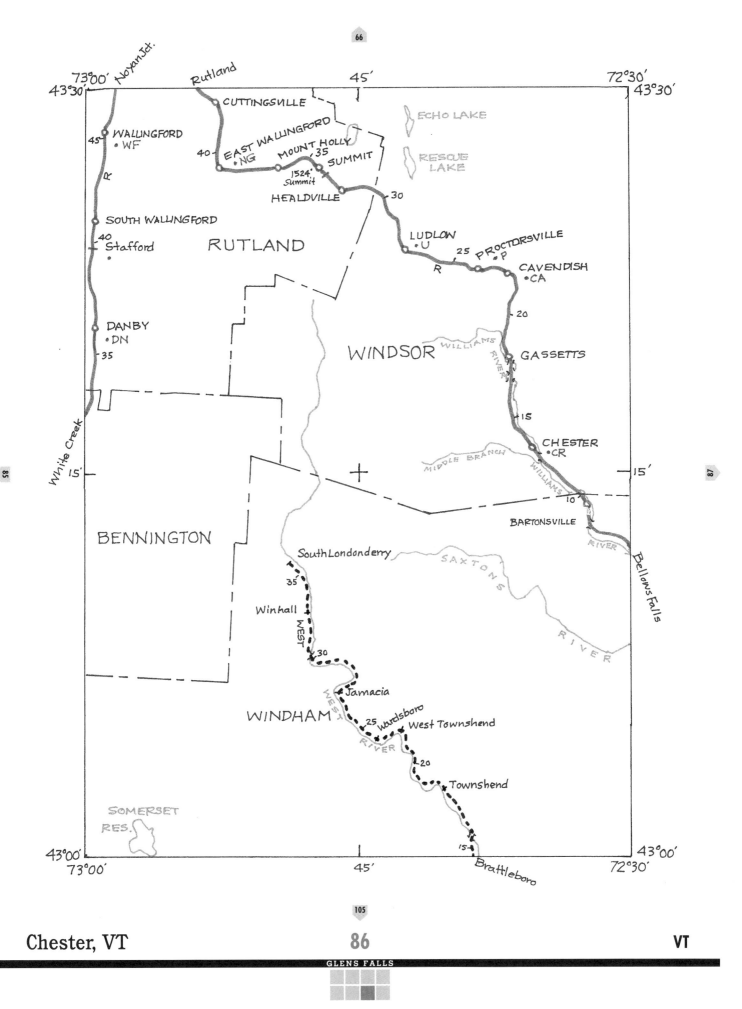

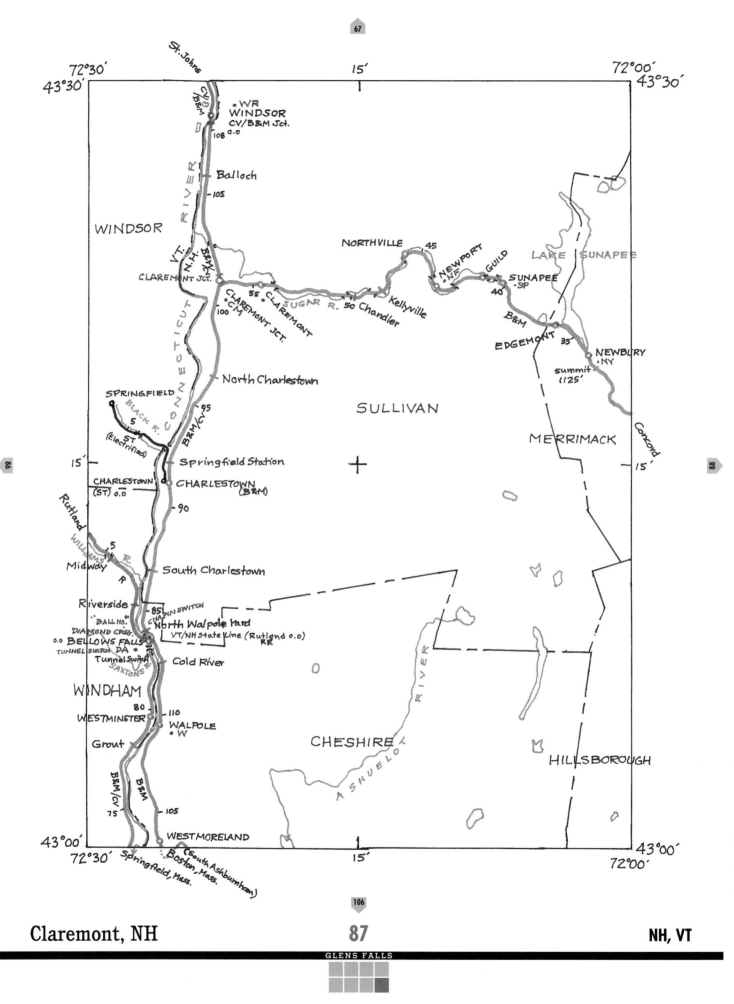

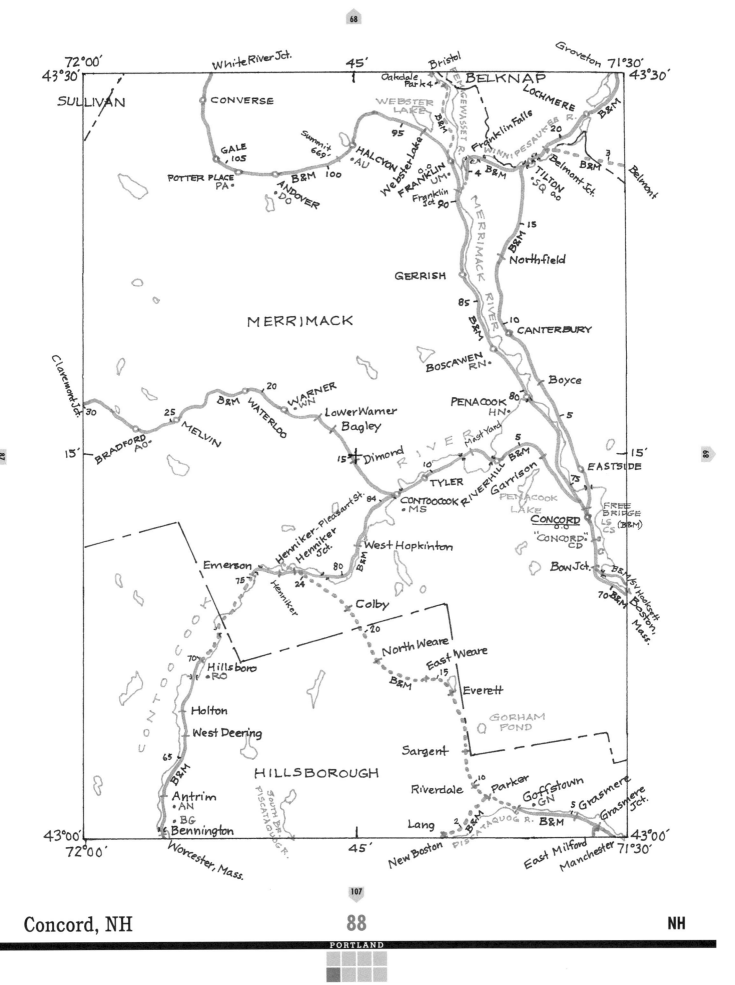

Concord, NH

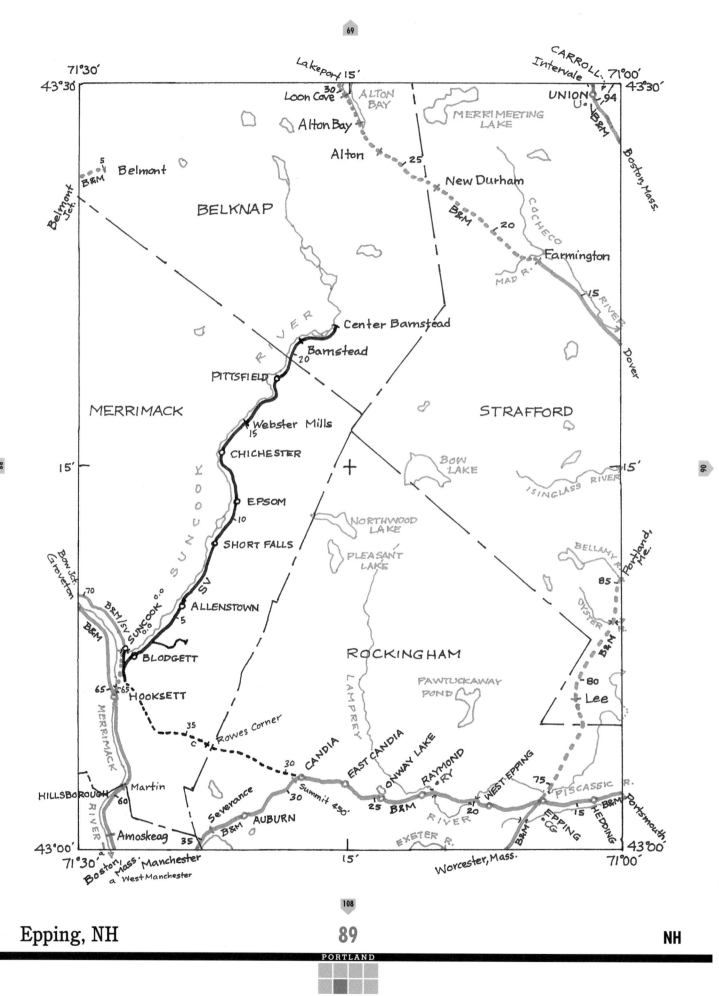

Epping, NH   89   NH

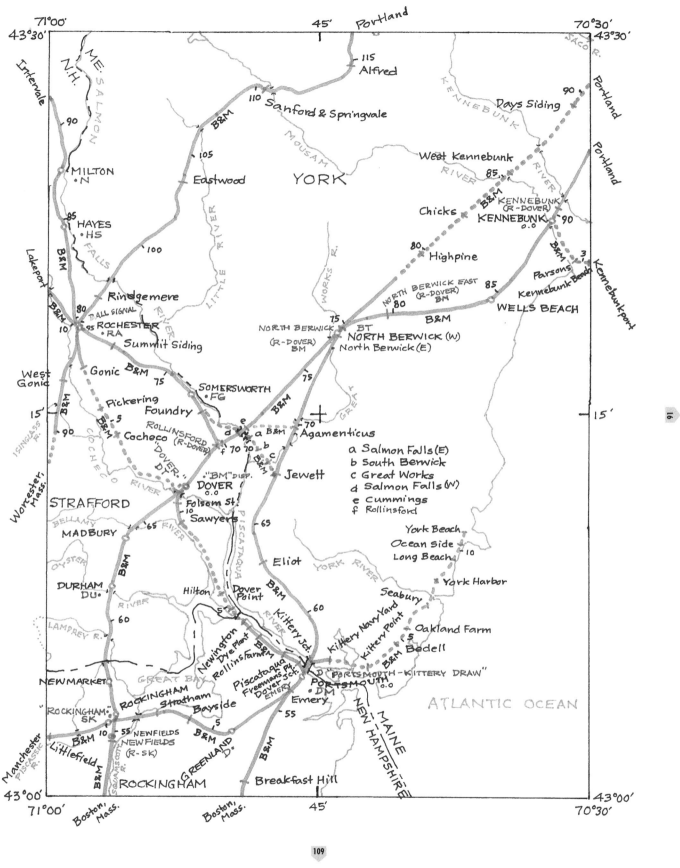

Dover, NH — ME, NH

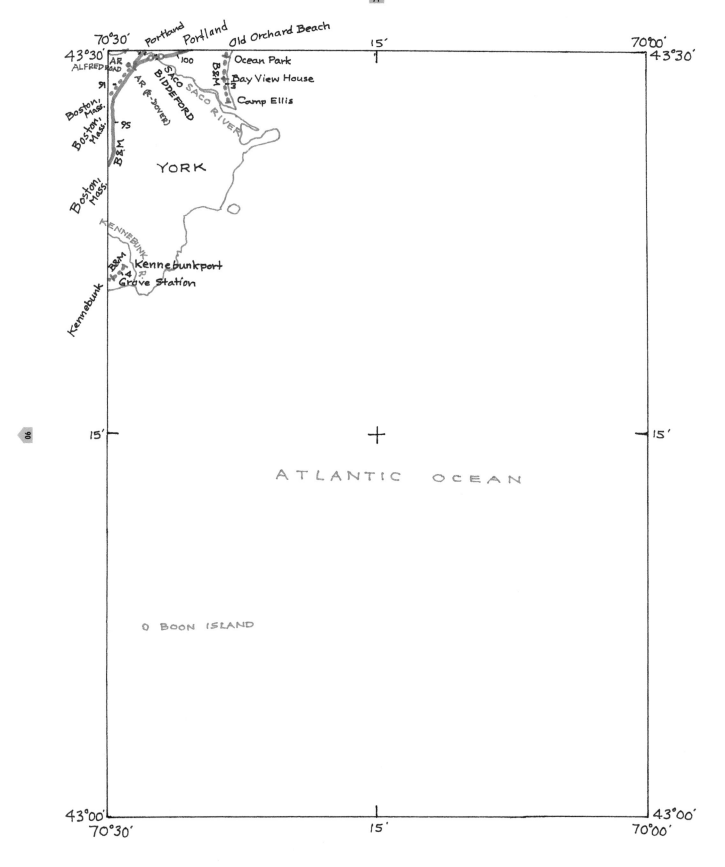

Biddeford, ME 91 ME

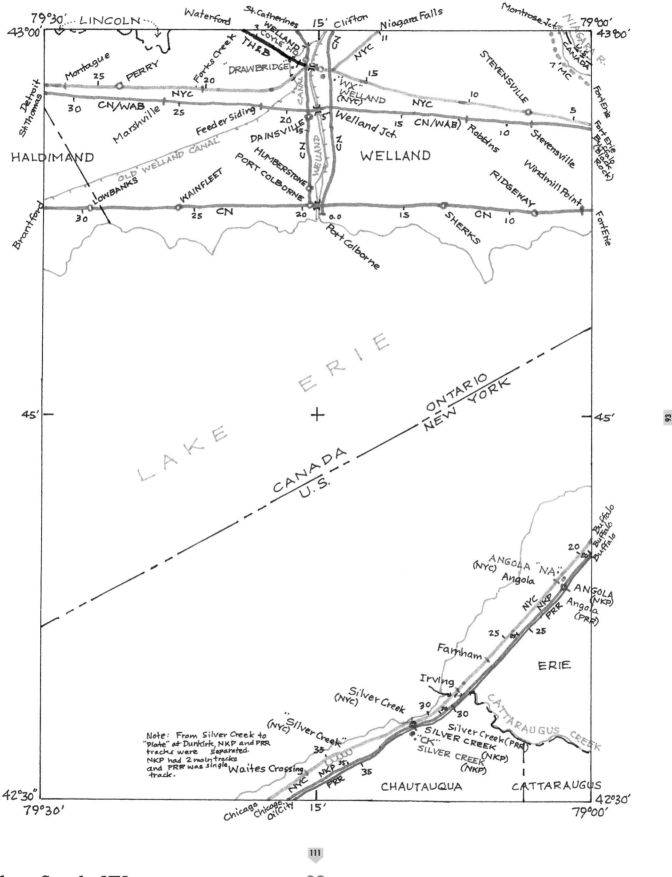

Silver Creek, NY   NY, ON

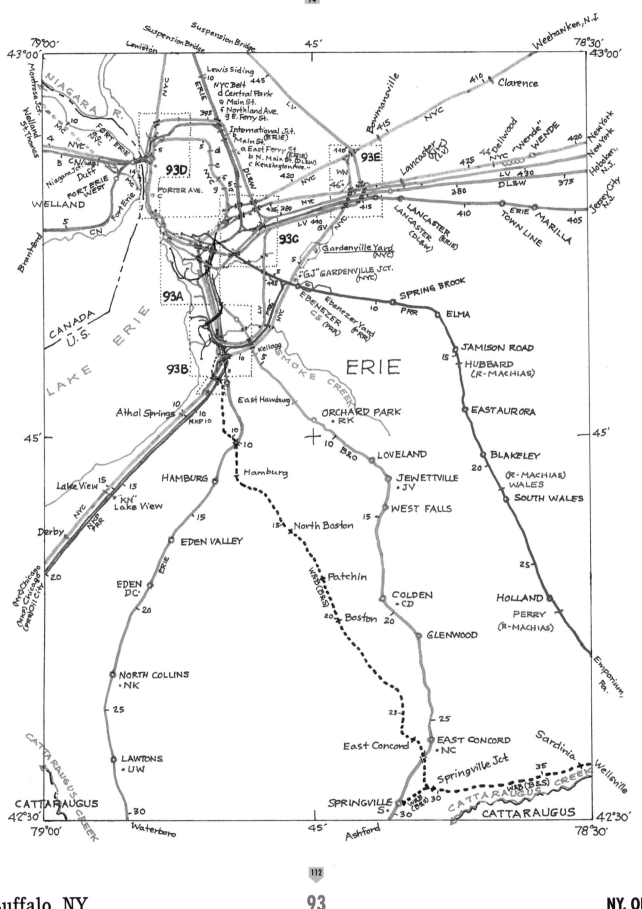

Buffalo, NY — 93 — NY, ON

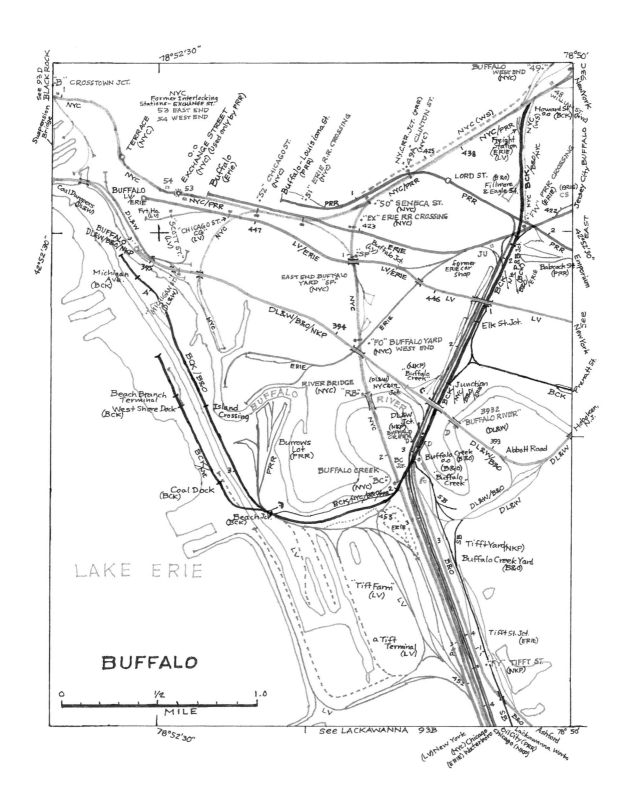

93A Buffalo, NY

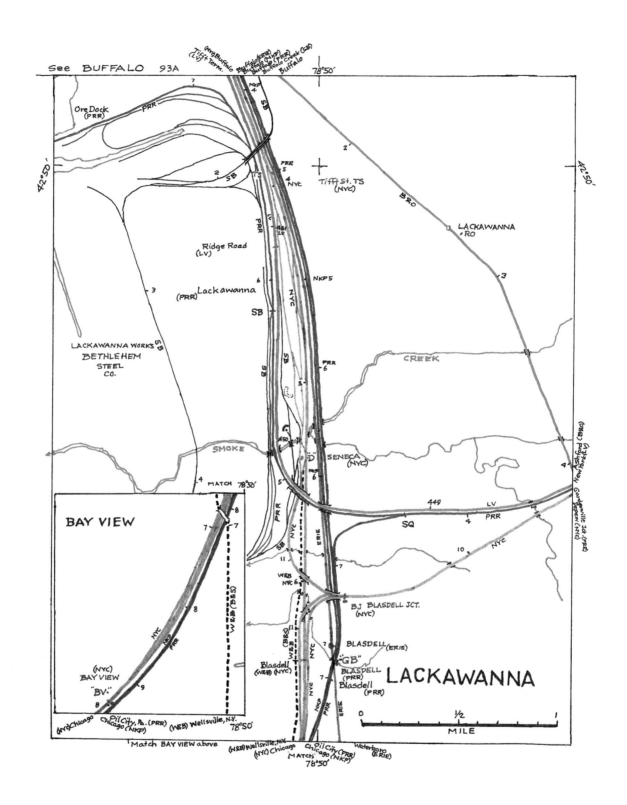

93B Lackawanna, NY

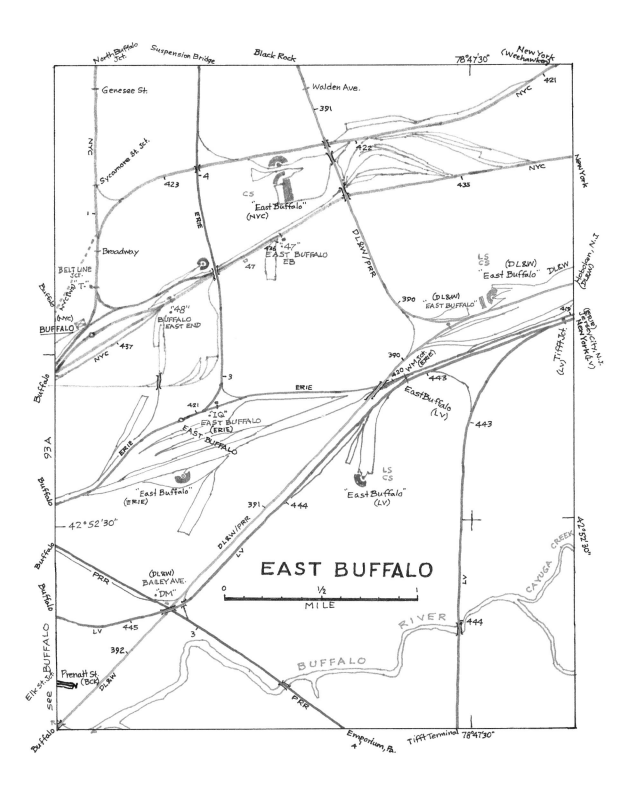

93C East Buffalo, NY

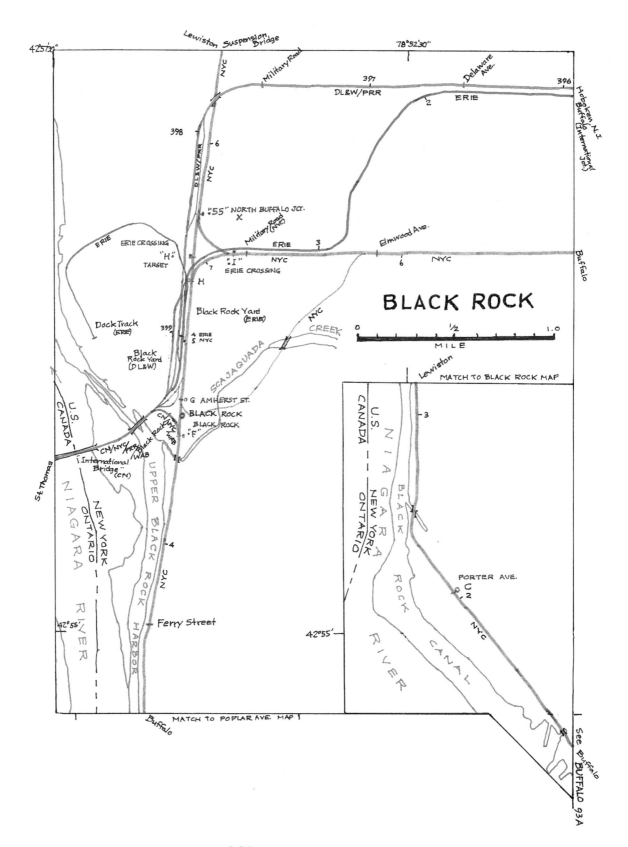

93D Black Rock, NY

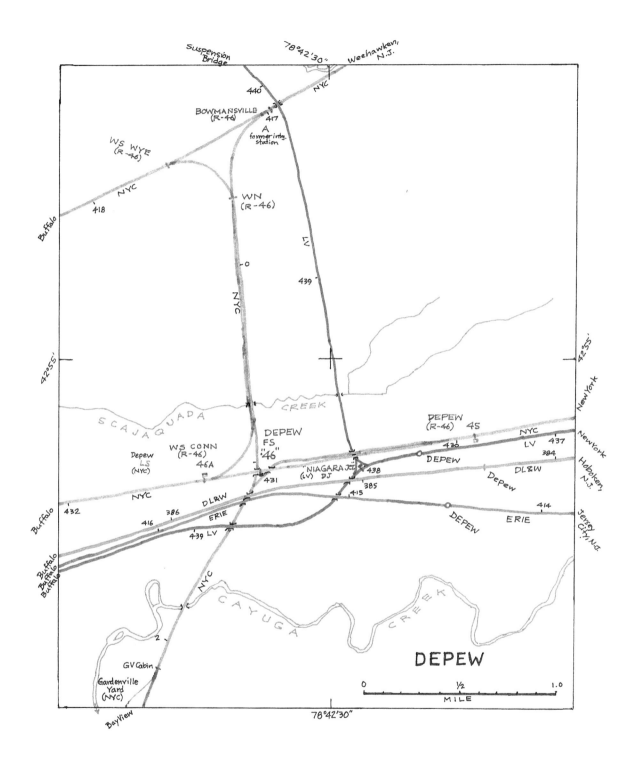

93E Depew, NY

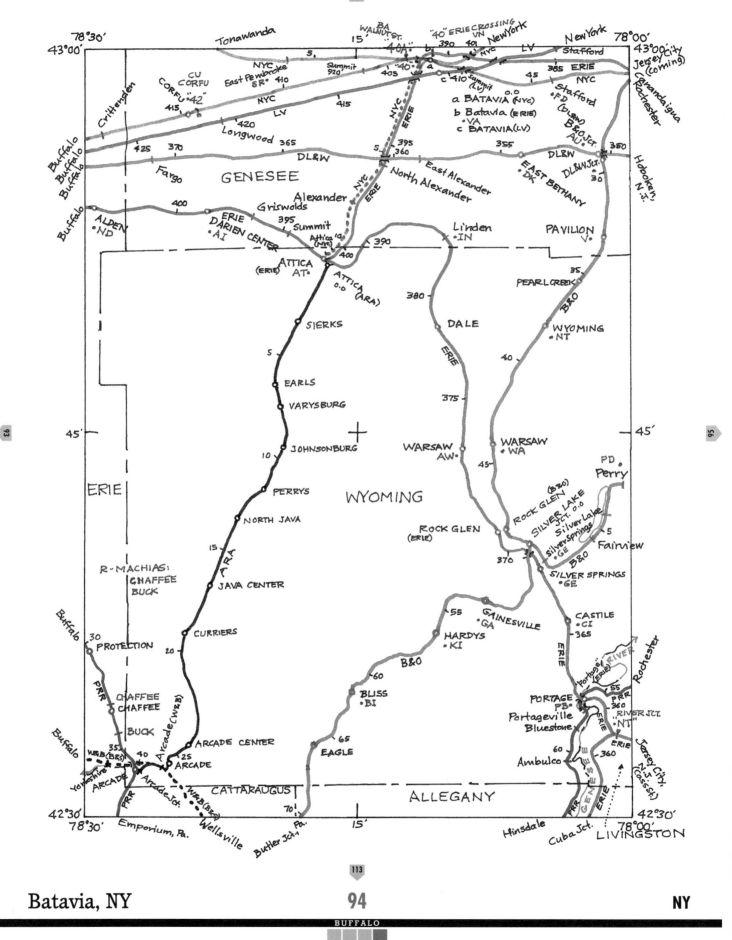

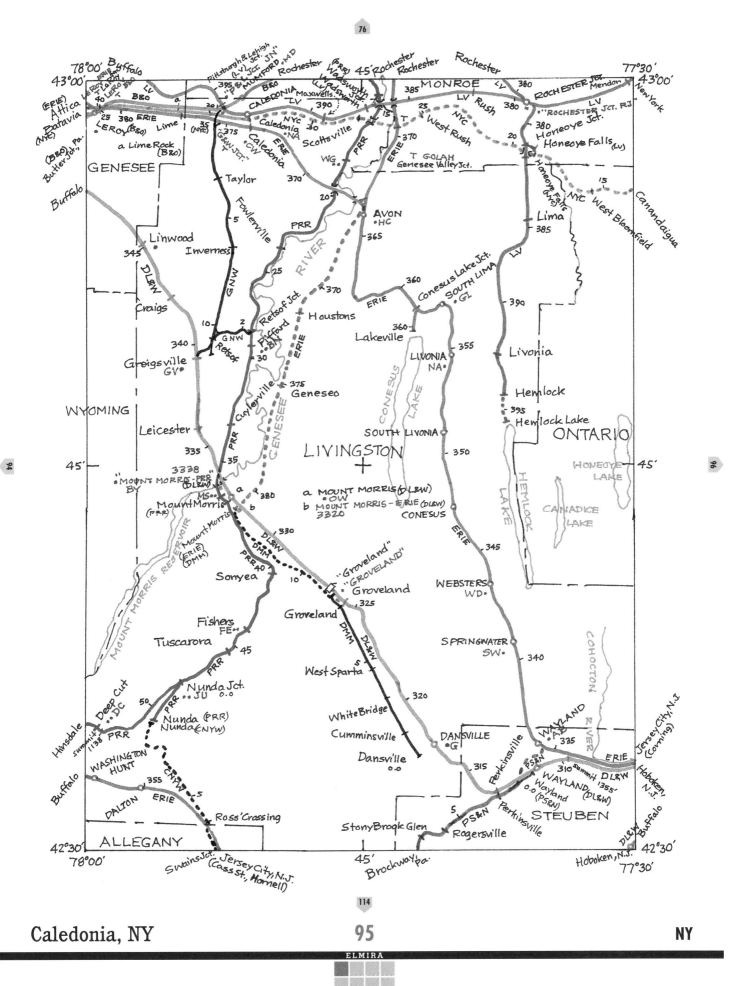

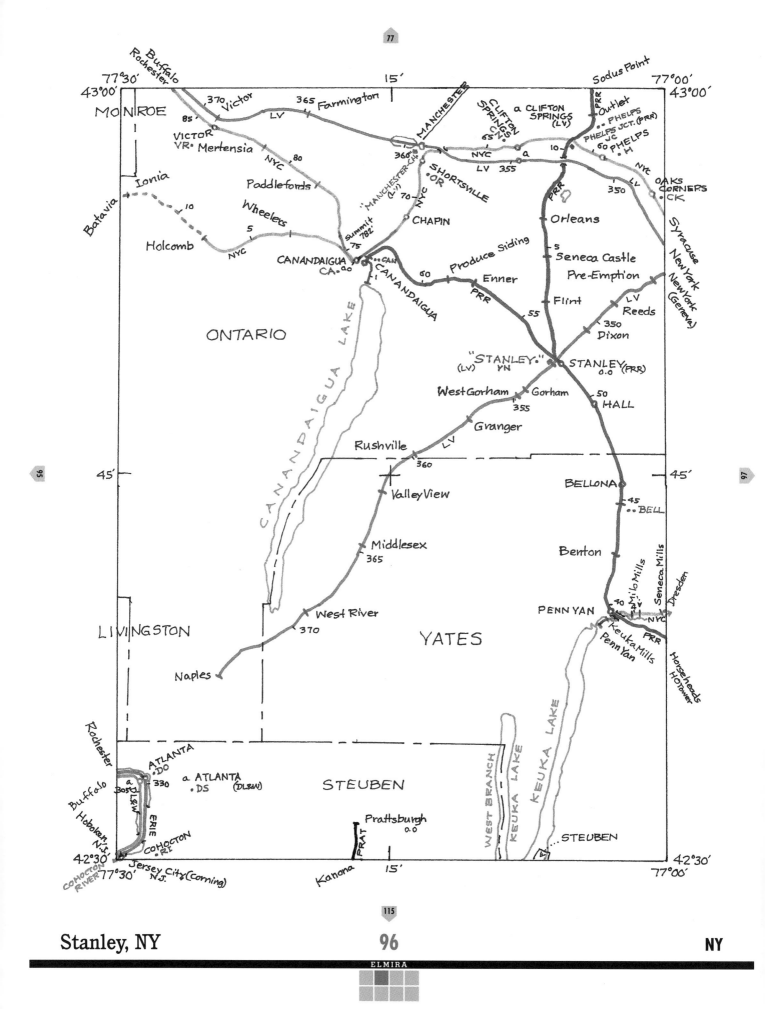

Stanley, NY

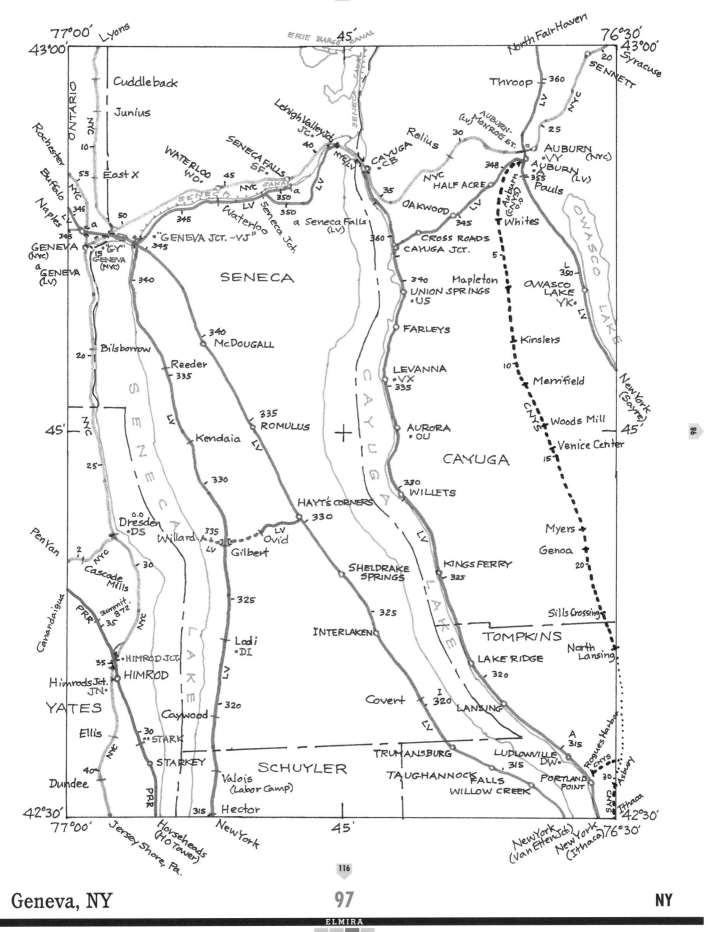

Geneva, NY

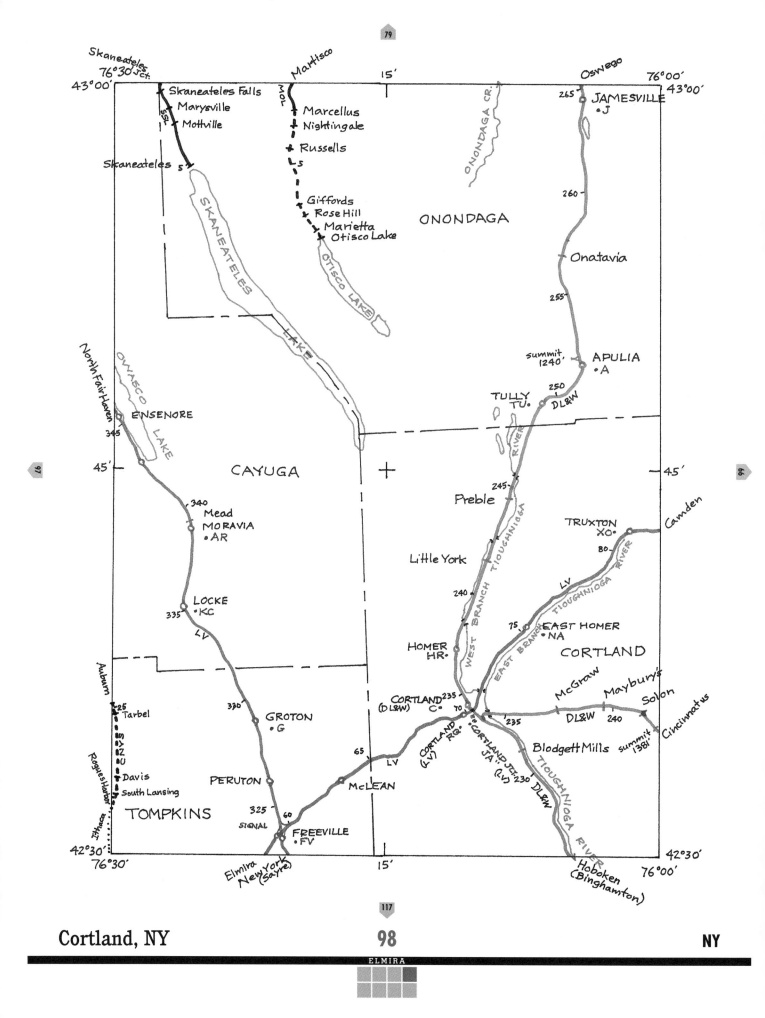

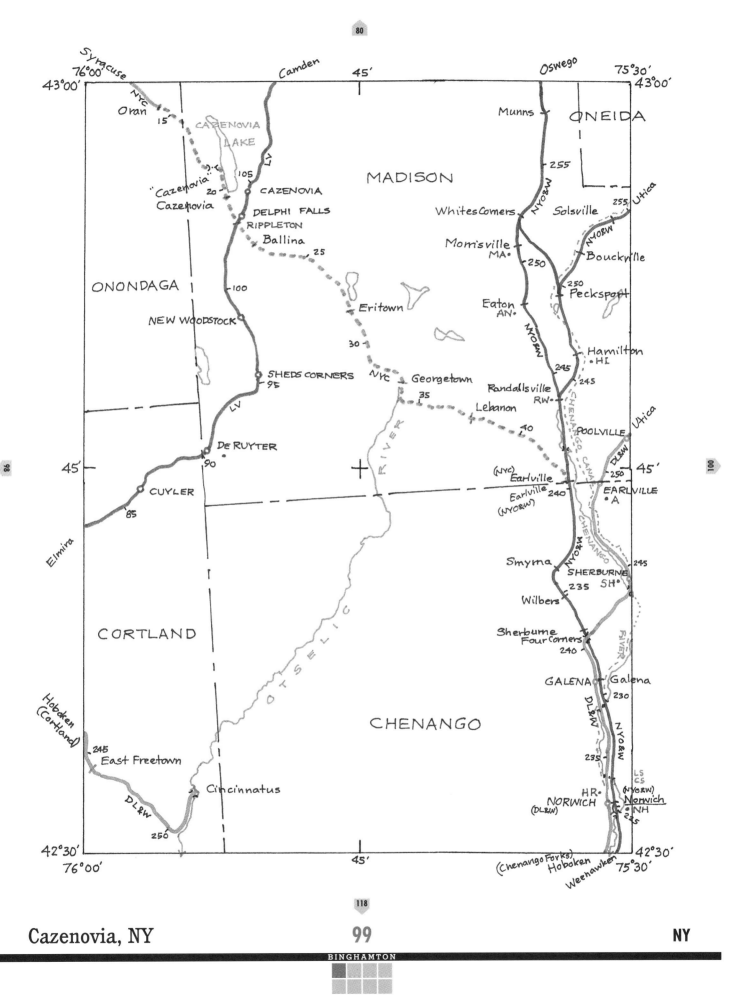

Cazenovia, NY

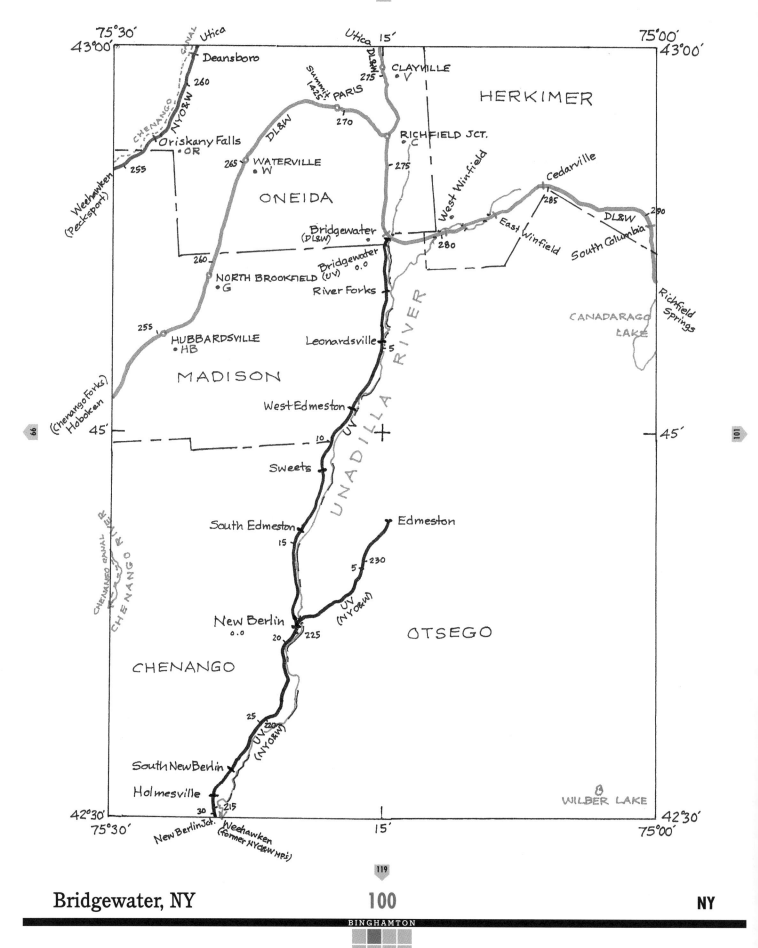

Bridgewater, NY

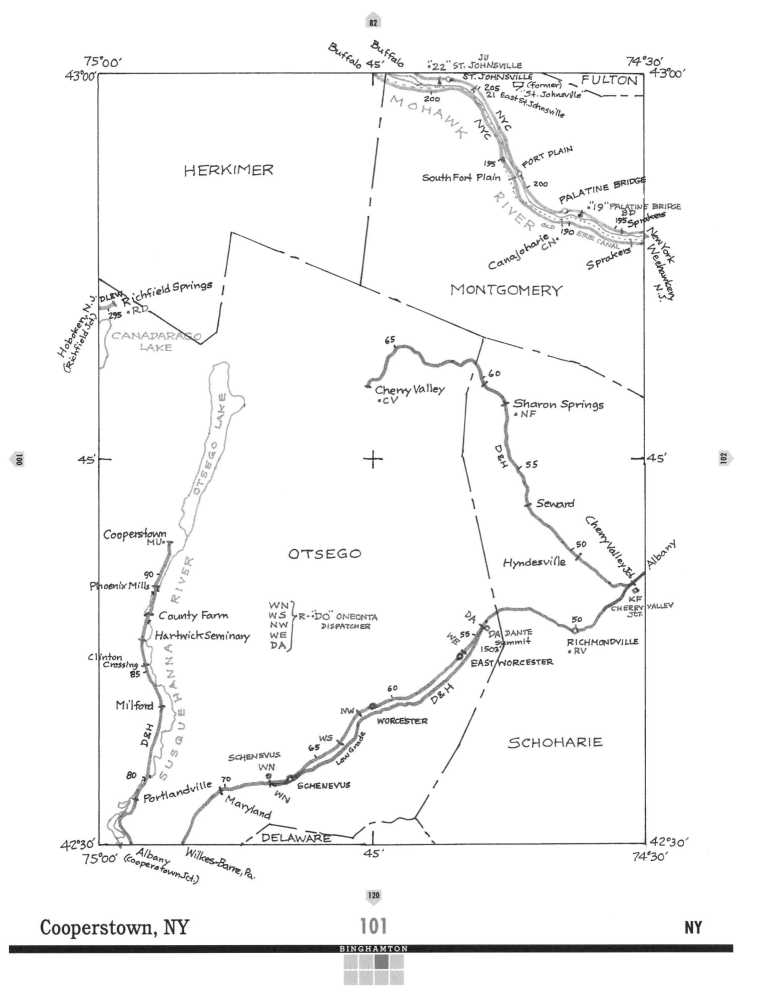

Cooperstown, NY

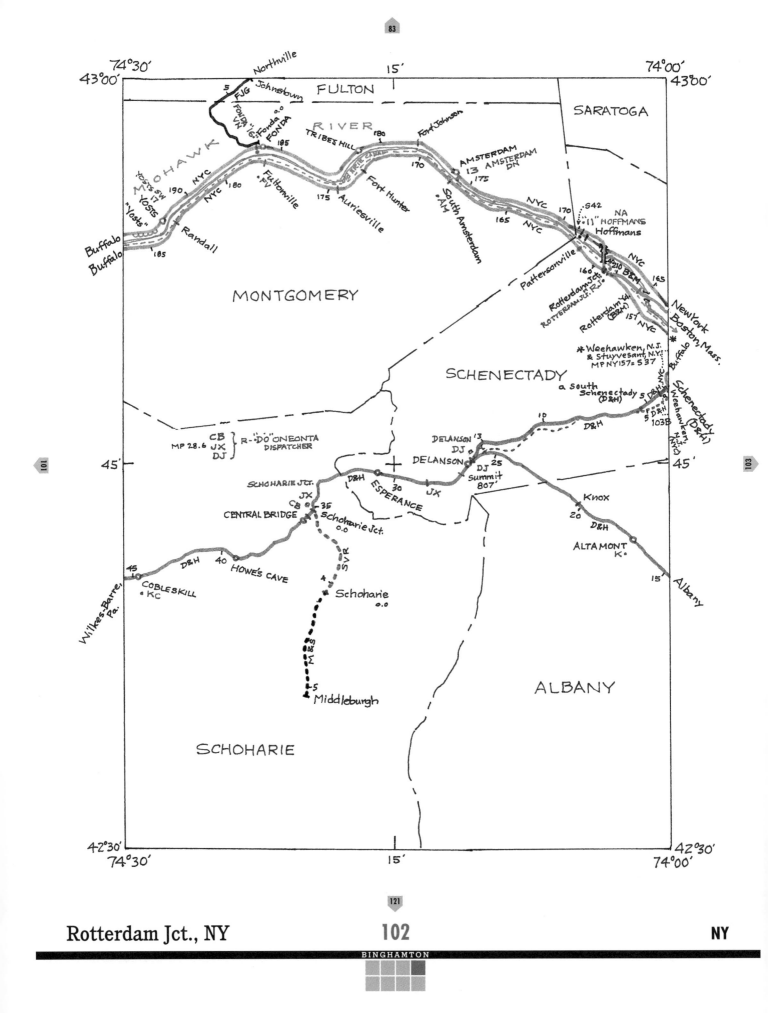

Rotterdam Jct., NY   102   NY

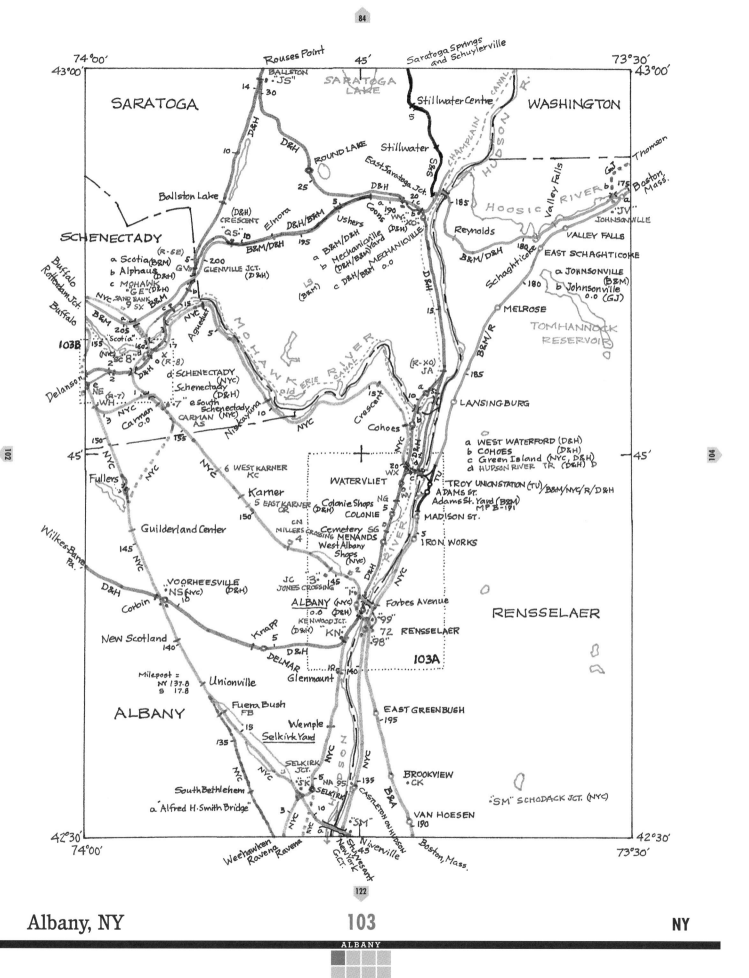

Albany, NY — 103 — NY

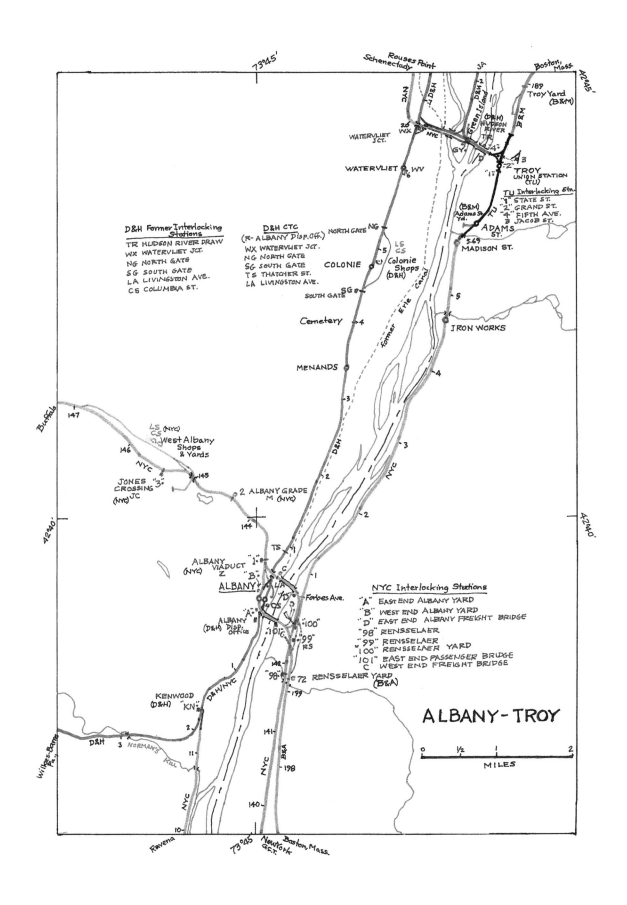

103A Albany–Troy, NY

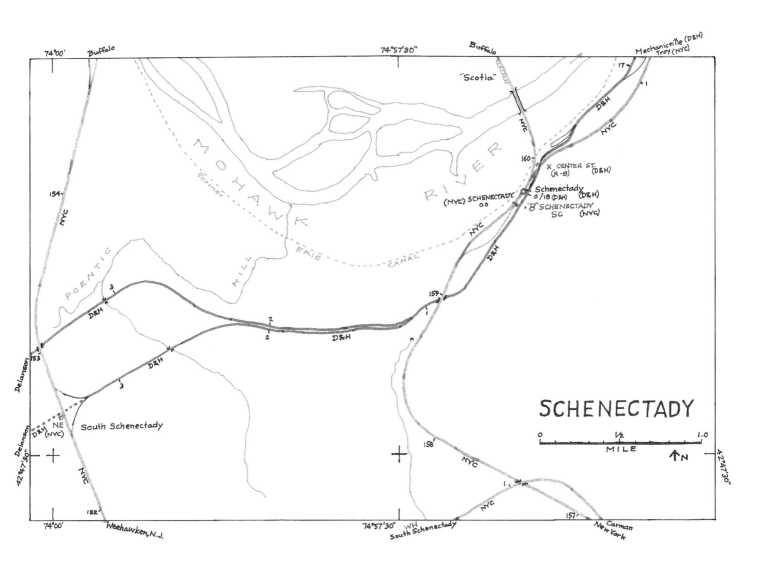

103B Schenectady, NY

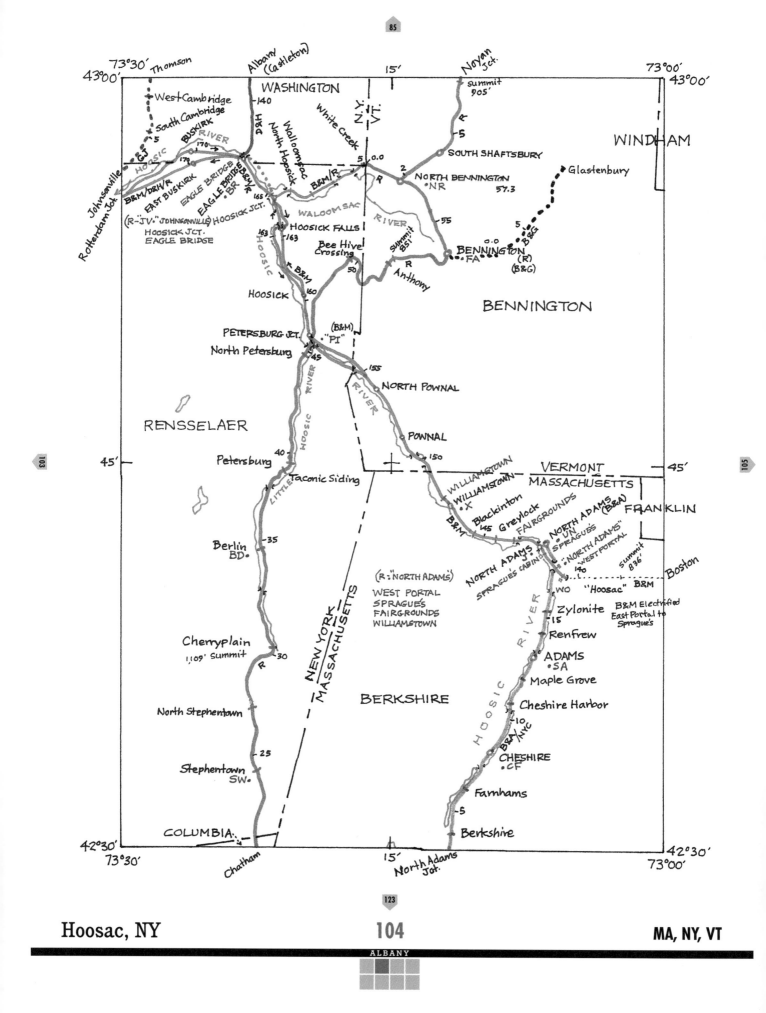

Hoosac, NY — 104 — MA, NY, VT

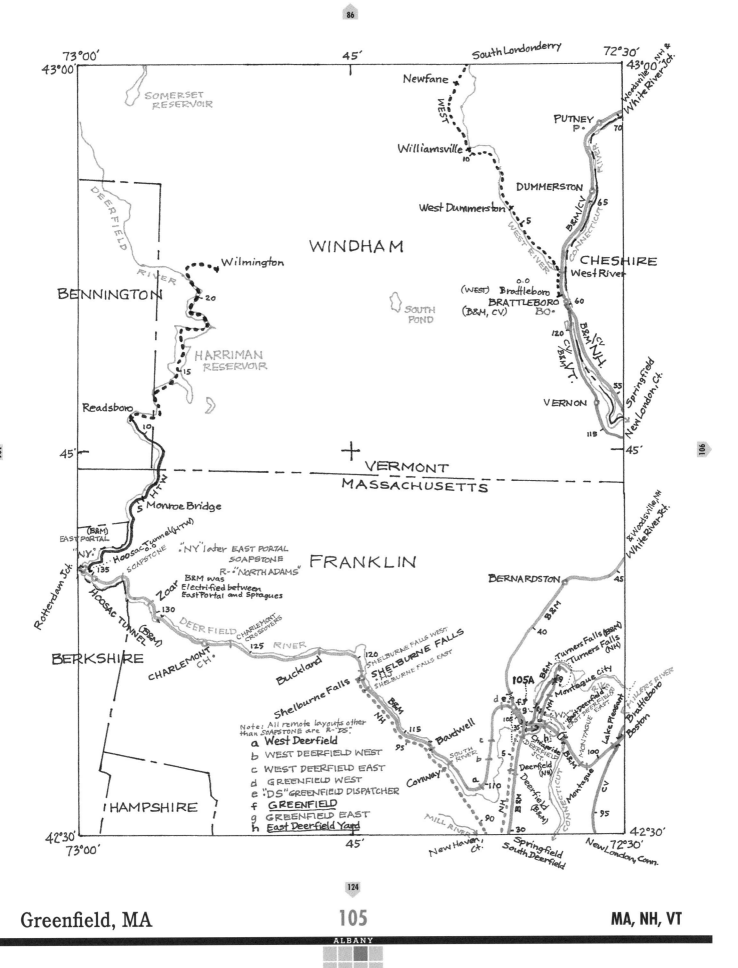

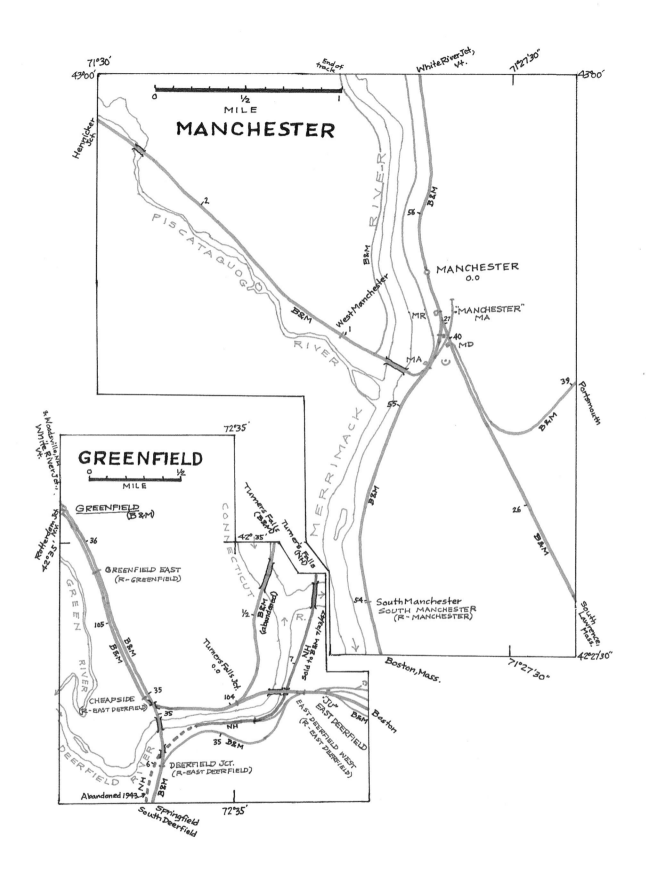

105A Greenfield, MA **108A** Manchester, NH

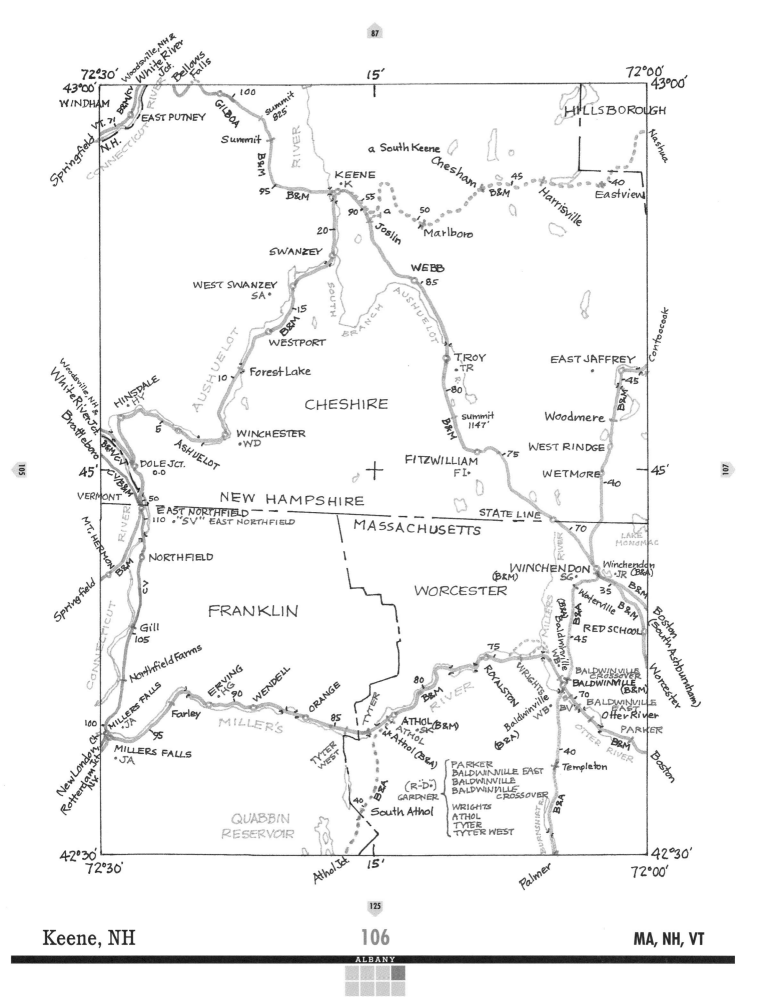

Keene, NH — 106 — MA, NH, VT

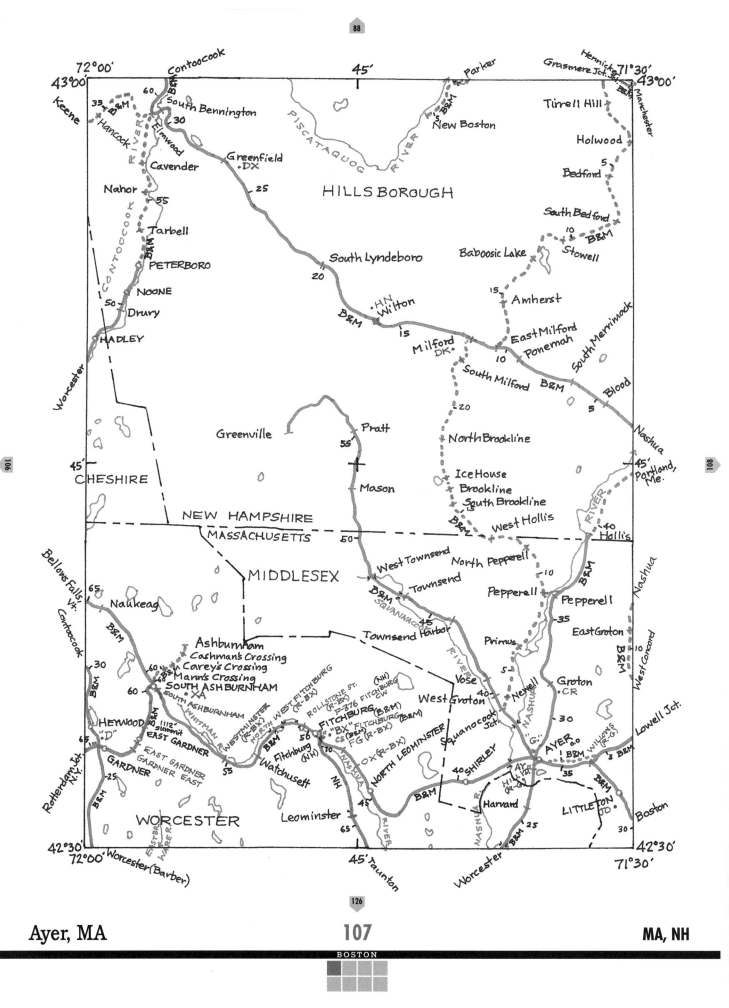

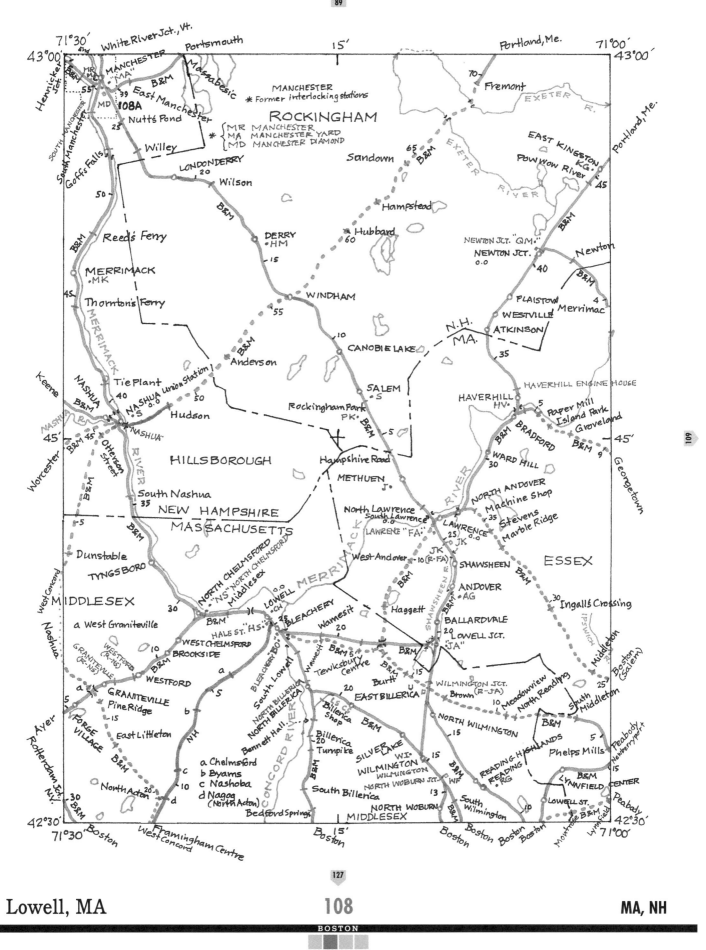

Lowell, MA — 108 — MA, NH

for detail, see map page 105A

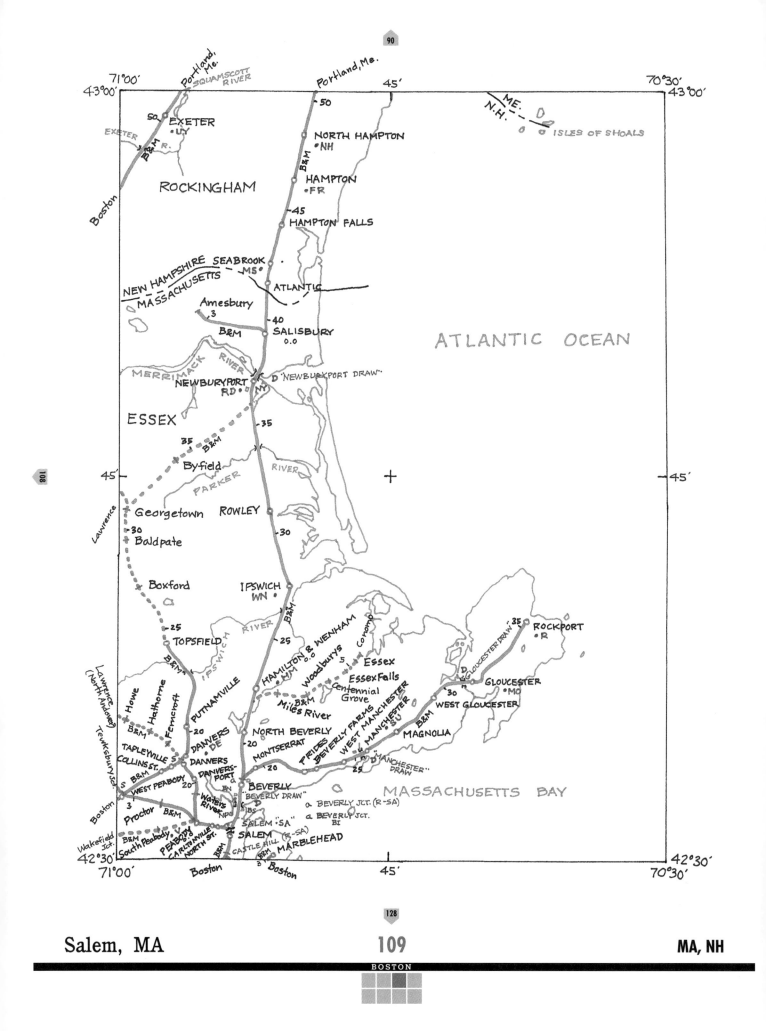

Salem, MA

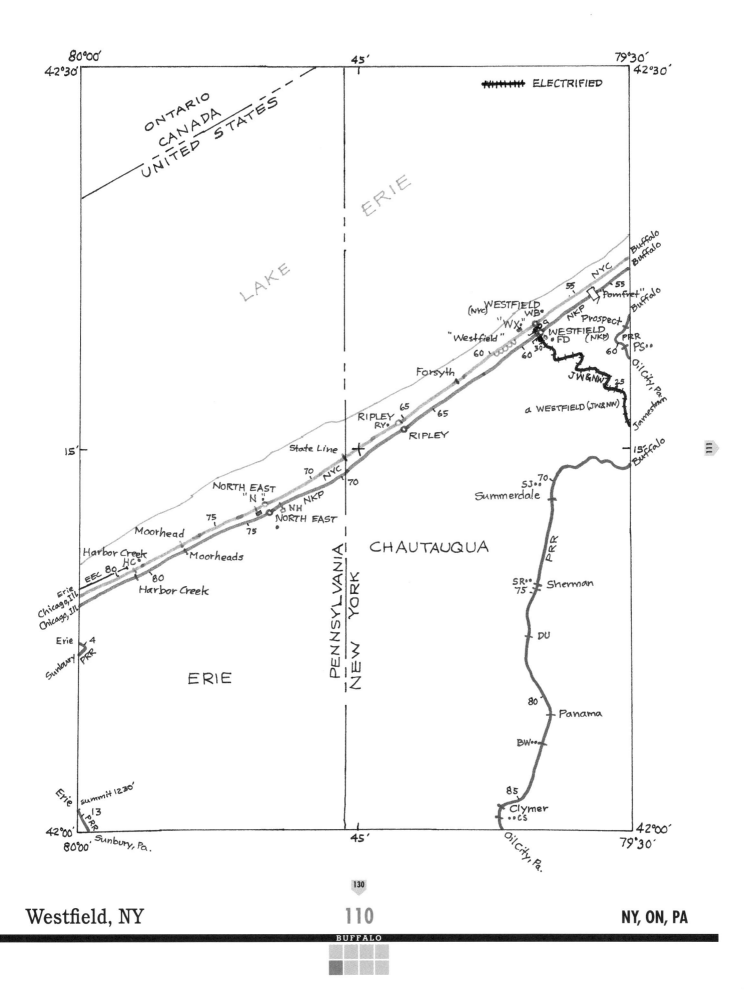

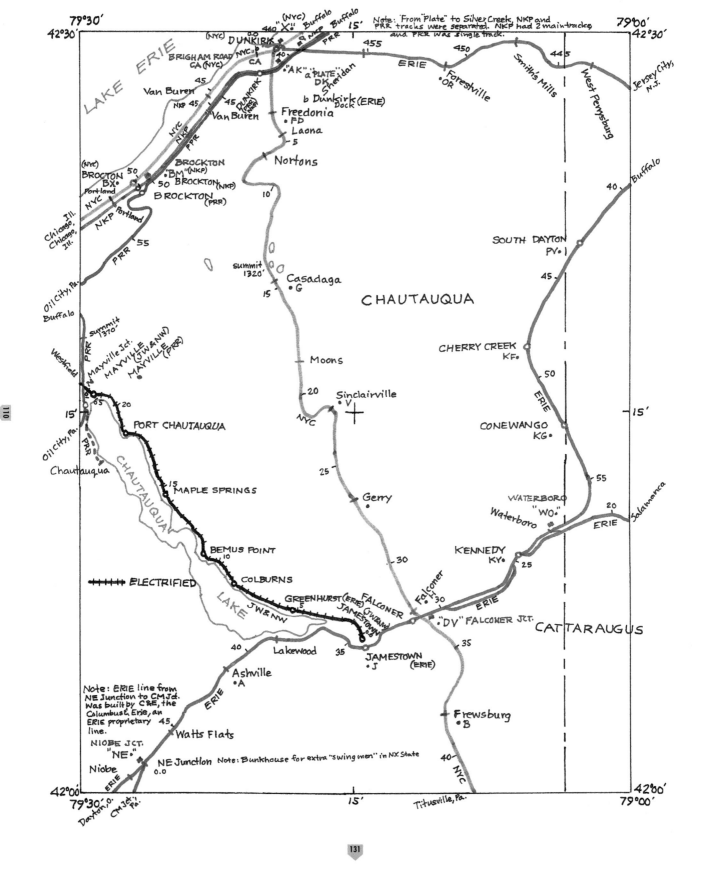

Dunkirk, NY   111   NY

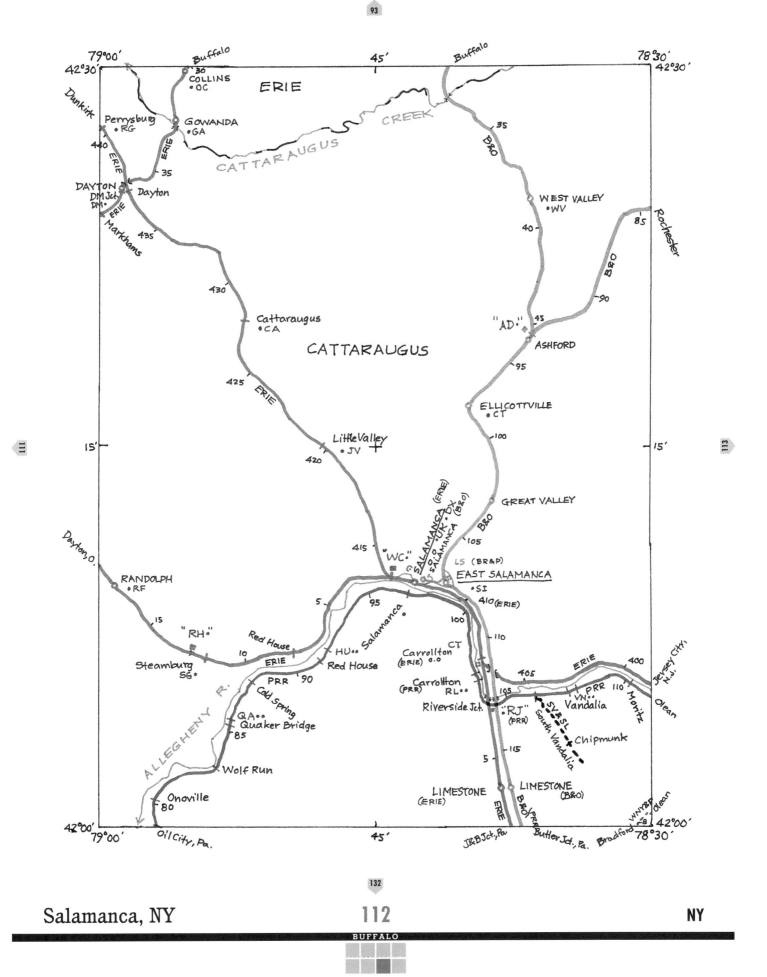

Salamanca, NY

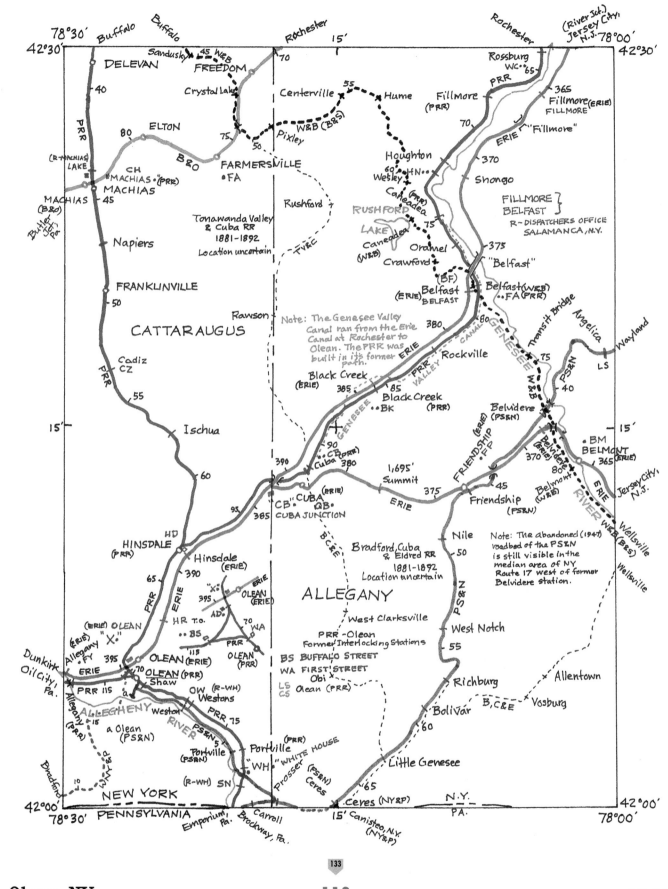

Olean, NY — 113 — NY, PA

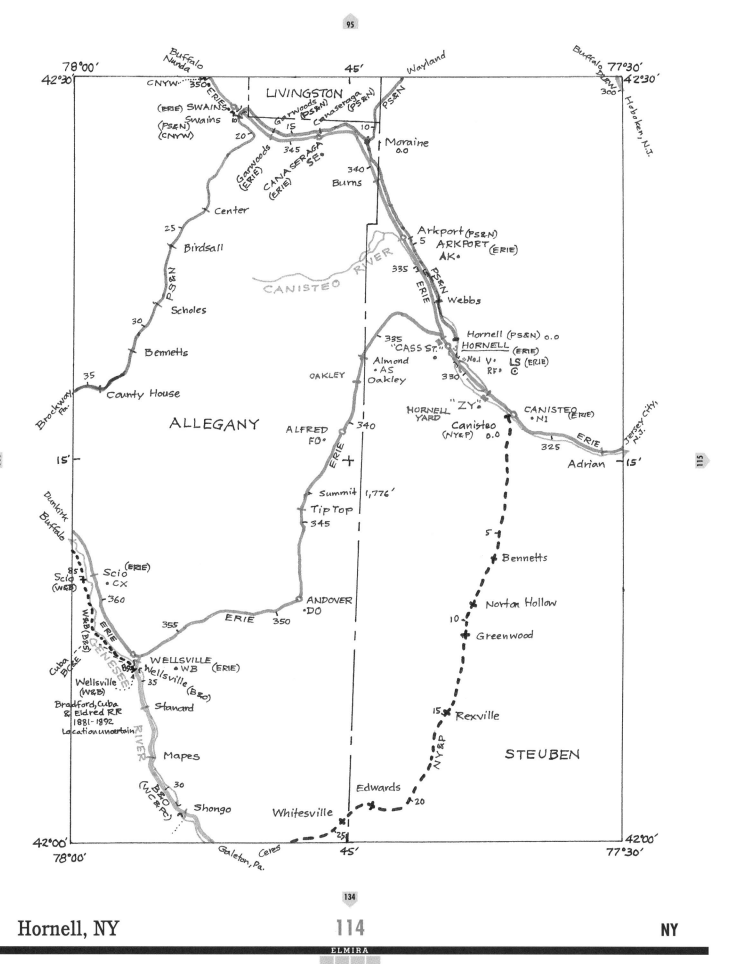

Hornell, NY

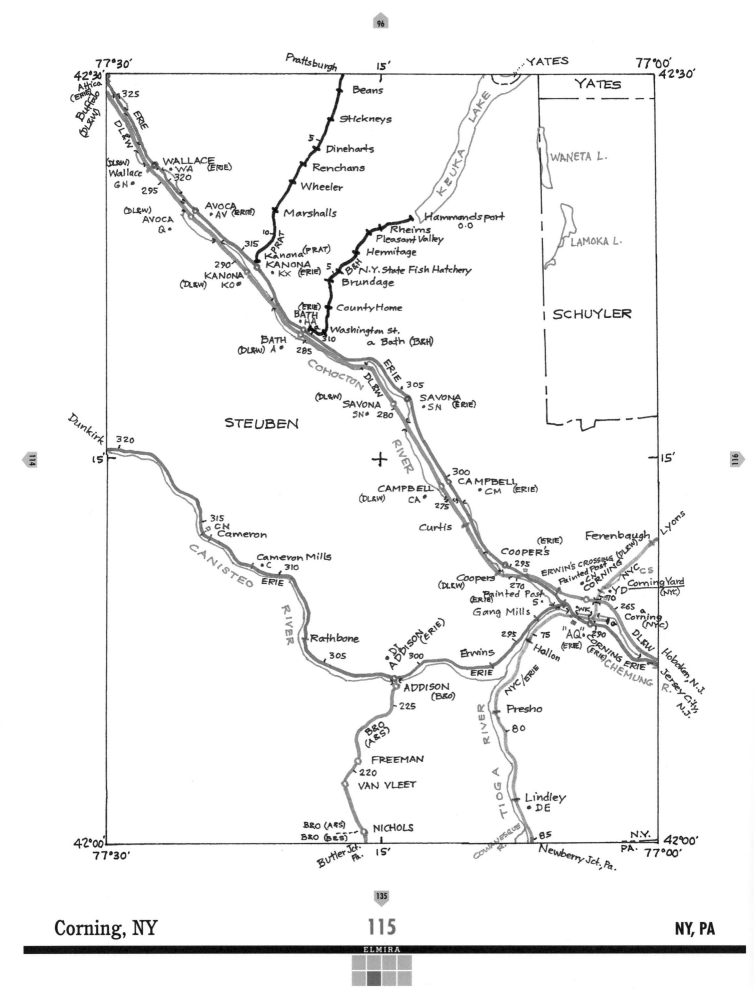

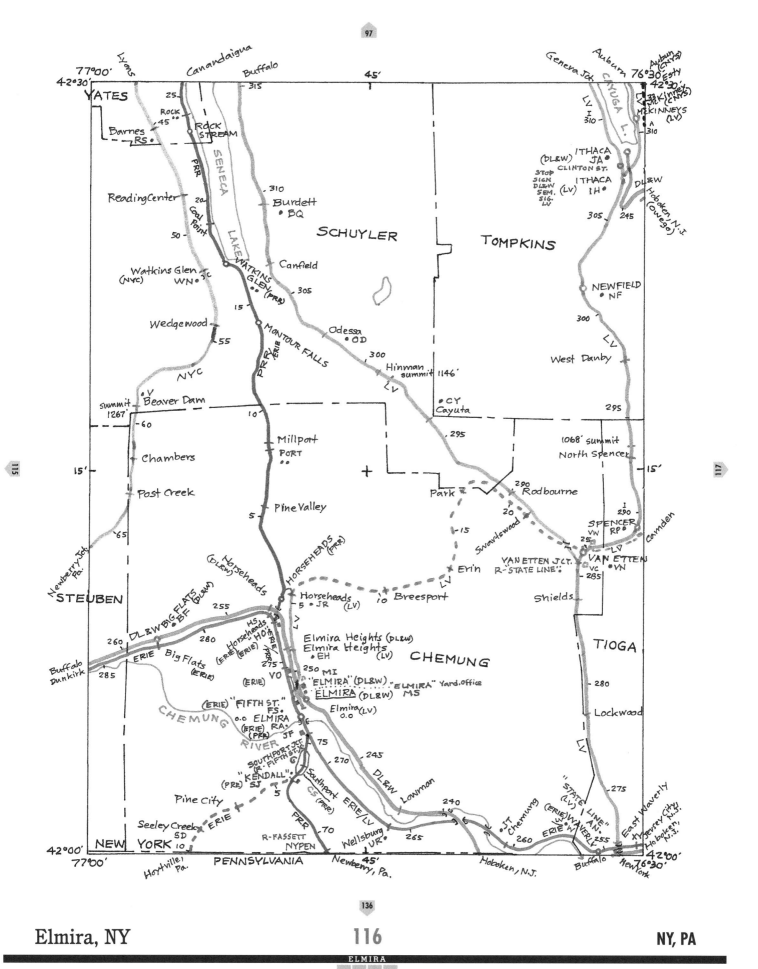

Elmira, NY

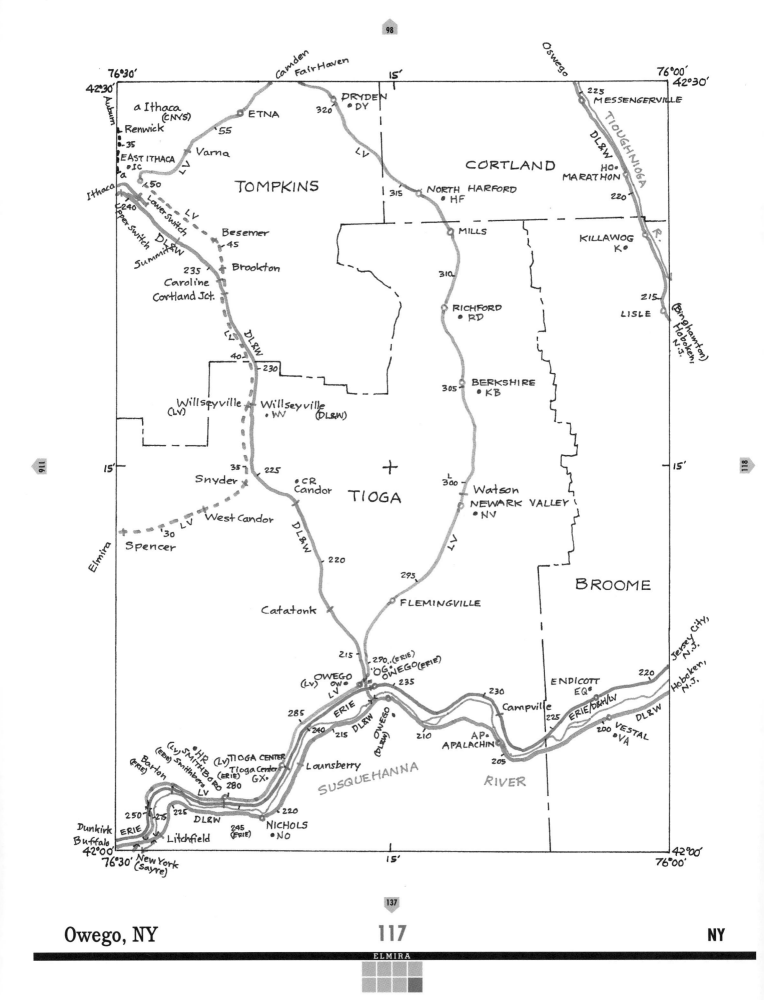

Owego, NY

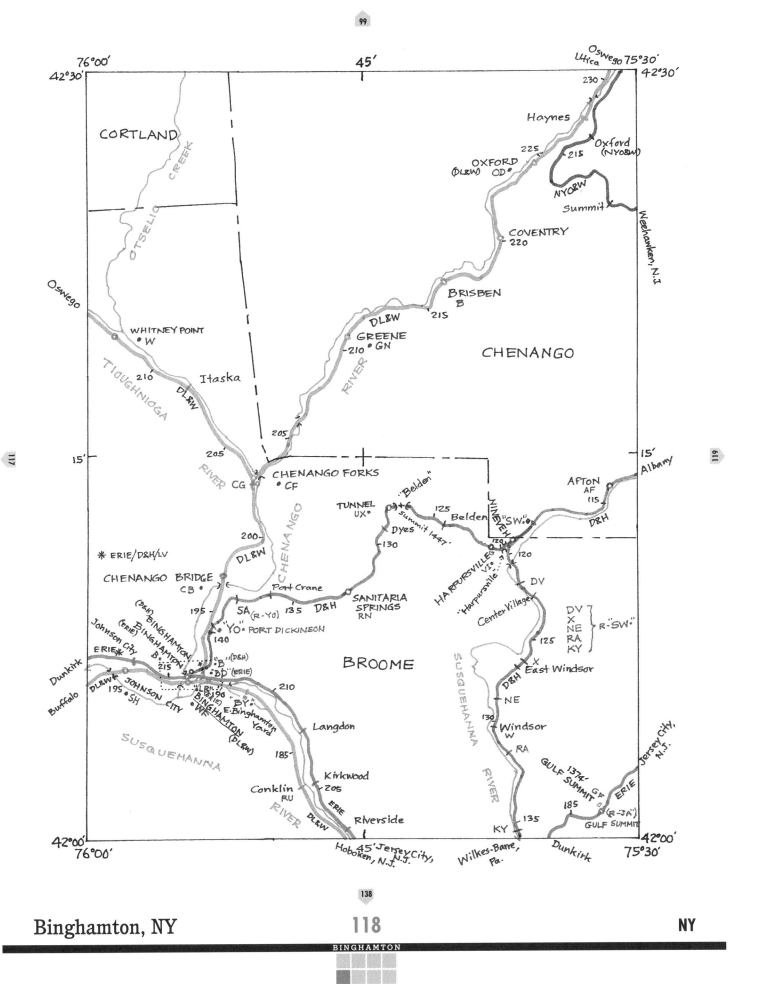

Binghamton, NY

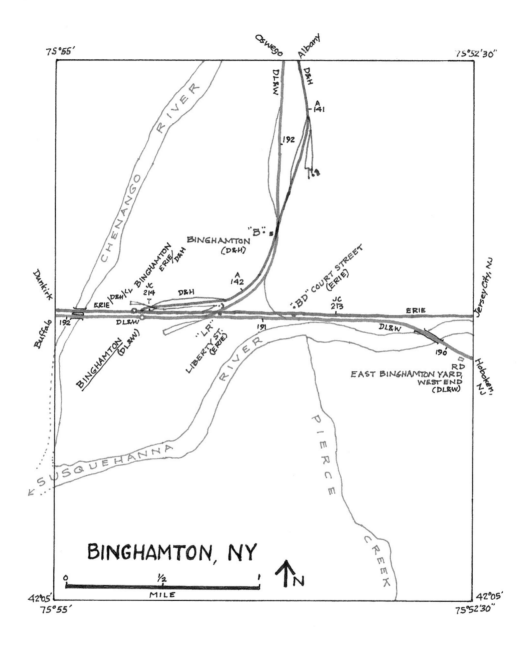

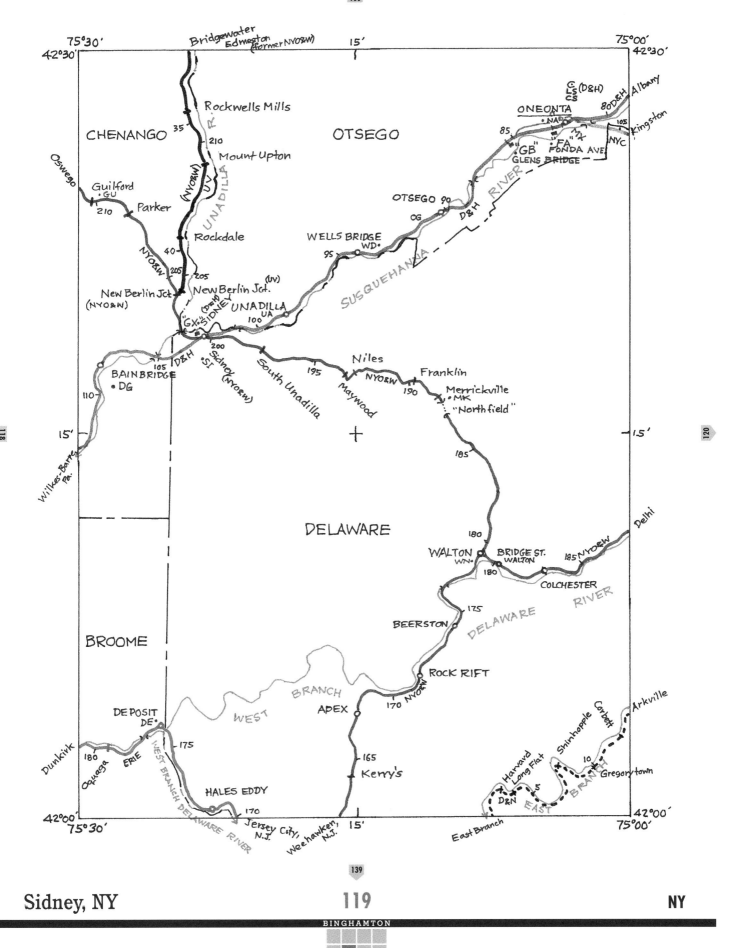

Sidney, NY   119   NY

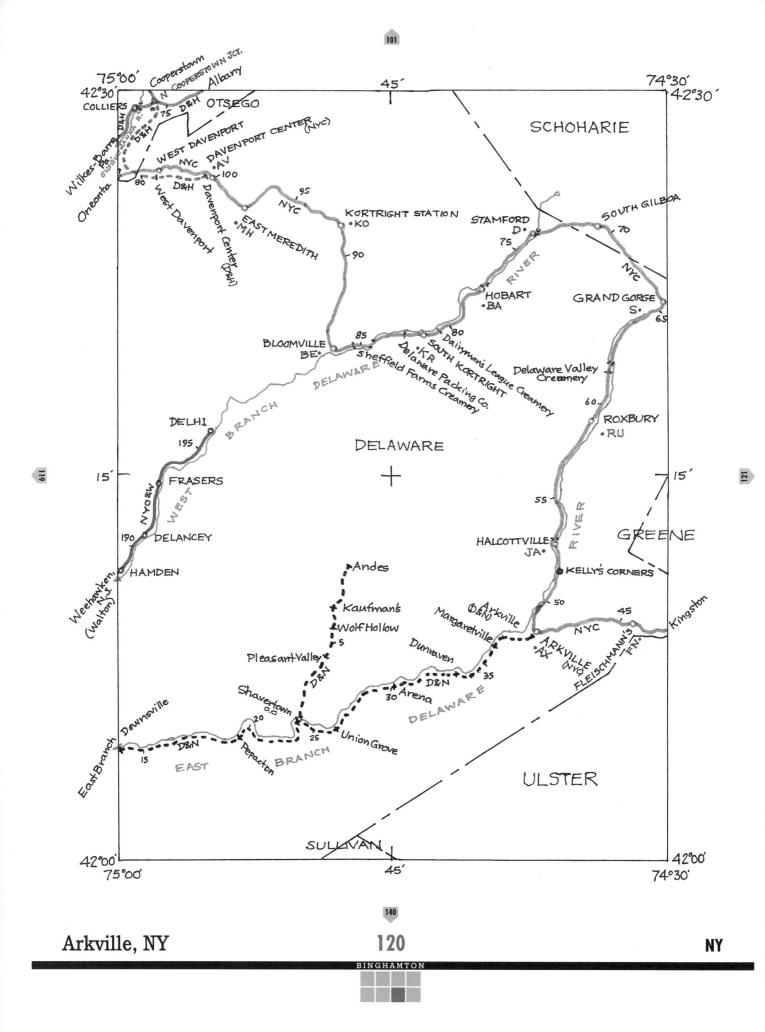

Arkville, NY

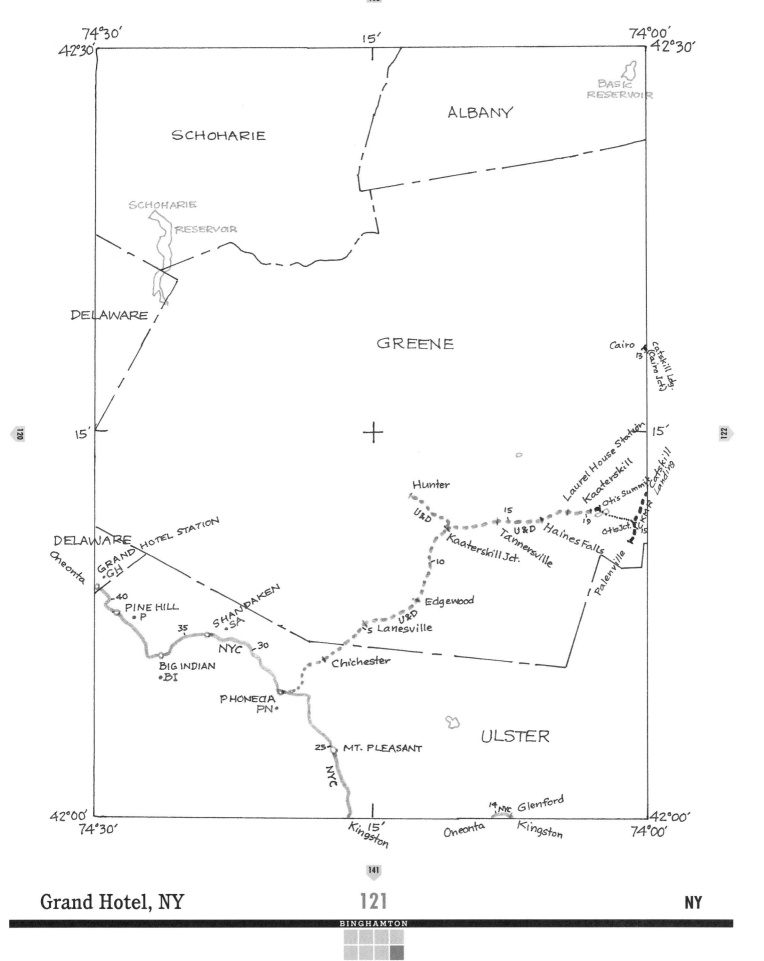

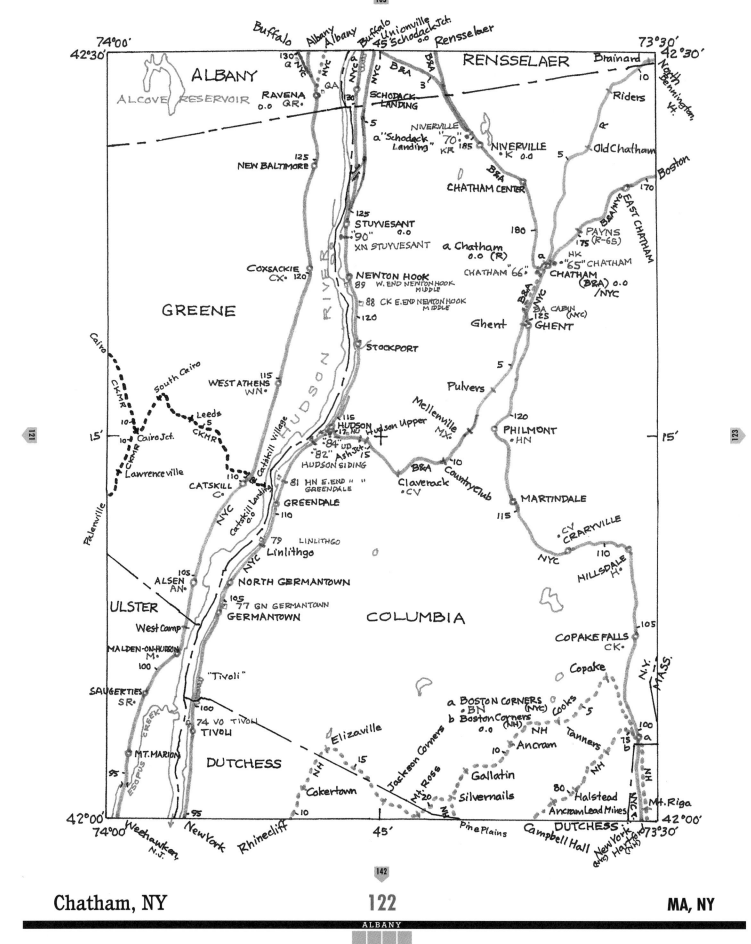

Chatham, NY — 122 — MA, NY

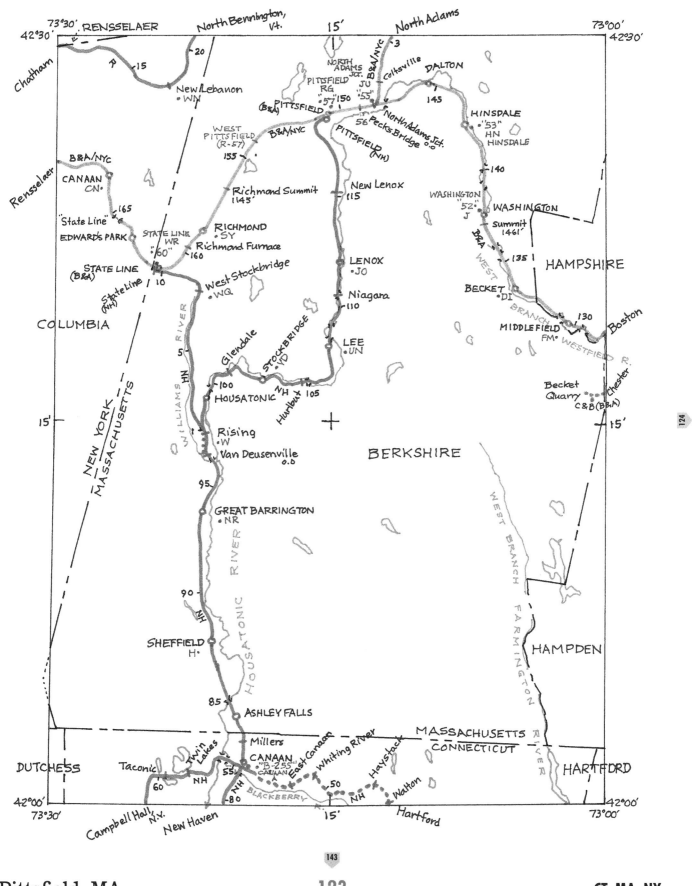

Pittsfield, MA

CT, MA, NY

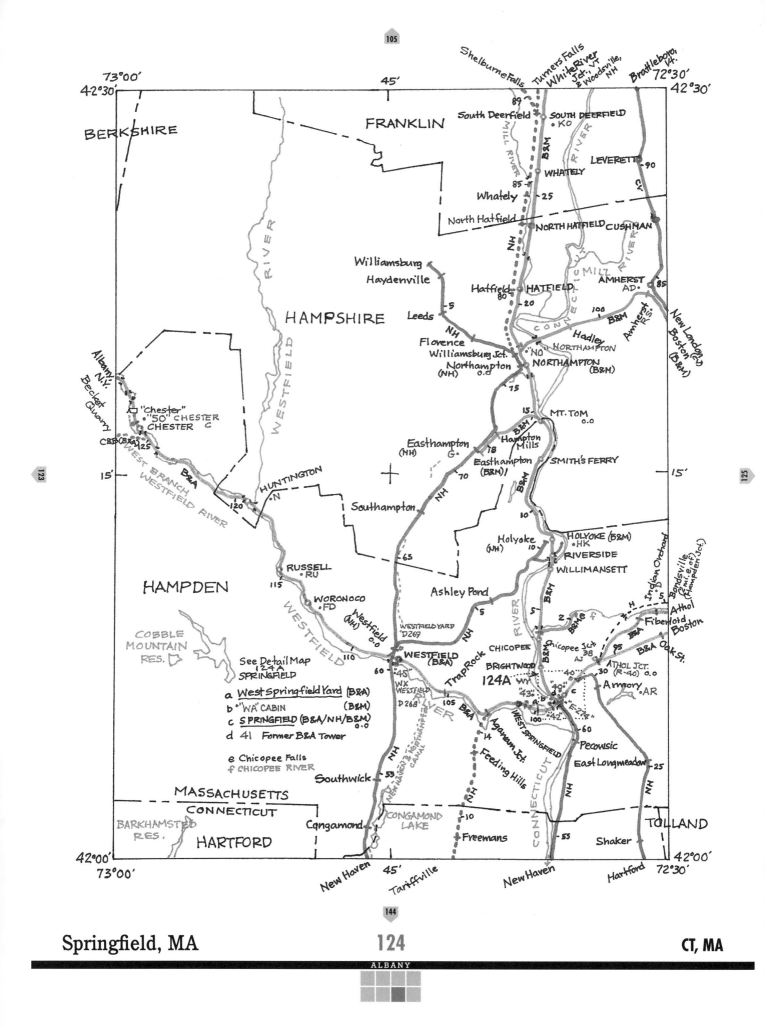

Springfield, MA — 124 — CT, MA

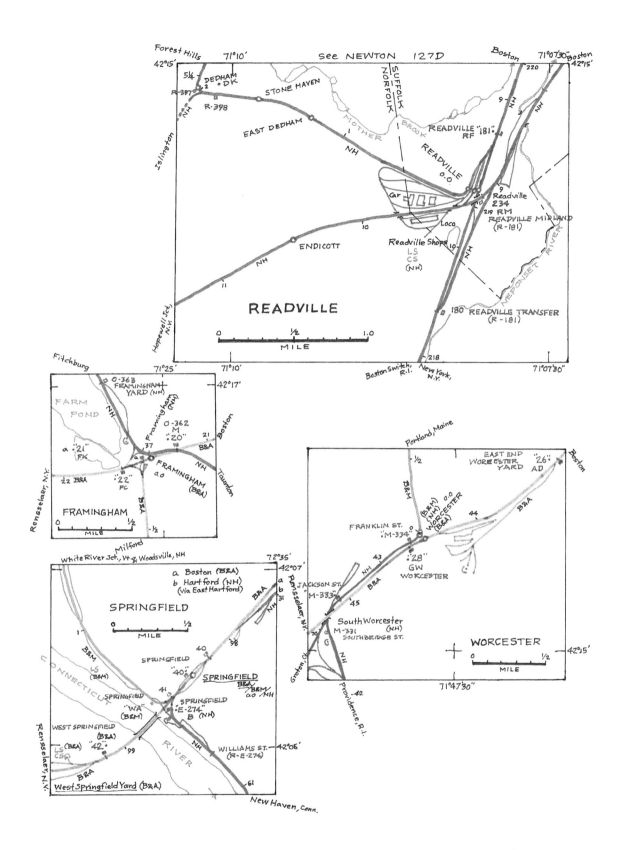

124A Springfield, MA 126A Worcester, MA

127F Readville, MA 127G Framingham, MA

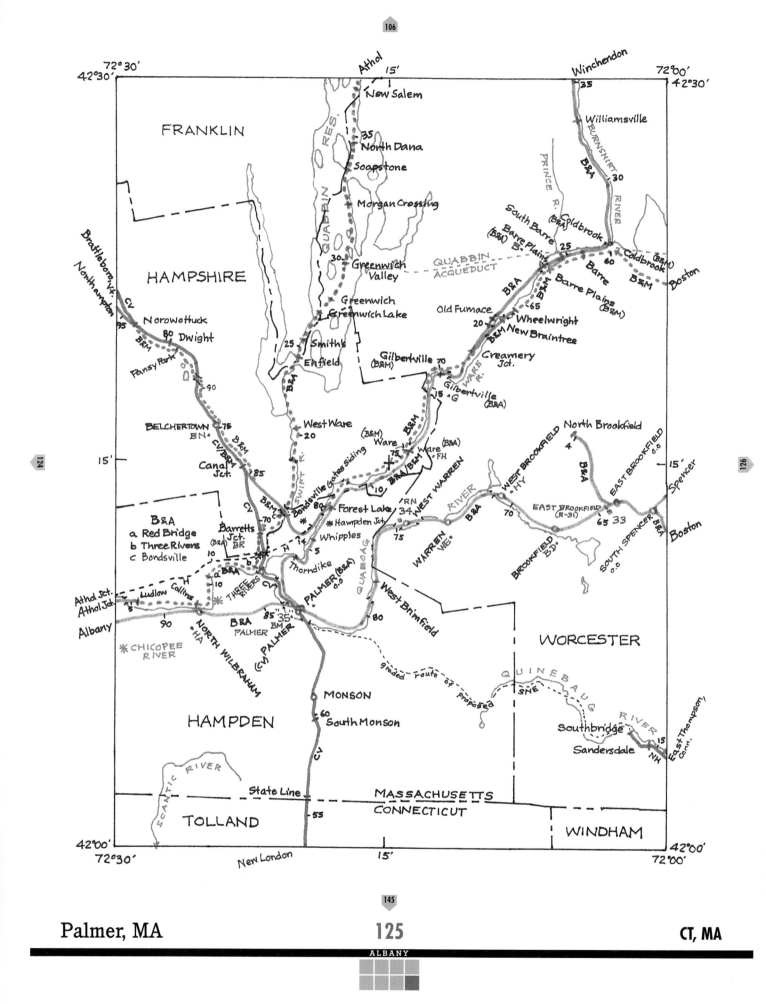
Palmer, MA

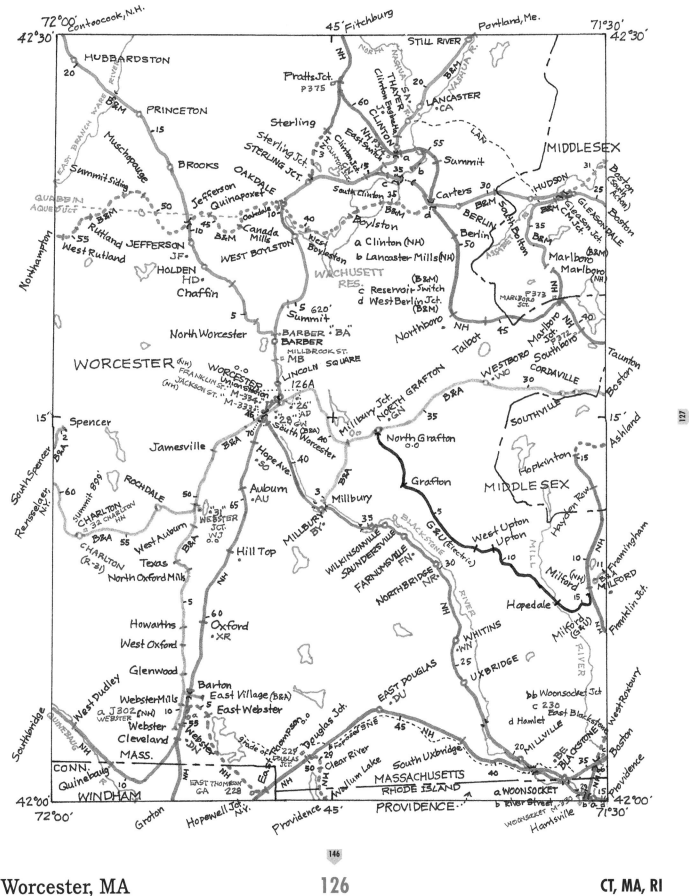

Worcester, MA — 126 — CT, MA, RI

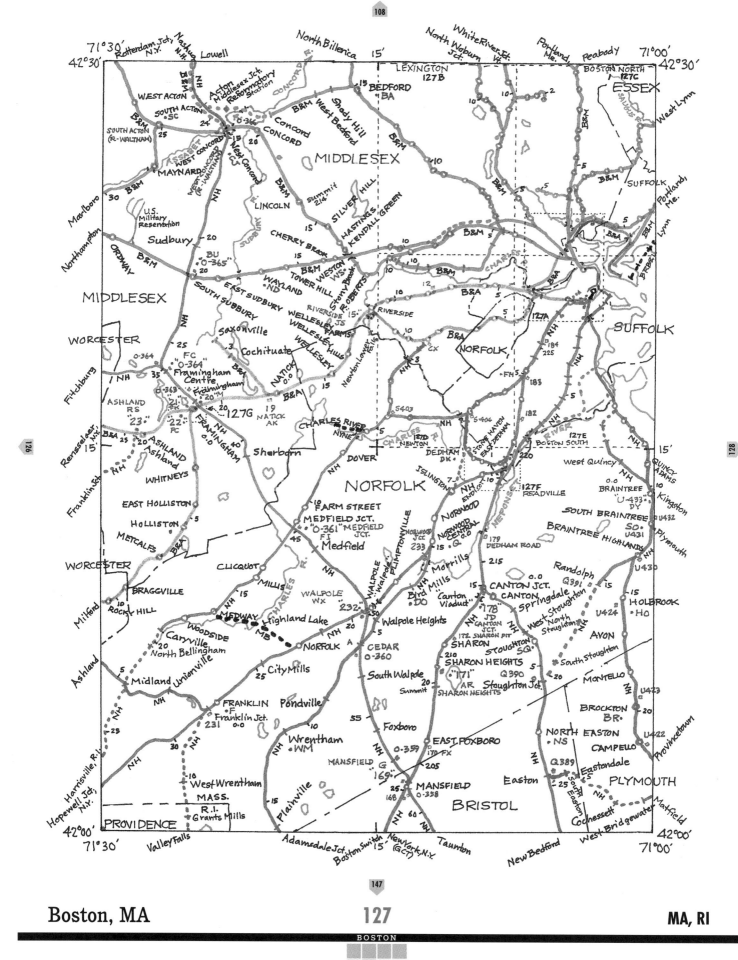

Boston, MA

127

MA, RI

for detail, see map pages 127A–E and 124A for 127F, G

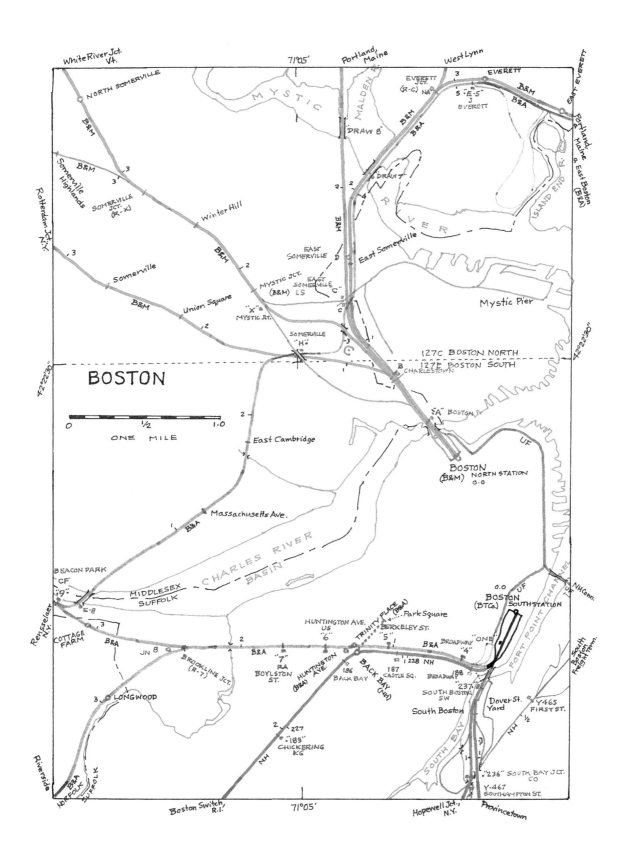

127A Boston, MA

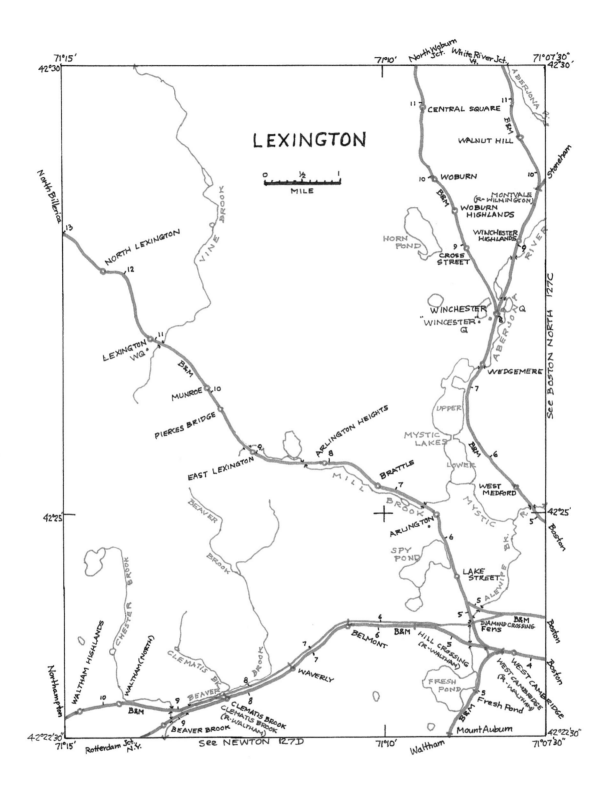

127B Lexington, MA

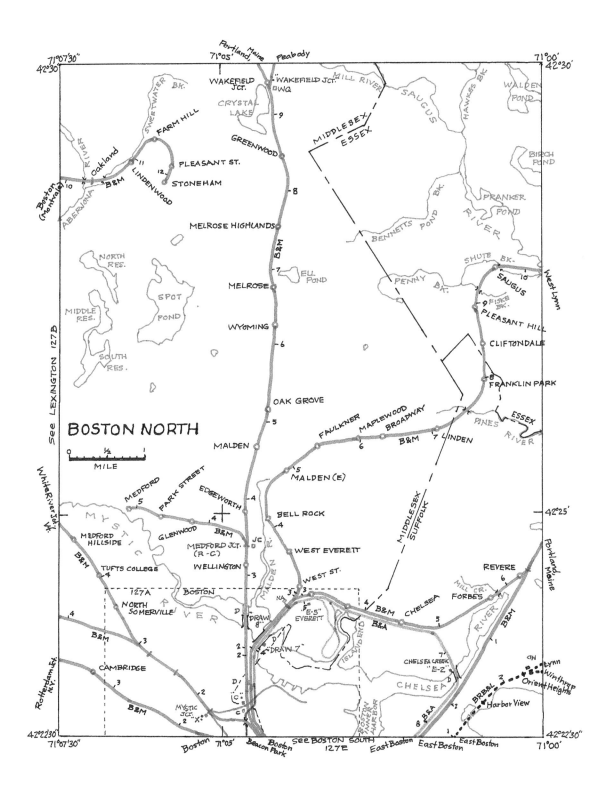

127C Boston North, MA

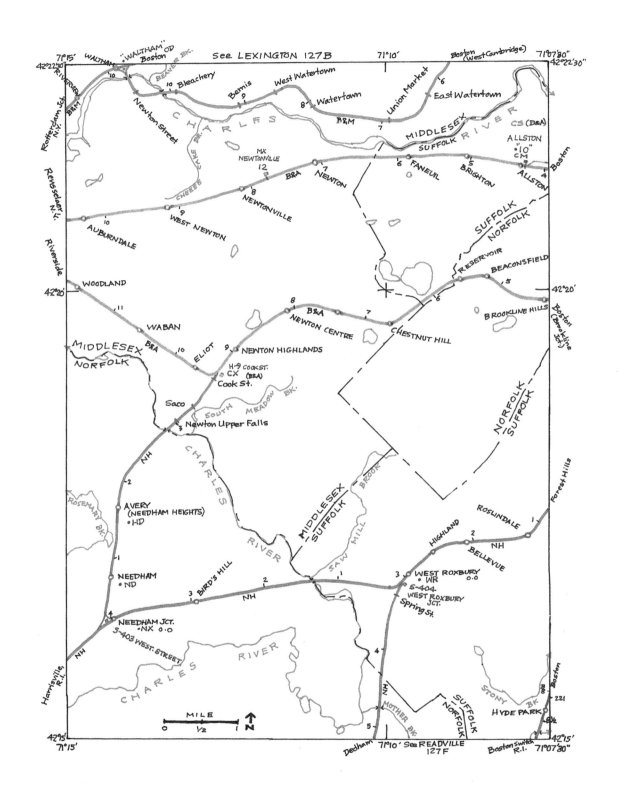

127D Newton, MA

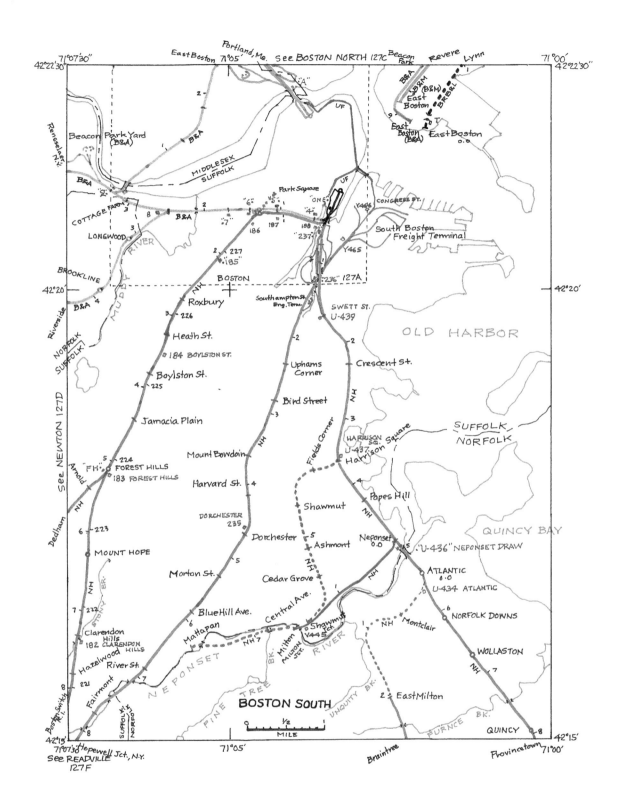

127E Boston South, MA

for 127F detail, see map page 124A

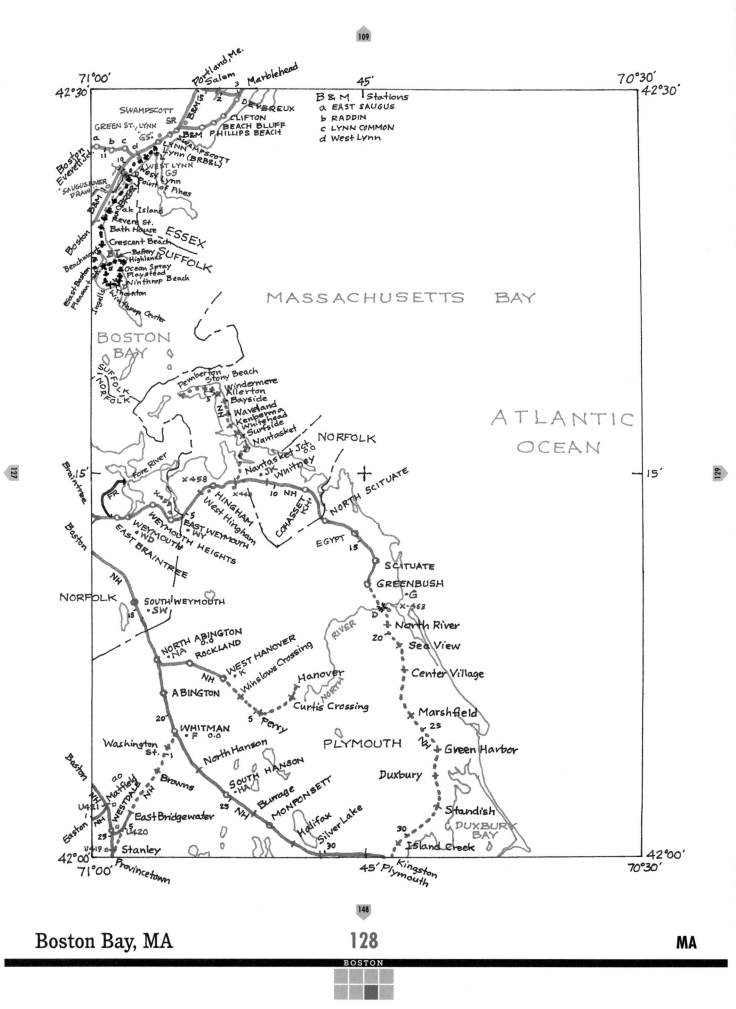

Boston Bay, MA

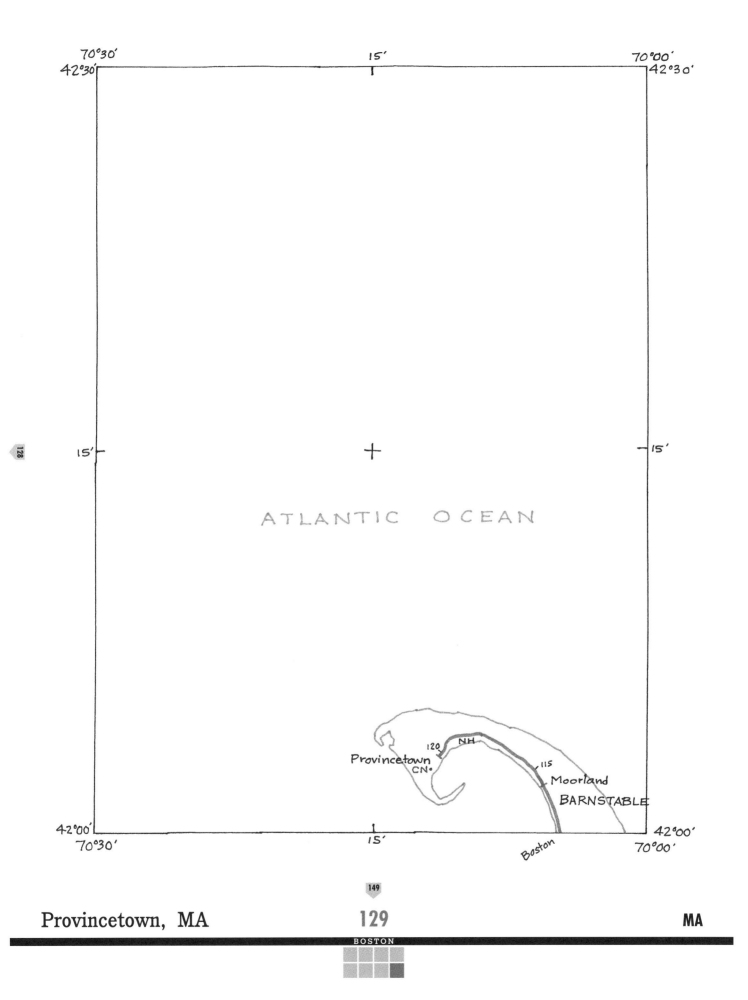

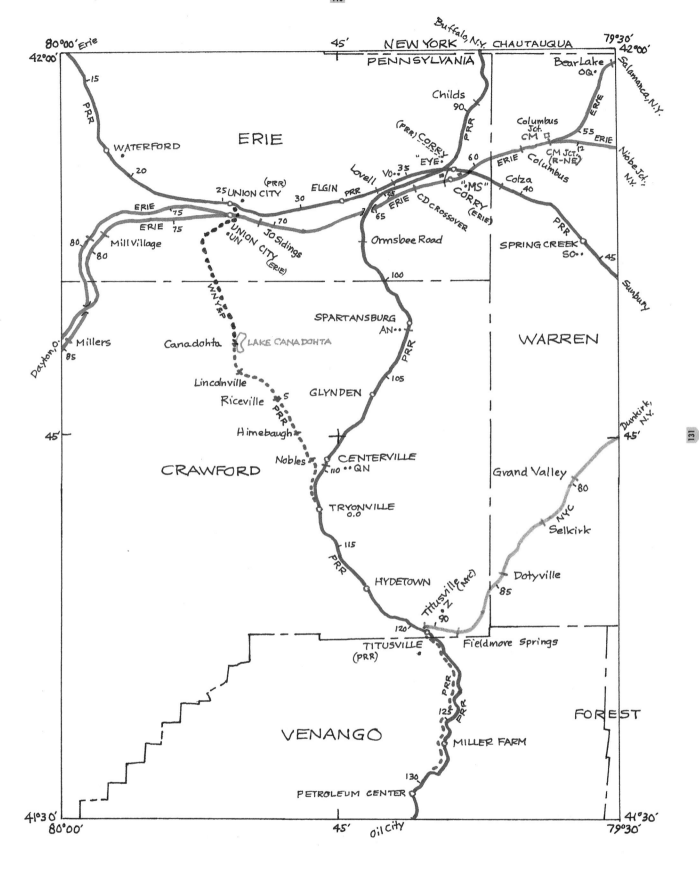

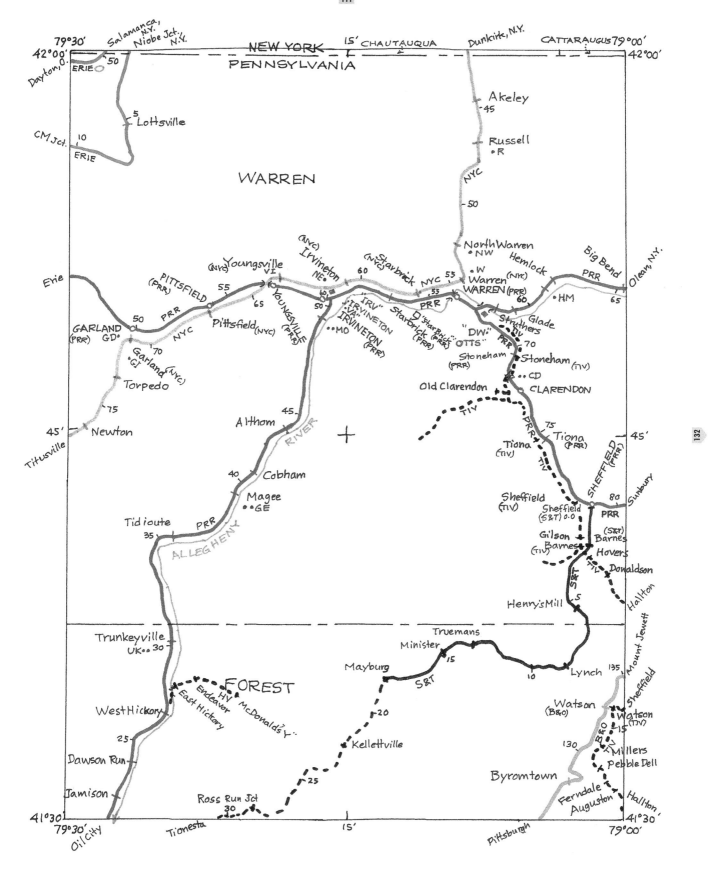

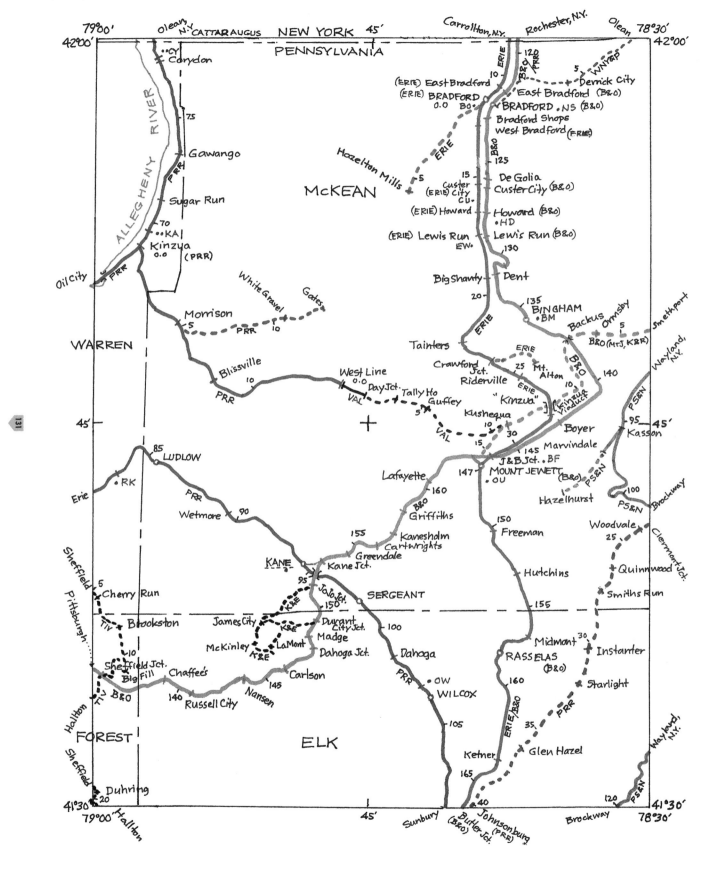

Kinzua, PA — 132 — NY, PA

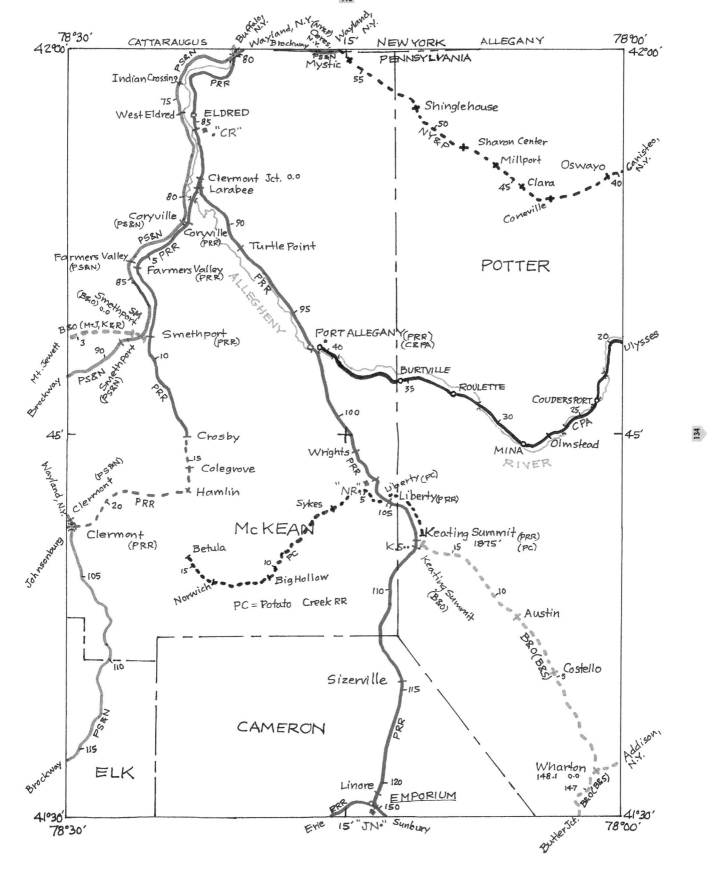

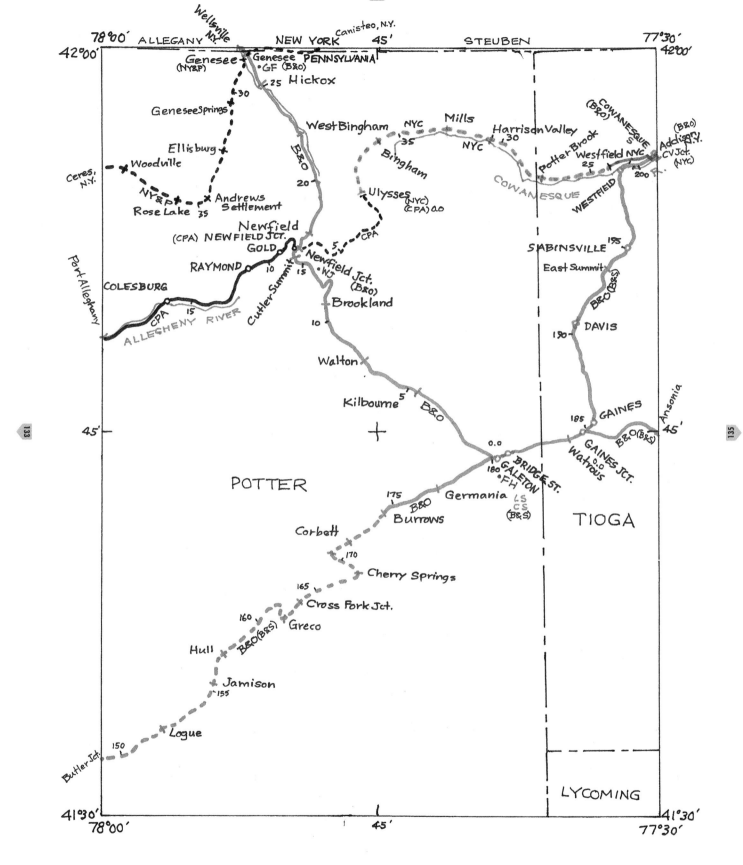

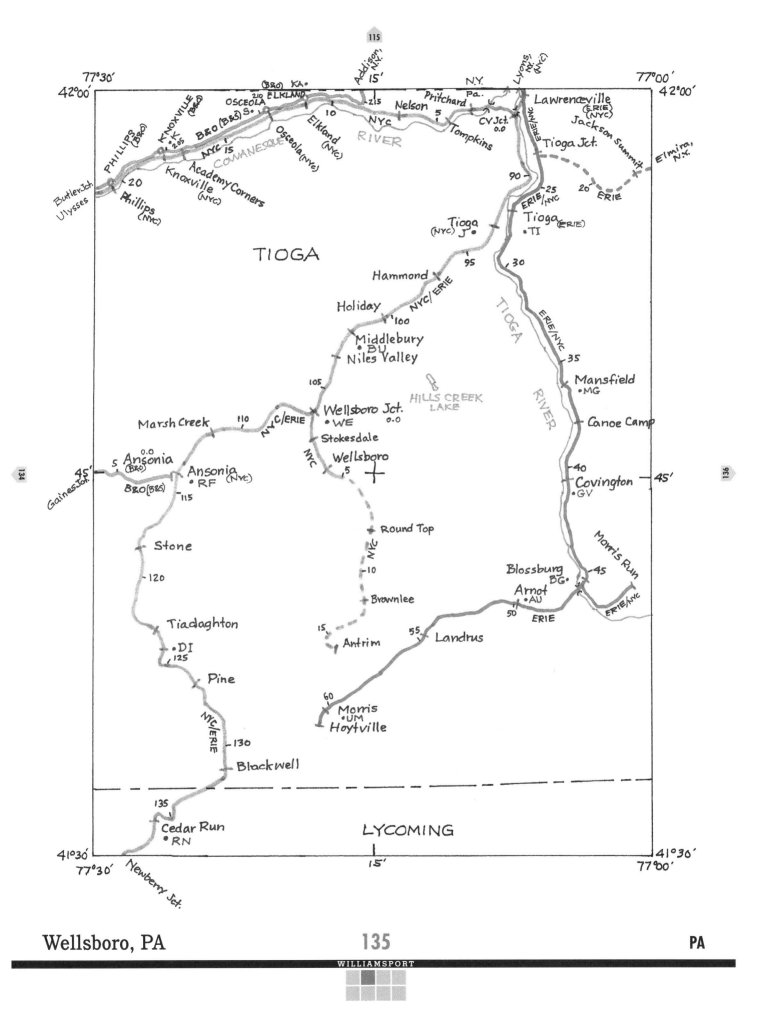

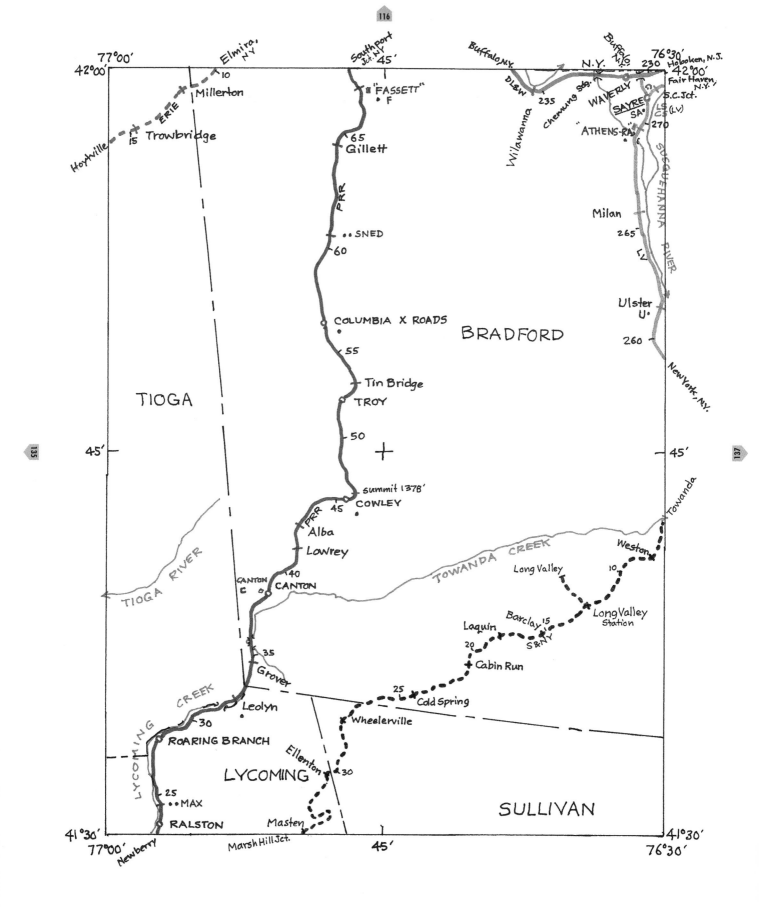

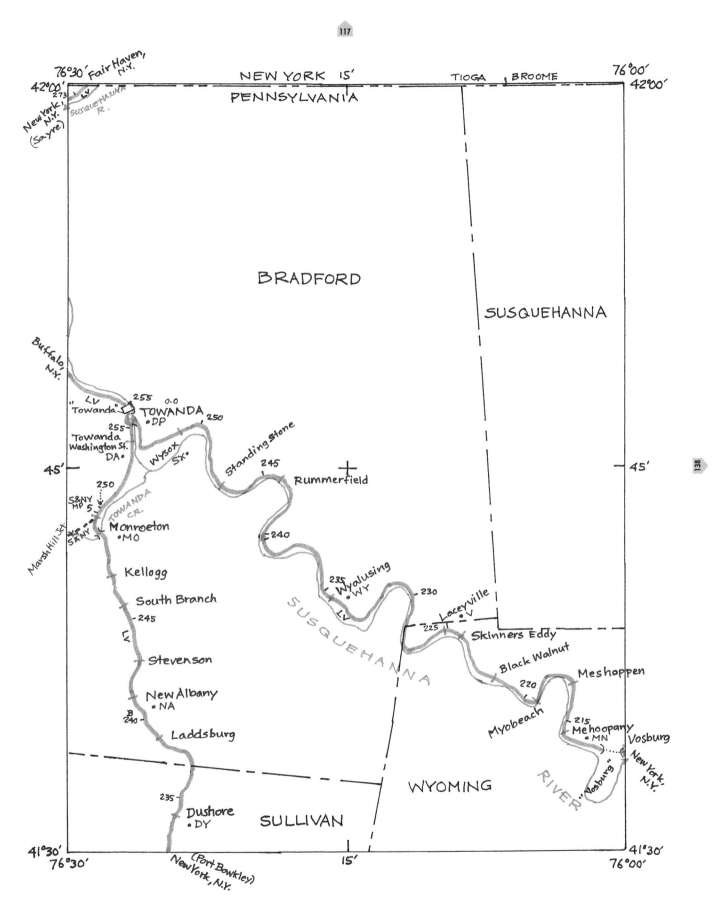

Towanda, PA — 137 — NY, PA

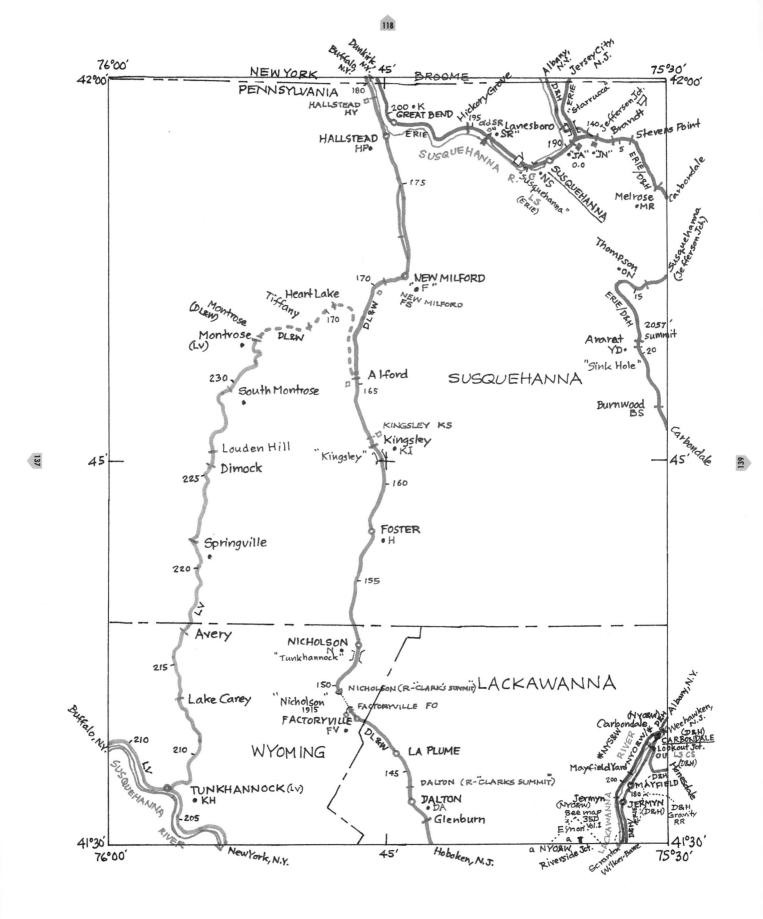

Susquehanna, PA — 138 — NY, PA
SCRANTON

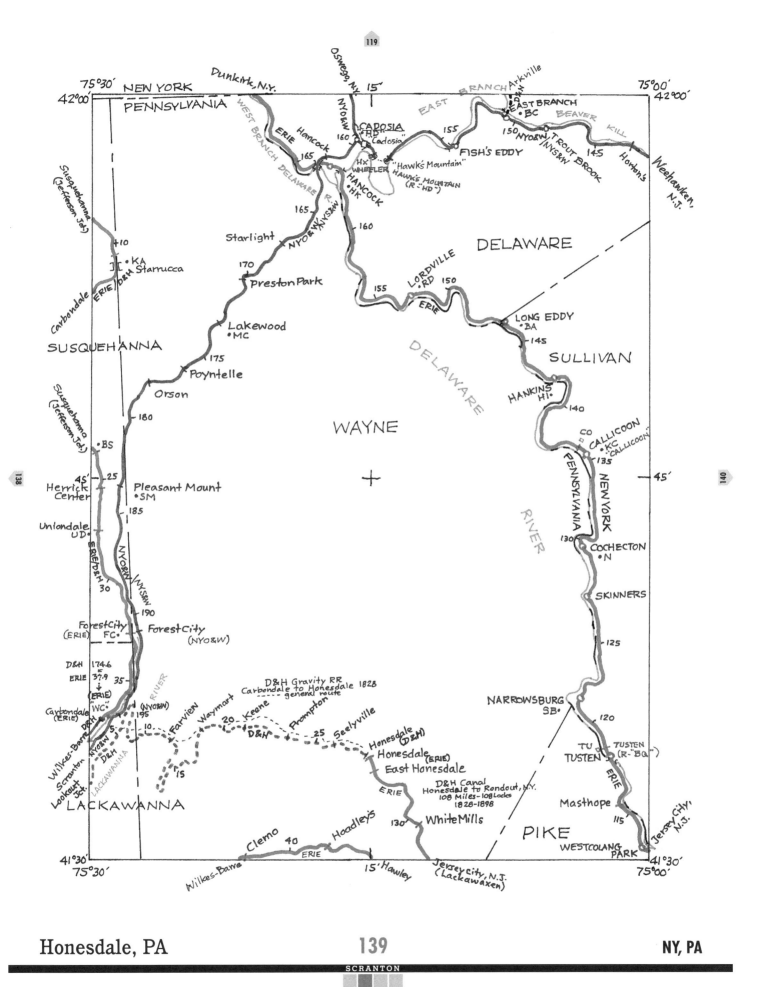

Honesdale, PA — 139 — NY, PA

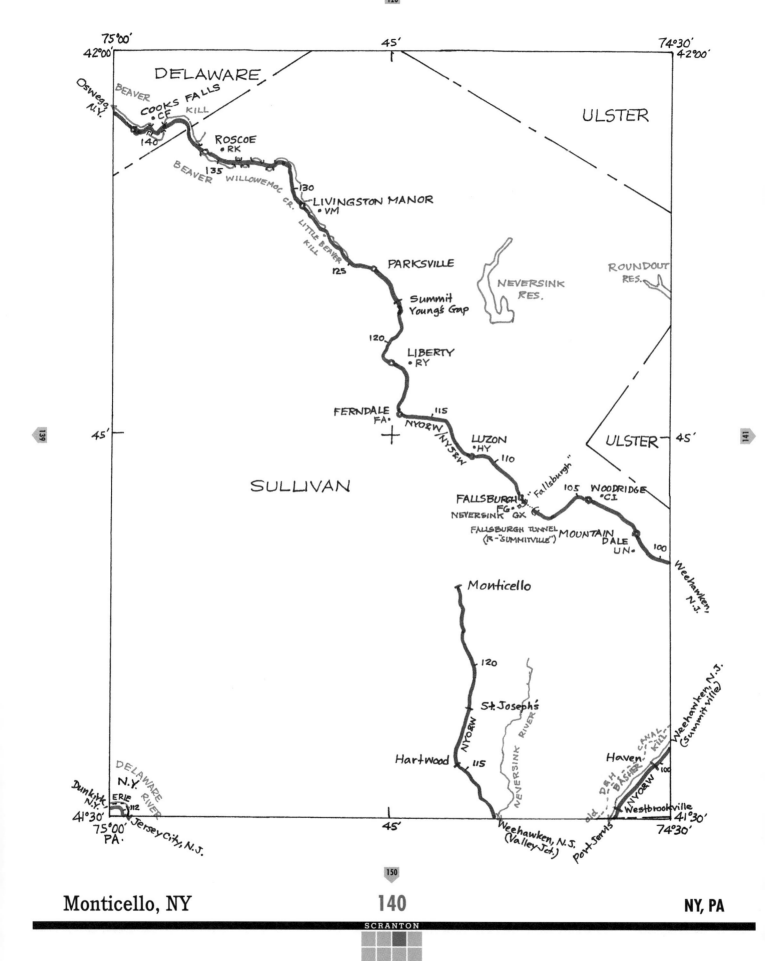

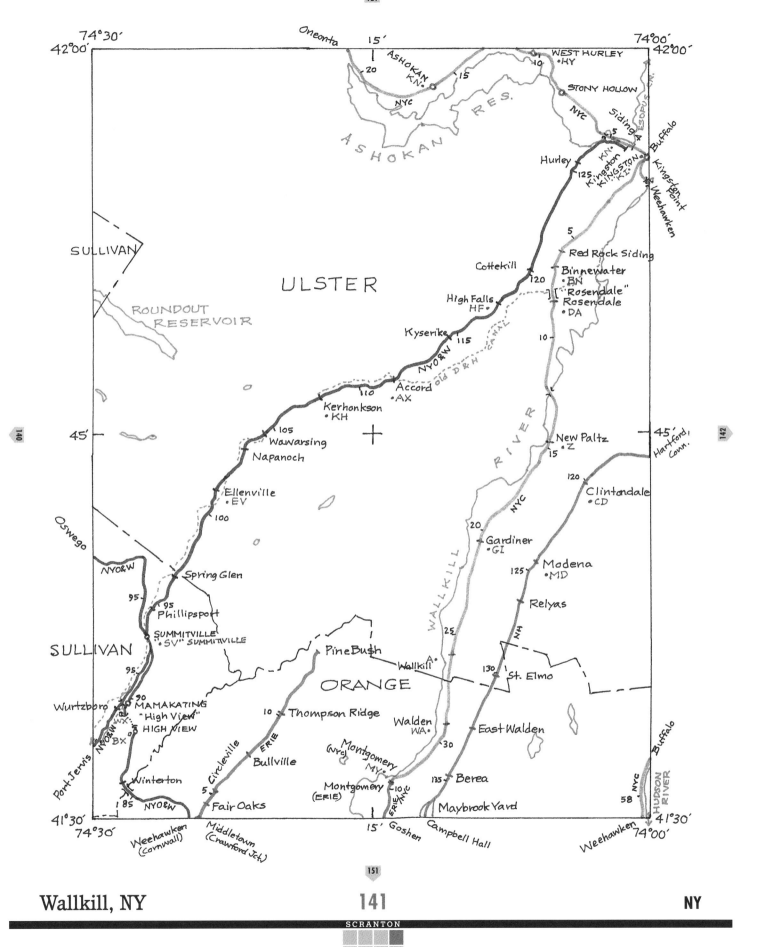

Wallkill, NY

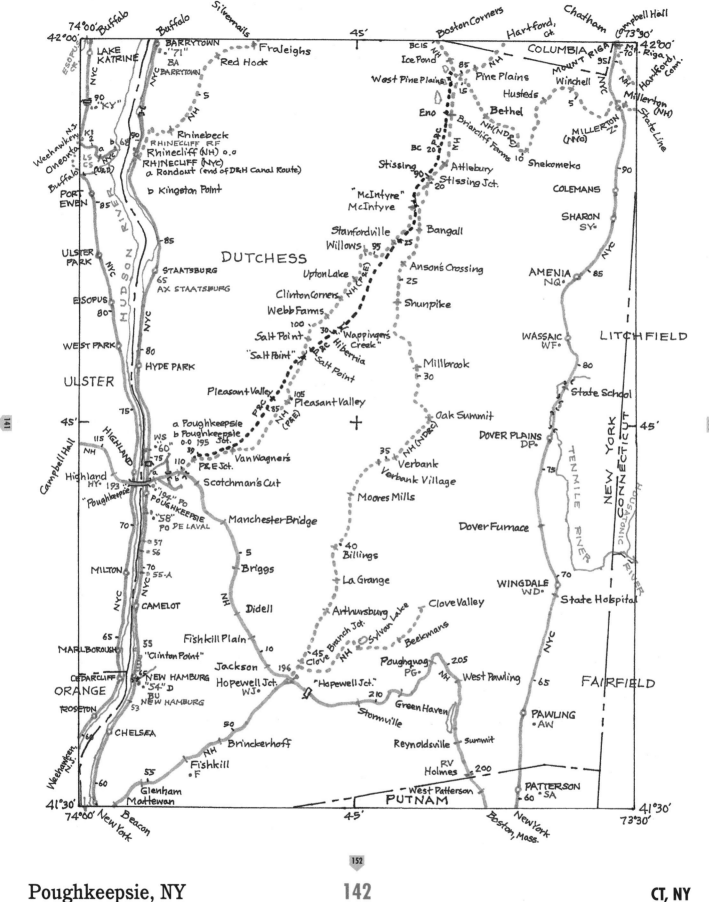

Poughkeepsie, NY   142   CT, NY

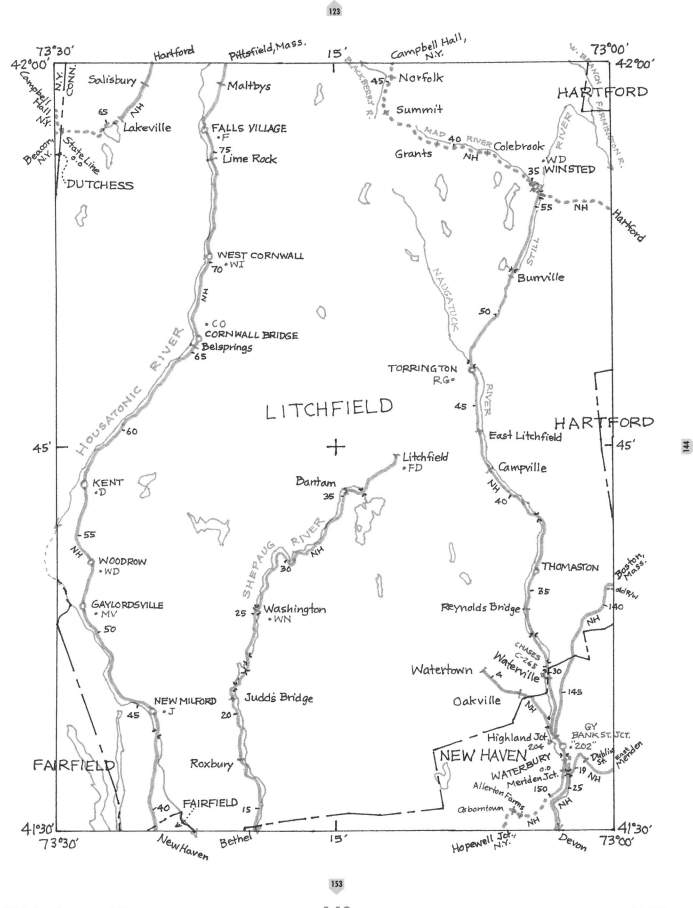

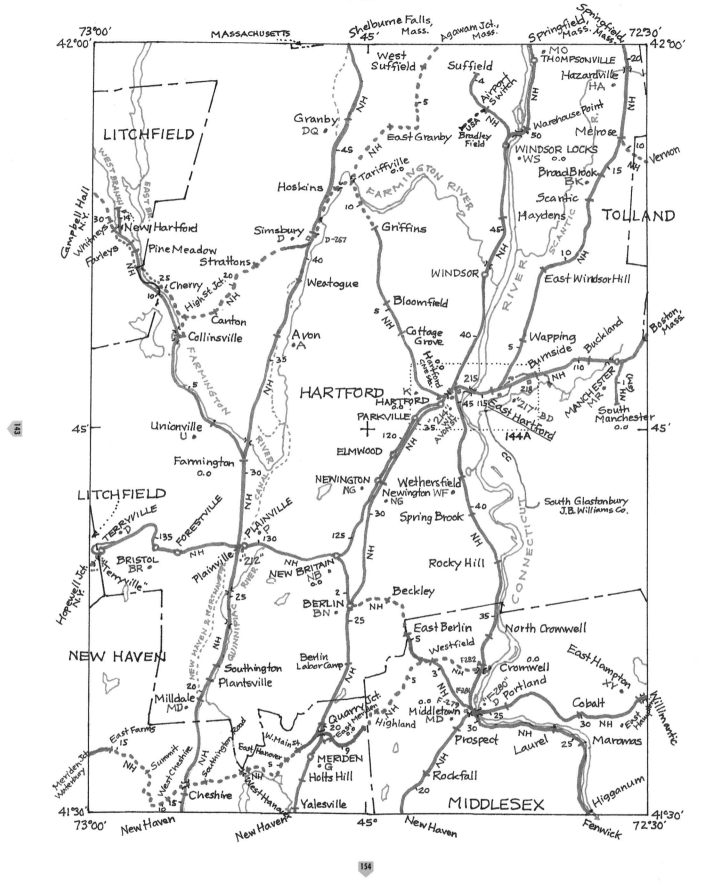

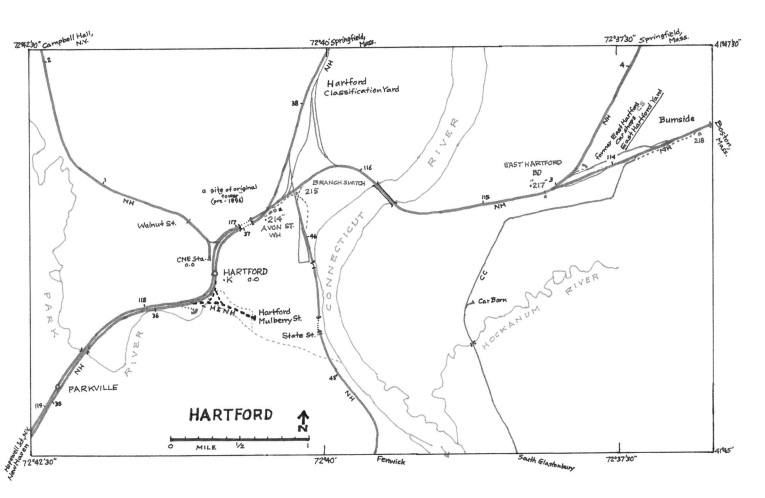

144A Hartford, CT

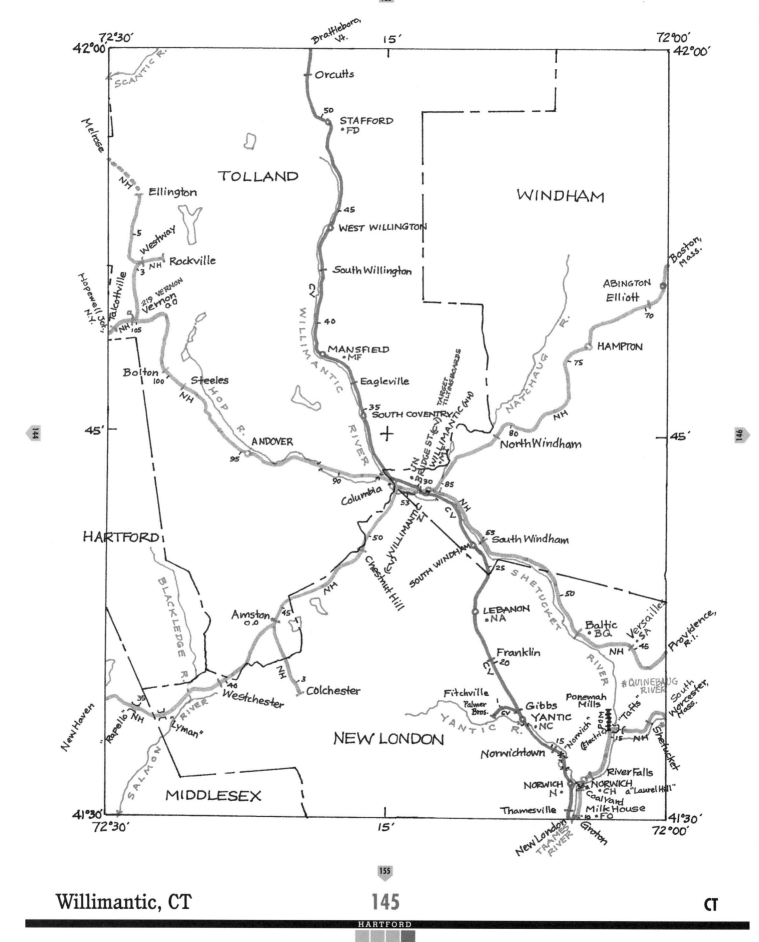

Willimantic, CT

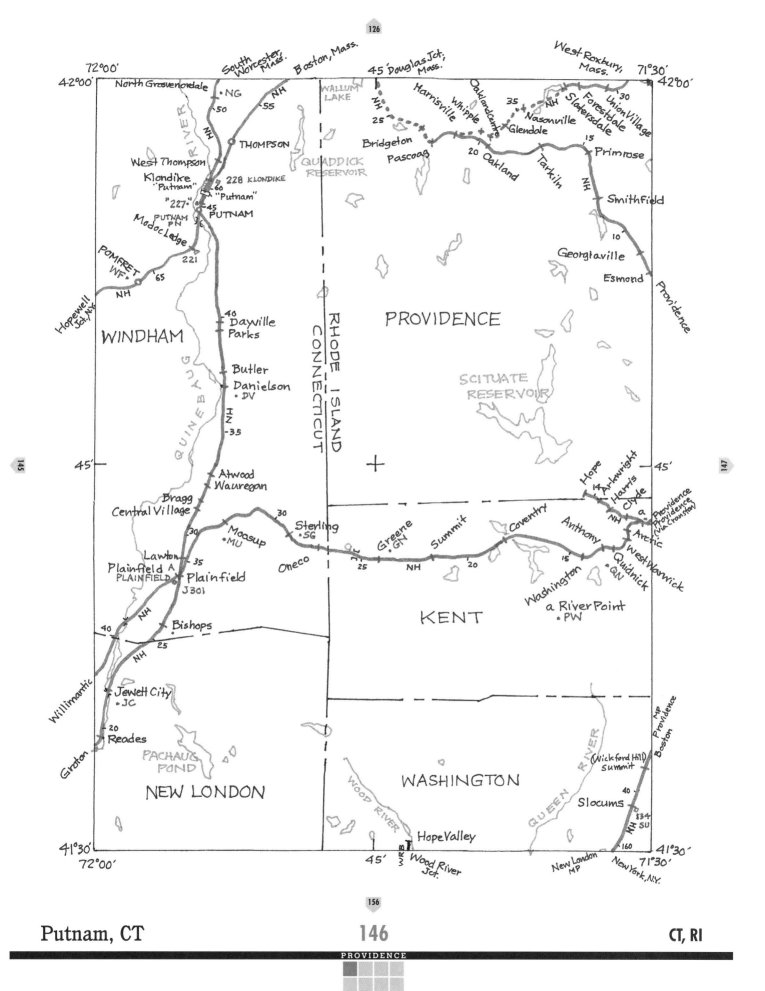

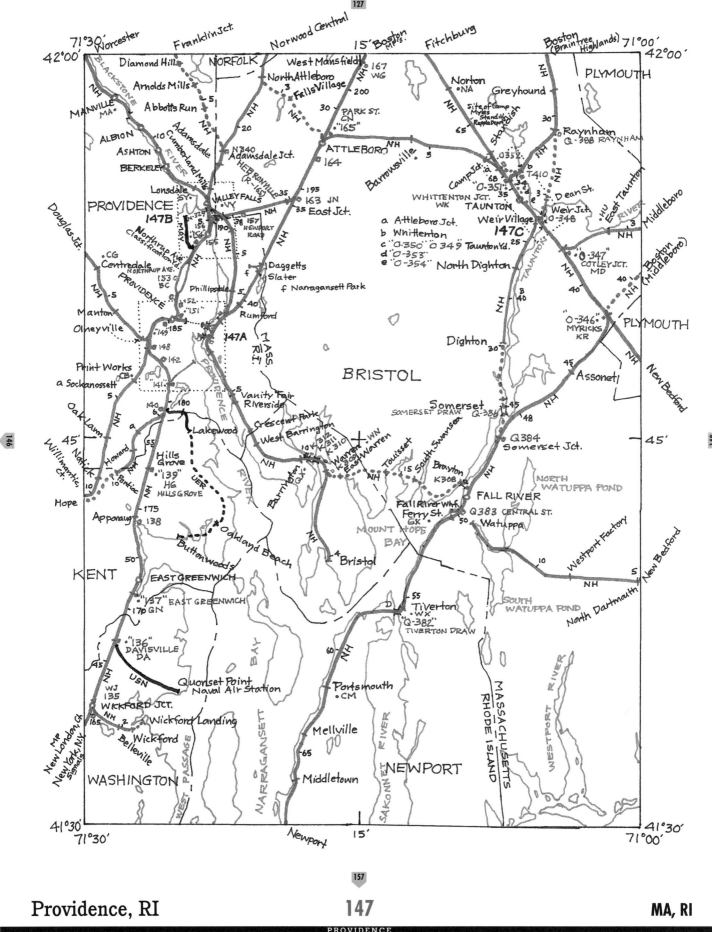

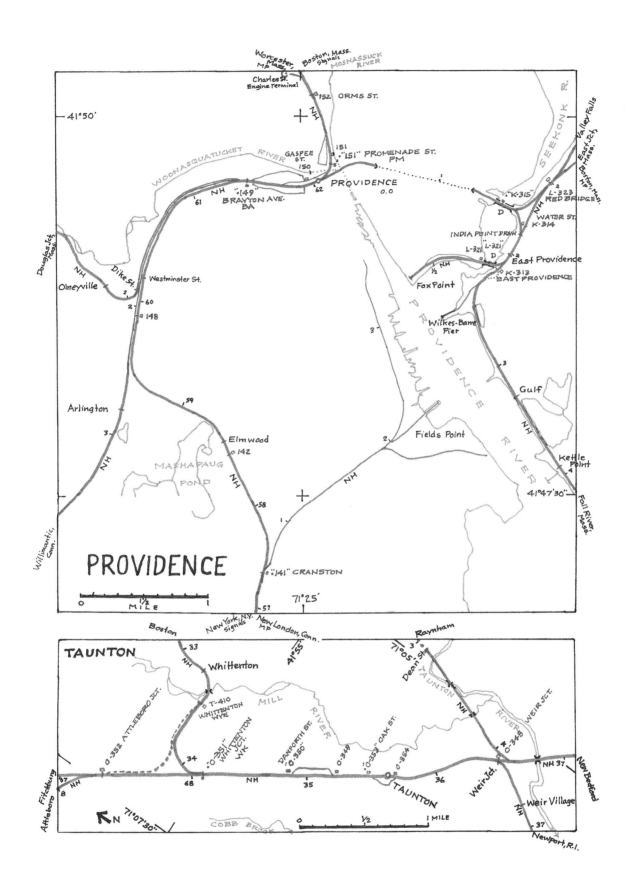

147A Providence, RI **147C** Taunton, MA

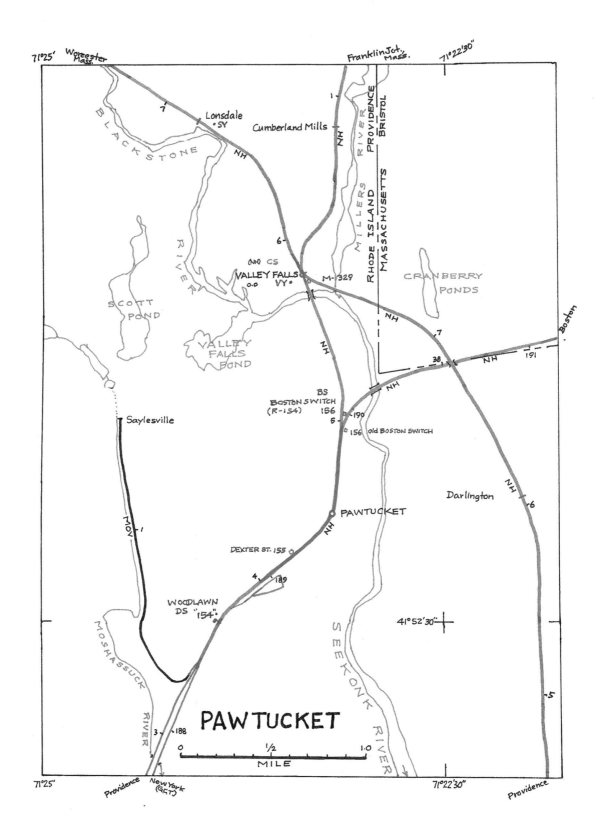

147B Pawtucket, RI

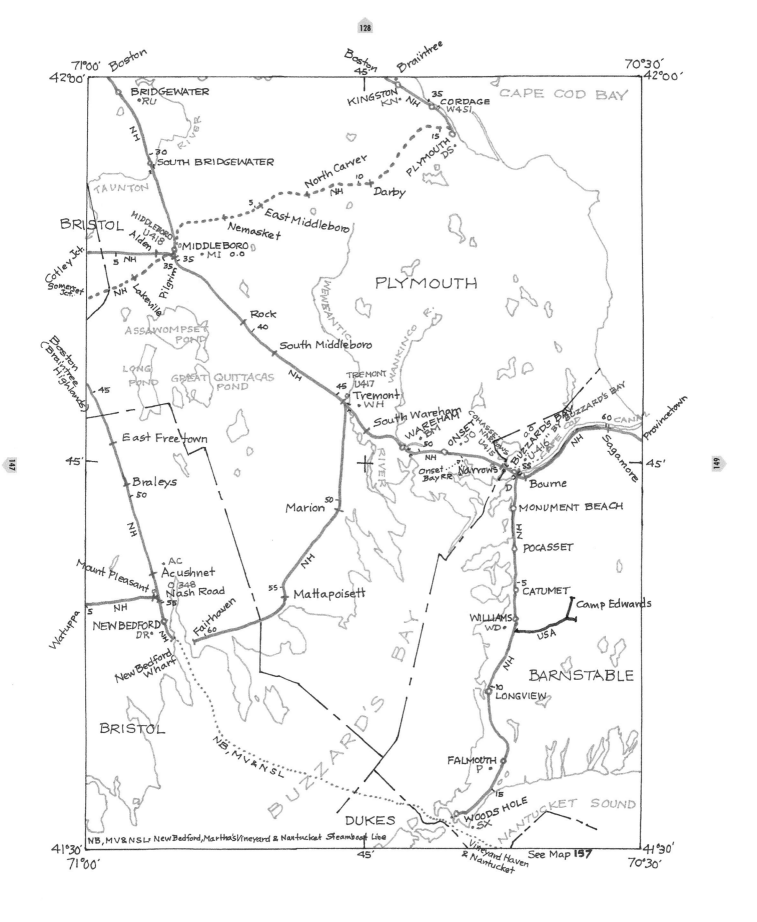

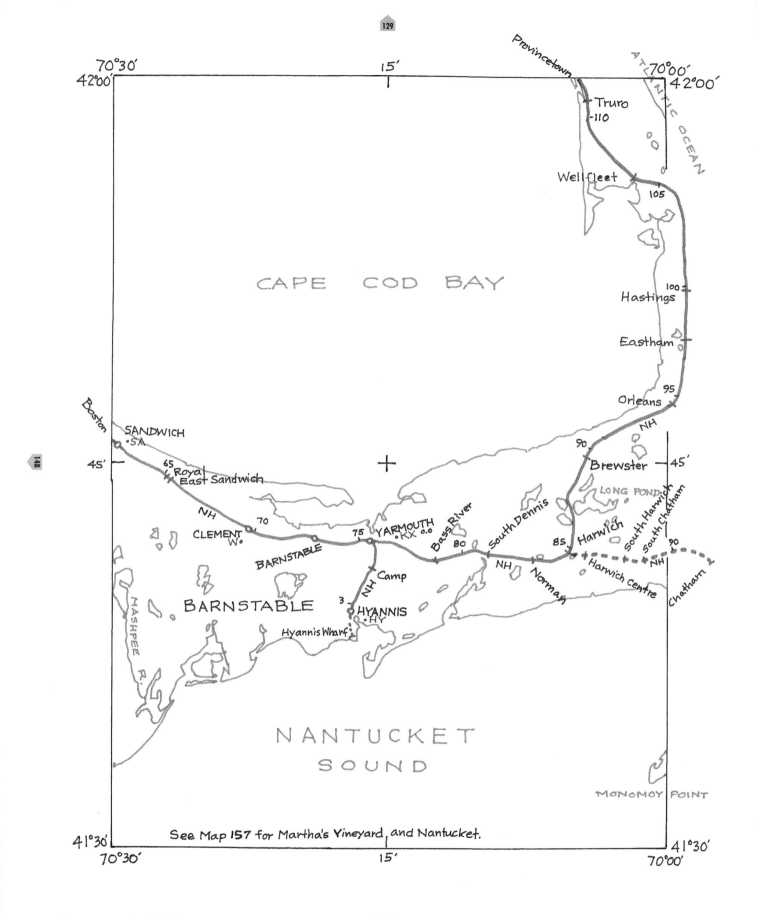
Yarmouth, MA

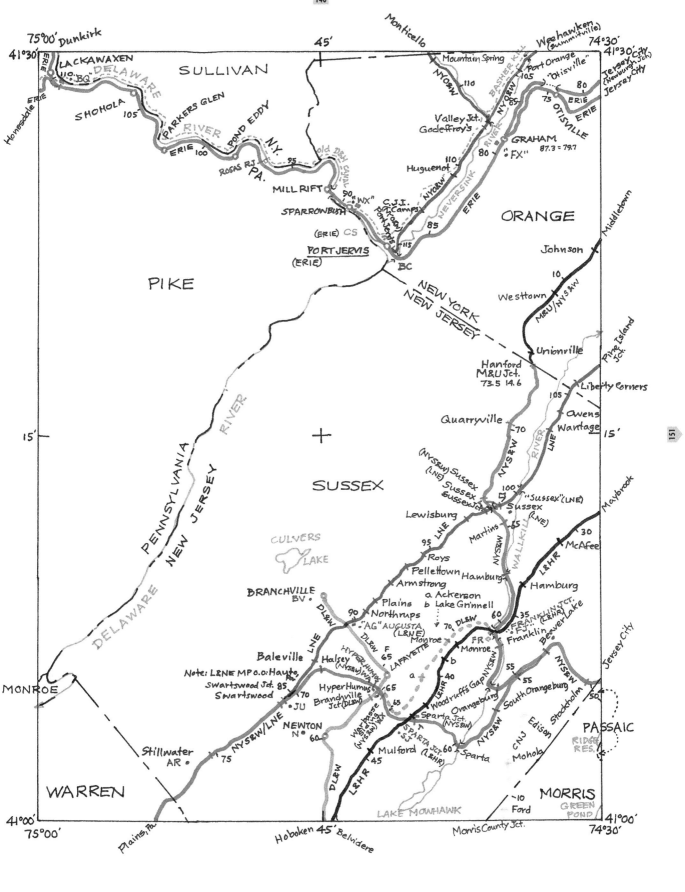

Port Jervis, NY — 150 — NJ, NY, PA

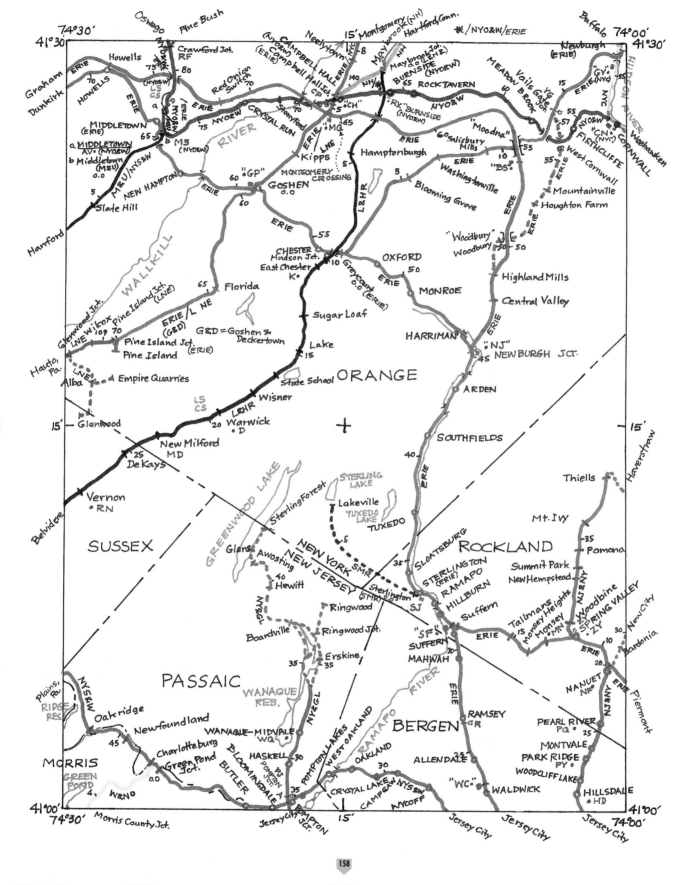

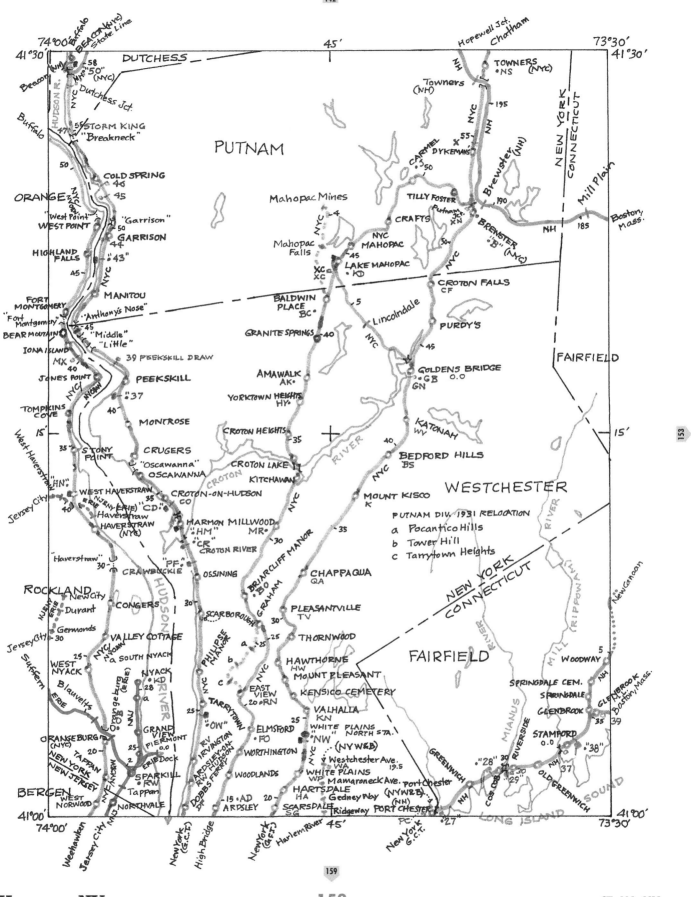

Harmon, NY

CT, NJ, NY

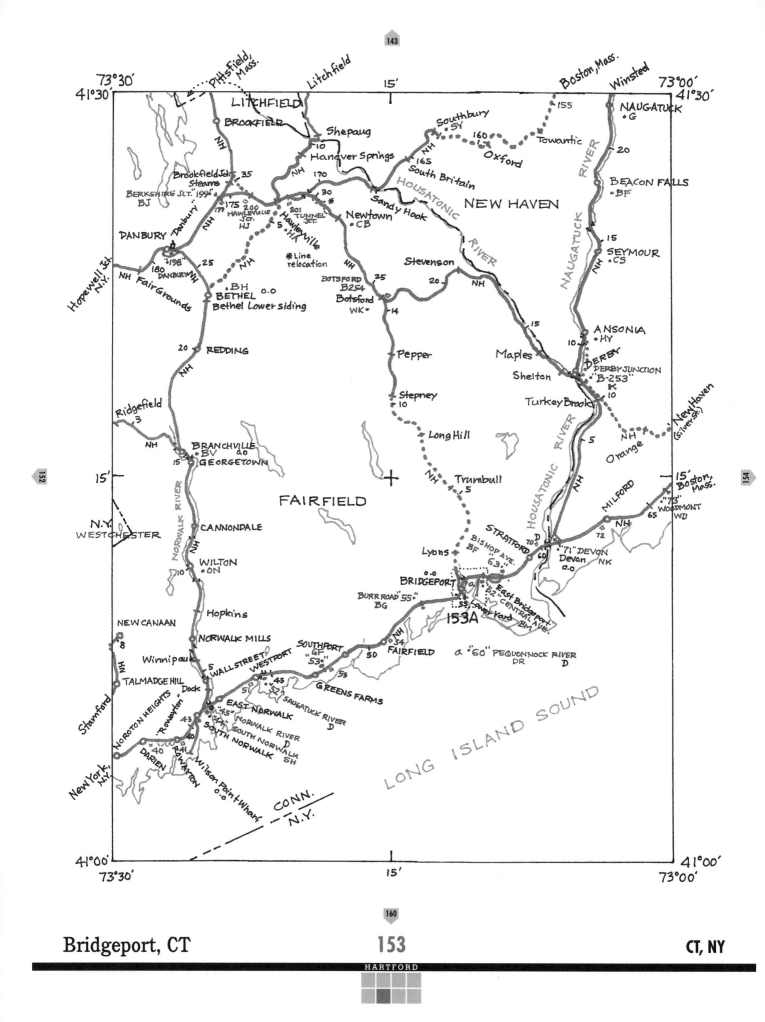

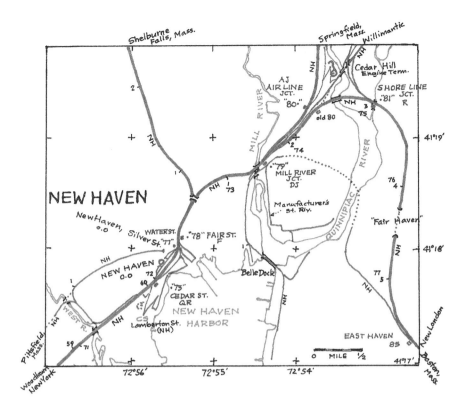

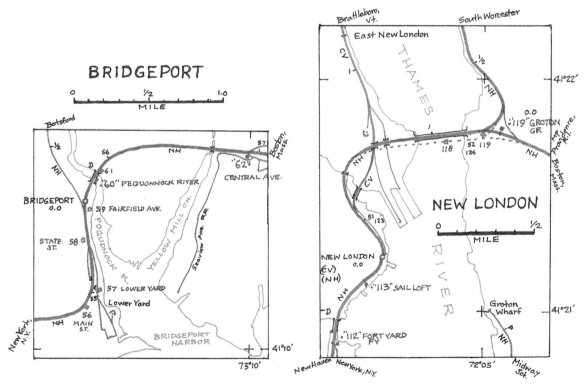

153A Bridgeport, CT **154A** New Haven, CT

155A New London, CT

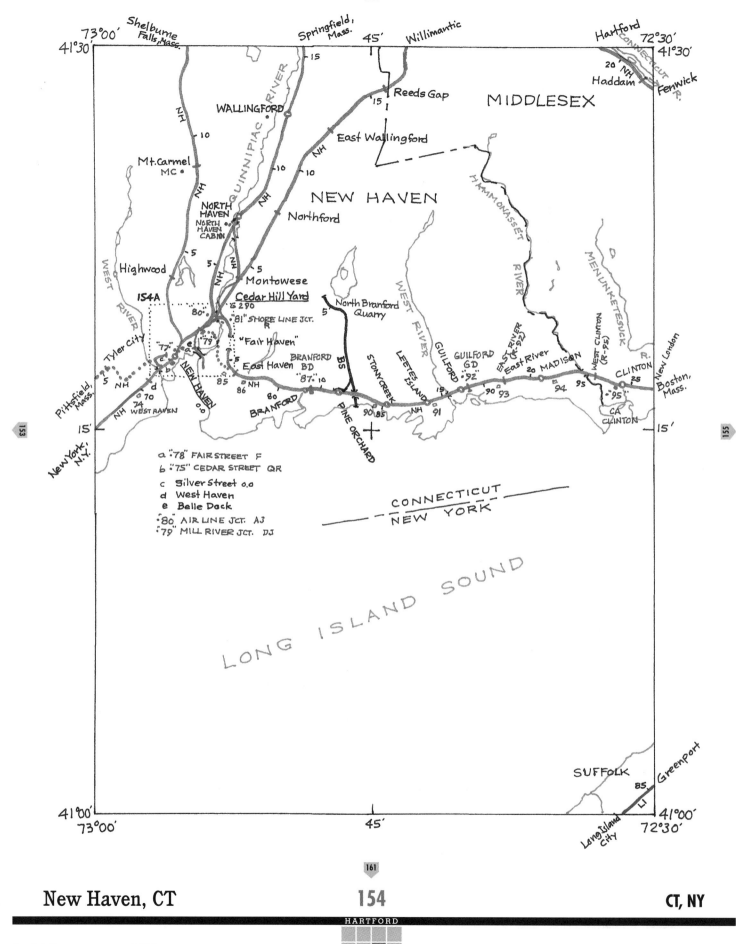

New Haven, CT

154

CT, NY

for detail, see map page 153A

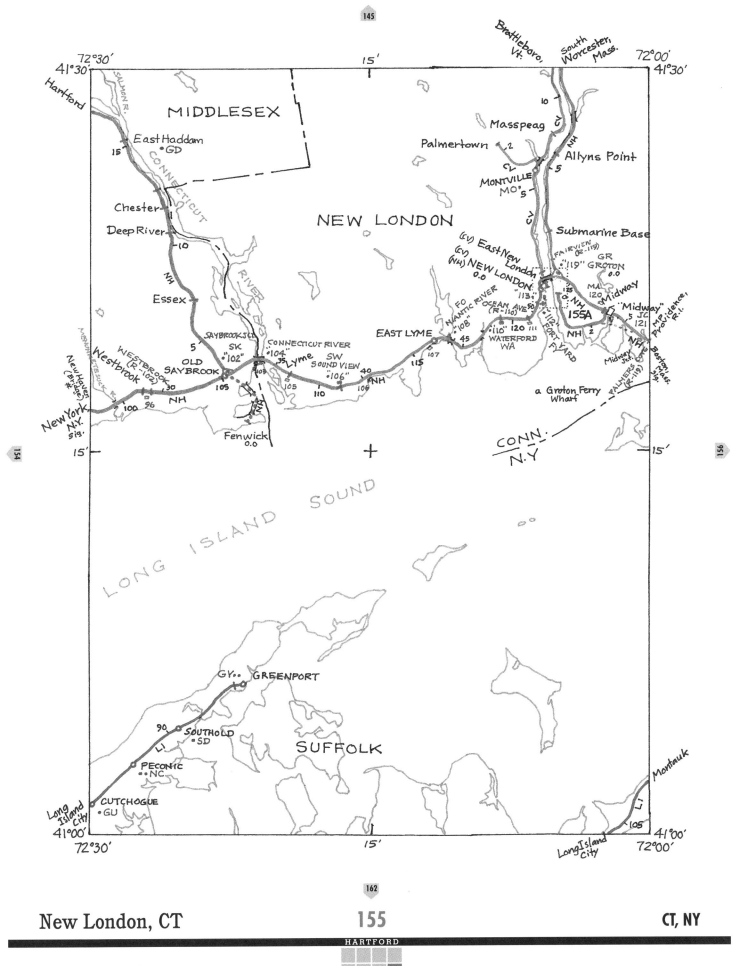

New London, CT

CT, NY

for detail, see map page 153A

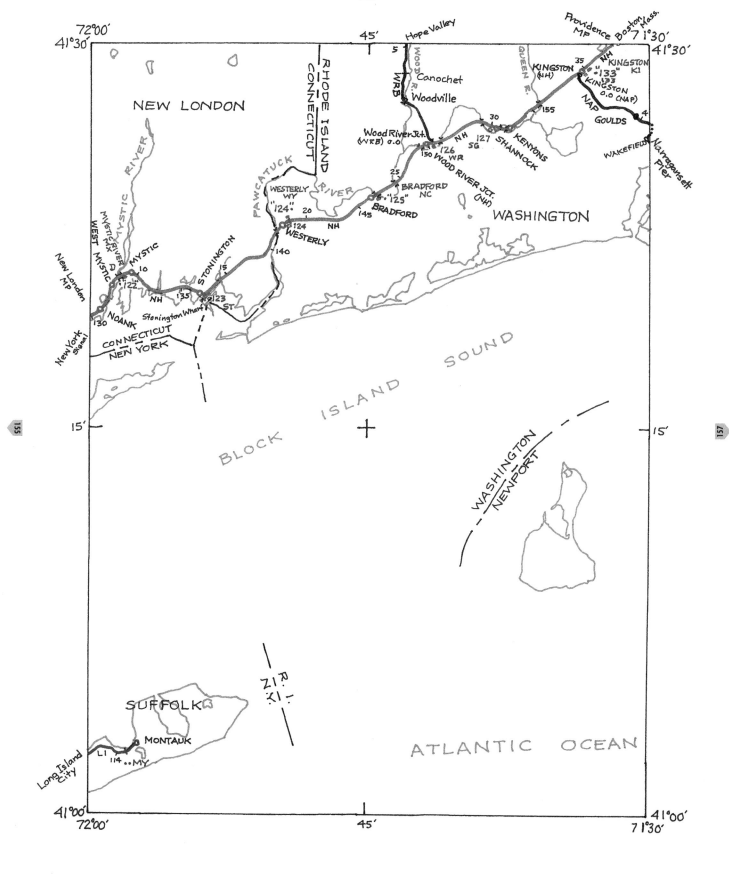

Kingston, RI — 156 — CT, NY, RI

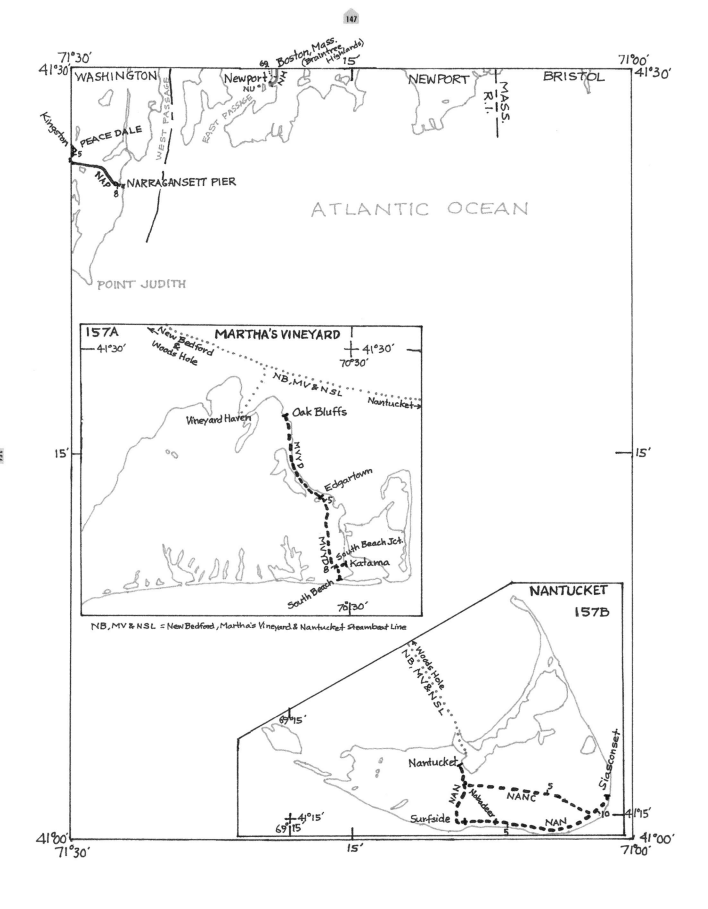

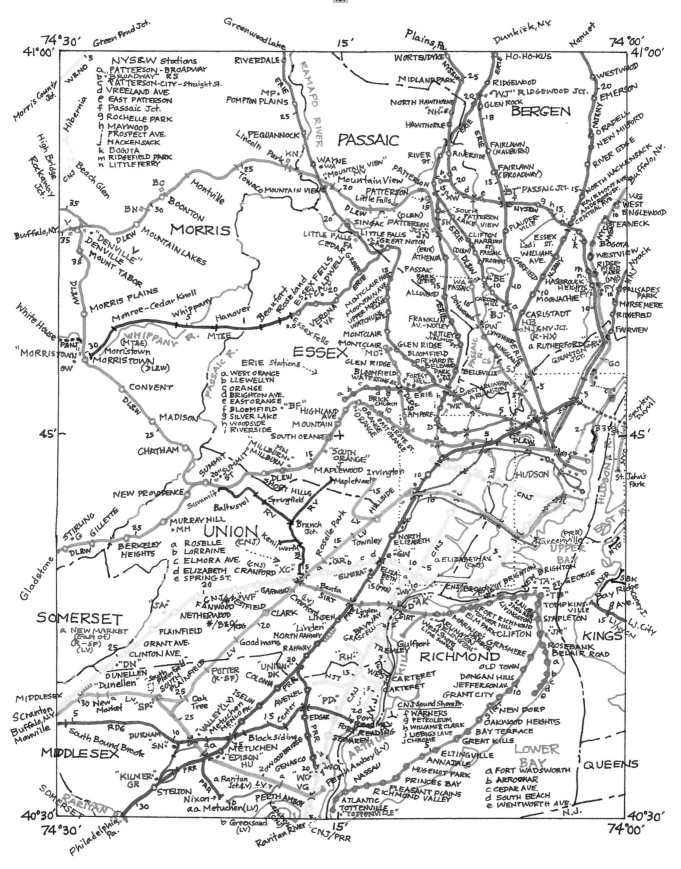

Newark, NJ — for Newark detail map, see map 53A, Volume 1: The Mid-Atlantic States

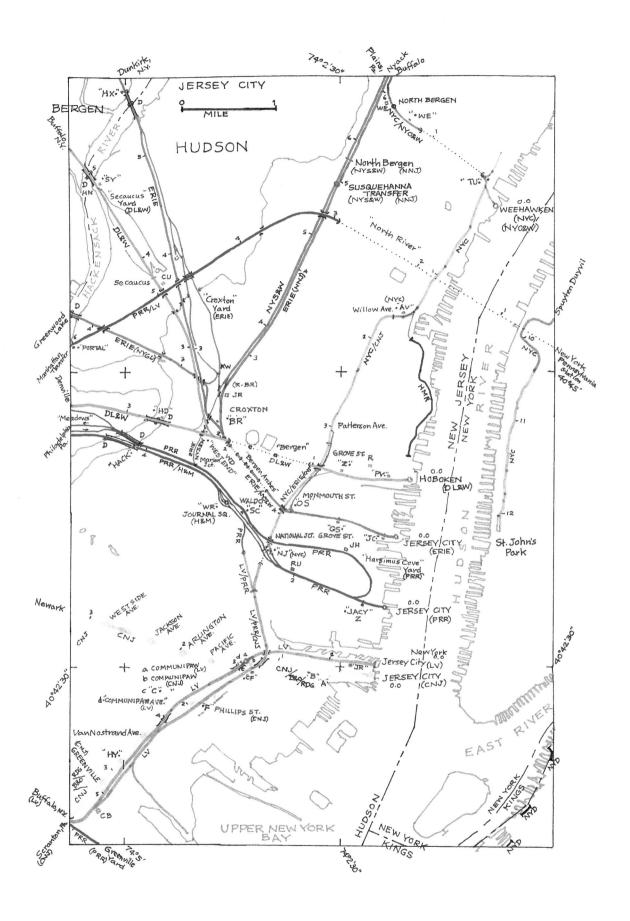

158A Jersey City, NJ

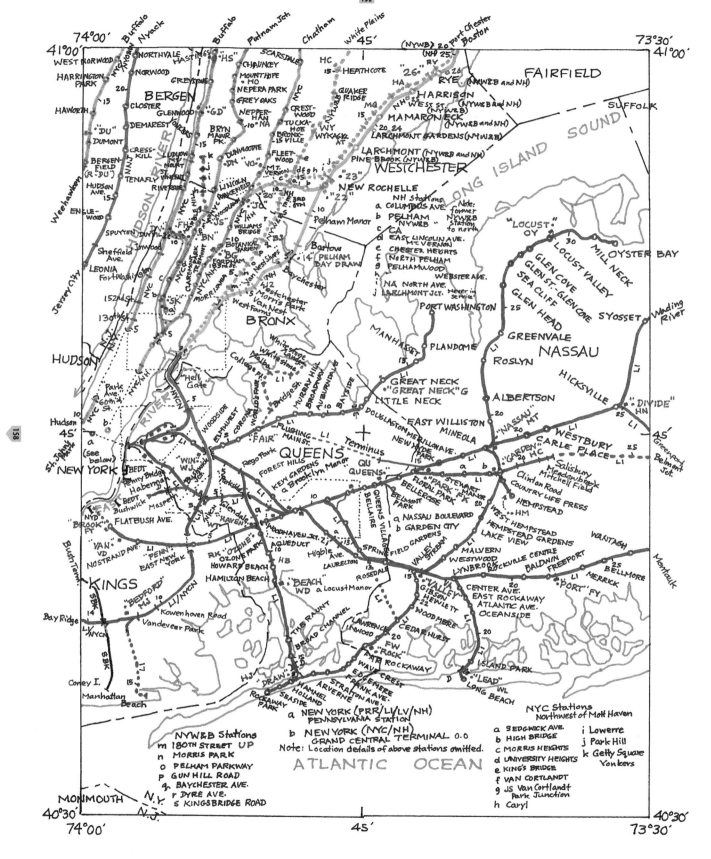

New York City, NY — 159 — CT, NJ, NY

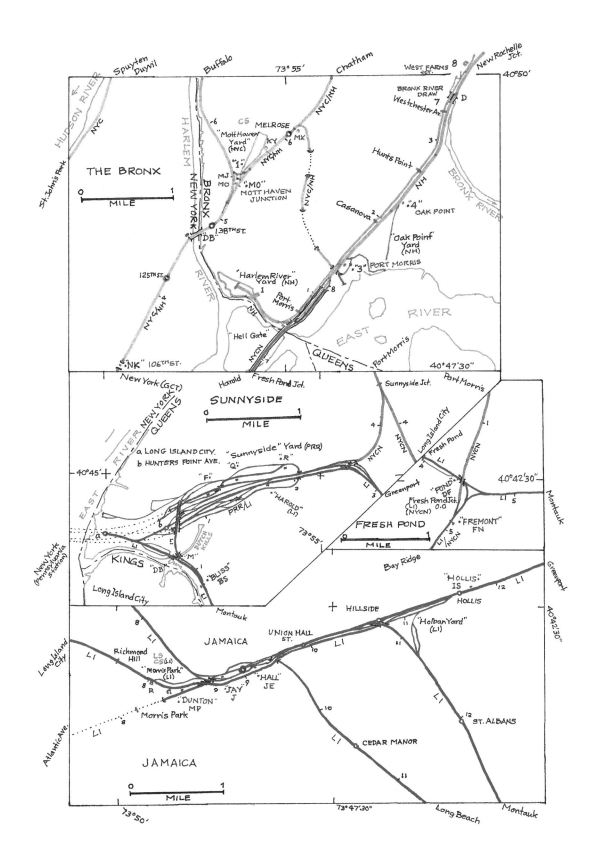

159A The Bronx, NY **159B** Sunnyside, NY

159C Fresh Pond, NY **159D** Jamaica, NY

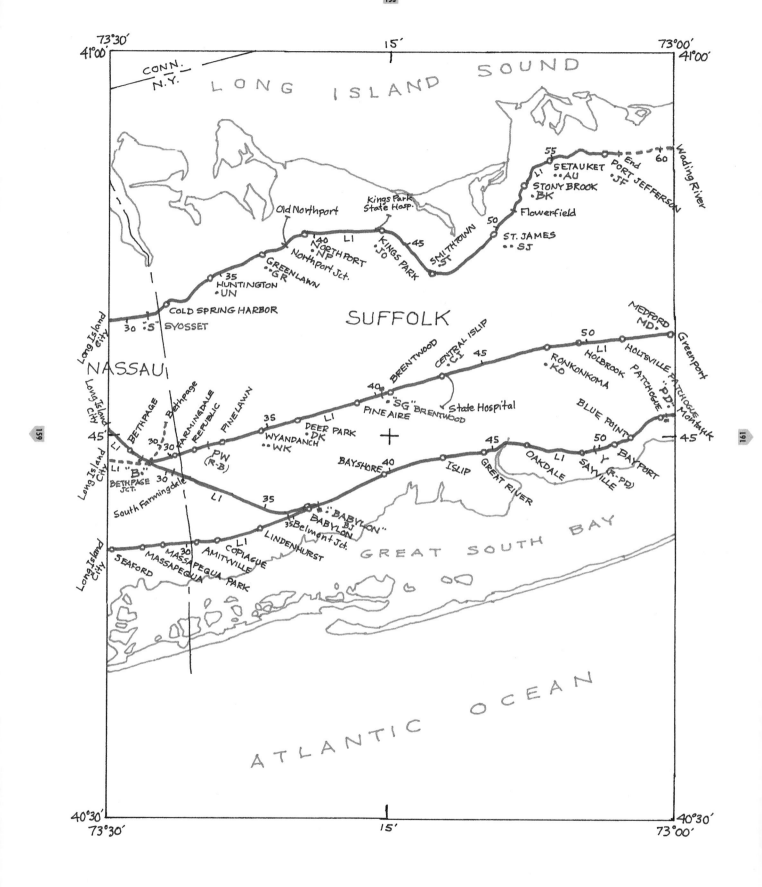

Babylon, NY

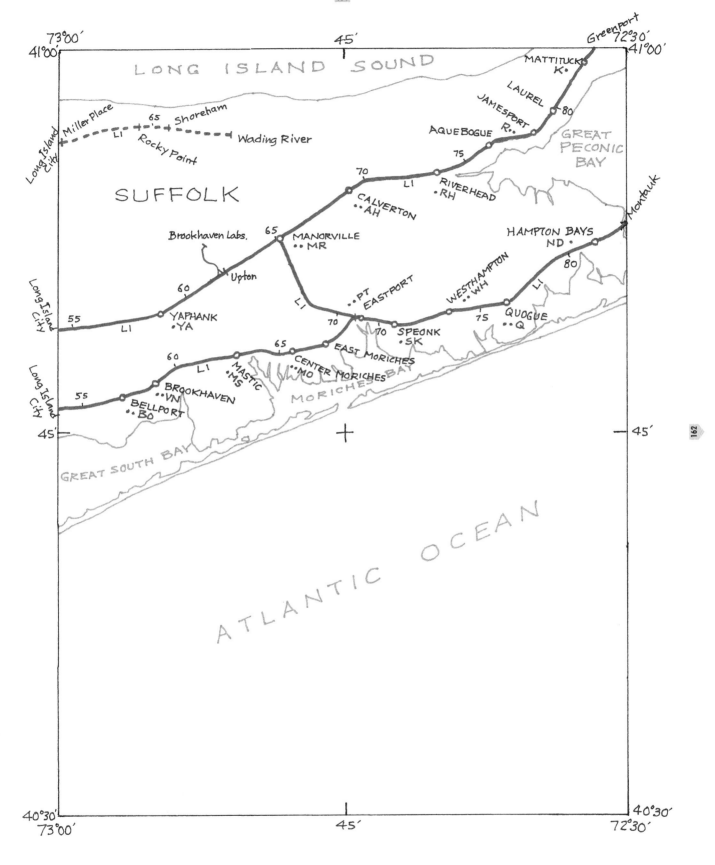

Manorville, NY

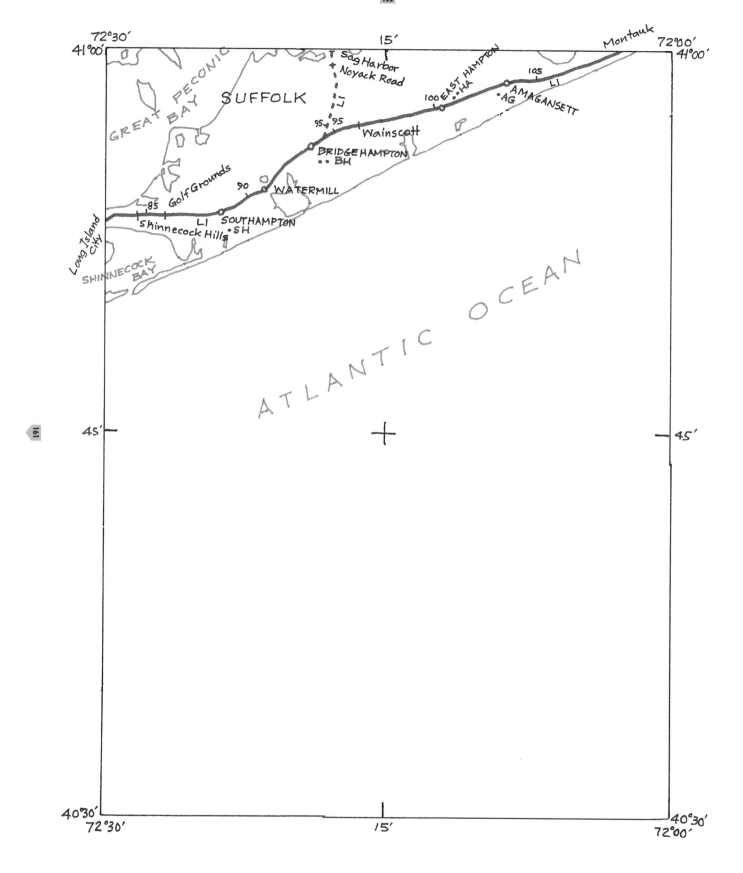

East Hampton, NY

Appendix

RAILROADS IN THE ATLAS

The abbreviations "Rly." and "RR" are used for "Railway" and "Railroad," respectively.

A&S: Addison & Susquehanna RR
A&STL: Adirondack & St. Lawrence RR
ARA: Arcade & Attica RR
AV: Aroostook Valley RR

B&A: Boston & Albany RR
B&G: Bennington & Glastenbury RR
B&H: Bath & Hammondsport RR
B&L: Burlington & Lamoille RR
B&M: Boston & Maine RR
B&O: Baltimore & Ohio RR
B&S: Buffalo & Susquehanna RR
BAR: Bangor & Aroostook RR
BC: Barre & Chelsea RR
BCC: Brooklyn Cooperage Co.
BCK: Buffalo Creek RR
BE&C: Bradford, Eldred & Cuba RR
BEDT: Brooklyn Eastern District Terminal RR
BG: Bethel Granite RR
BH: Bridgton & Harrison Rly.
BML: Belfast & Moosehead Lake RR
BMRR: Bombay & Moira RR
BMS: Berlin Mills Rly.
BO&M: Bangor, Oldtown & Milford RR
BRB&L: Boston, Revere Beach & Lynn RR
BRIS: Bristol RR
BS: Branford Steam RR
BTCo.: Boston Terminal Co.

C&B: Chester & Becket RR
C&E: Columbus & Erie RR

C: Concord RR
CC: Connecticut Company
CKMR: Catskill Mountain RR
CLP: Clarendon & Pittsford RR
CLR: Cranberry Lake RR
CN: Canadian National Rlys.
CNJ: Central Railroad Co. of New Jersey
CNYS: Central New York Southern RR
CNYW: Central New York & Western RR
CPA: Coudersport & Port Allegheny RR
CPI: Crown Point Iron Co. RR
CPR: Canadian Pacific Rly.
CV: Central Vermont Rly.

D&H: Delaware & Hudson RR
D&N: Delaware & Northern RR
DL&W: Delaware, Lackawanna & Western RR
DMM: Dansville & Mount Morris RR
DR: Deer River RR

ERIE: Erie RR

FC: Fulton Chain Rly.
FJG: Fonda, Johnstown & Gloversville RR
FR: Fore River RR

G&D: Goshen & Deckertown Rly.
G&U: Grafton & Upton RR
G&W: Glenfield & Western RR
GJ: Greenwich & Johnsonville Rly.
GNW: Genesee & Wyoming RR

GR: Grassy River RR
GT: Grand Trunk Rly.

H&M: Hudson & Manhattan RR
H&NH: Hartford & New Haven RR
H&W: Hardwick & Woodbury RR
H: Hampden RR
HMR: Hoboken Manufacturers RR
HR: Hannawa RR
HTW: Hoosac Tunnel & Wilmington RR
HV: Hickory Valley RR

INT: International Rly.

JW&NW: Jamestown, Westfield & Northwestern RR

K&E: Kane & Elk RR
K&P: Kennebec & Portland RR
KACL: Keesville, Ausable Chasm & Lake Champlain RR
KCL: Kennebec Central RR
KNOX: Knox RR

L&HR: Lehigh & Hudson River RR
LAN: Lancaster RR
LBR: Lowville & Beaver River RR
LC&M: Lake Champlain & Moriah RR
LI: Long Island RR
LNE: Lehigh & New England RR
LR: Lime Rock RR
LV: Lehigh Valley RR

M&S: Middleburgh & Schoharie RR
M&U: Middletown & Unionville RR
MB: Medway Branch RR
MC: Michigan Central RR
MDG: Manchester, Dorset & Granville RR
MEC: Maine Central RR
MOL: Marcellus & Otisco Co.'s RR
MONSON: Monson RR
MOV: Moshassuck Valley RR
MT&E: Morristown & Erie RR
MT: Massena Terminal RR
MTJK&R: Mt. Jewett, Kinzua & Riterville RR
MVYD: Martha's Vineyard RR
MWCR: Mount Washington Rly.

N&STL: Norwood & St. Lawrence RR
NAN: Nantucket RR
NANC: Nantucket Central RR
NAP: Narragansett Pier RR
ND&C: Newburgh, Dutchess & Connecticut RR
NH: New York, New Haven & Hartford RR
NIJ: Niagara Junction Rly.
NJ&NY: New Jersey & New York RR
NJ: Napierville Junction Rly.

NJT: New Jersey Terminal RR
NKP: New York, Chicago & St. Louis RR
NLI: National Lead Co.
NNJ: Northern RR of New Jersey
NSTC&T: Niagara, St. Catherines & Toronto Rly.
NY&GL: New York & Greenwood Lake Rly.
NY&P: New York & Pennsylvania Rly.
NYC: New York Central RR
NYCN: New York Connecting RR
NYD: New York Dock Rly.
NYNE: New York & New England RR
NYO&W: New York, Ontario & Western Rly.
NYS&W: New York, Susquehanna & Western RR
NYW&B: New York, Westchester & Boston Rly.

P&C: Poughkeepsie & Connecticut RR
P&E: Poughkeepsie & Eastern RR
PC: Potato Creek RR
POM: Ponemah Mills RR
PRAT: Prattsburgh Rly.
PRR: Pennsylvania RR
PS&N: Pittsburgh, Shawmut & Northern RR
PTM: Portland Terminal Co.

QC: Quebec Central Rly.

R: Rutland RR
RDG: Reading Co.
RL&M: Rangley Lakes & Megantic RR
RL: Raquette Lake Rly.
RPT: Rockport RR
RV: Rahway Valley RR

S&NY: Susquehanna & New York RR
S&S: Saratoga & Schuylerville RR
S&T: Sheffield & Tionesta Rly.
SA: Seaview Ave. Rly.
SBK: South Brooklyn Rly.
SIRT: Staten Island Rapid Transit Rly.
SJL: St. Johnsbury & Lake Champlain RR
SM: South Manchester RR
SMR: Sterling Mountain Rly.
SNE: Southern New England RR
SRRL: Sandy River & Rangeley Lakes RR
SSL: Skaneateles Short Line RR
ST: Springfield Terminal Rly.
STJL: St. Joe Lead RR
SV: Suncook Valley RR
SV&SL: South Vandalia & State Line RR
SVR: Schoharie Valley Rly.

TH&B: Toronto, Hamilton & Buffalo Rly.
TIV: Tionesta Valley RR
TMC: Temiscouata Rly.
TU: Troy Union RR
TV&C: Tonawanda Valley & Cuba RR

U&D: Ulster & Delaware RR
UER: United Electric Rlys.
UF: Union Freight RR
US: United States Government
USA: United States Army
USN: United States Navy
UV: Unadilla Valley Rly.

VAL: Valley RR
VB: Van Buren Bridge Co.

W&B: Wellsville & Buffalo RR
W&NO: Wharton & Northern RR
WAB: Wabash
WC&PC: Wellsville, Coudersport & Pine Creek RR
WEST: West River RR
WNY&P: Western New York & Pennsylvania Rly.
WR: Woodstock Rly.
WRB: Wood River Branch RR
WRR: White River RR
WWF: Wiscasset, Waterville & Farmington Rly.

Notes on the Maps

1. **Fort Kent, Maine:** The northern end of the historic federal highway designated U.S. 1, Fort Kent is located on the south bank of the St. John River on the international border with Canada.

2. **Madawaska, Maine:** The northernmost point on the national railroad network of the United States is at Madawaska (NW), on the Bangor & Aroostook Railroad.

3. **Van Buren, Maine:** The north end of the Bangor & Aroostook Railroad main line is at Van Buren (SW), 259.4 miles north of Searsport, which is located at the head of Penobscot Bay on the Maine coast.

4. **Winterville, Maine:** The Bangor & Aroostook branch through Winterville (NE) provided a sixty-nine-mile, shorter and more direct route to Fort Kent than the main line to Van Buren.

5. **Presque Isle, Maine:** The Canadian Pacific Railway crossed the Canadian border in order to directly serve Presque Isle, Maine (SE).

6. **Caribou, Maine:** Despite close proximity, there was no physical connection at Fort Fairfield (NW) between the Bangor & Aroostook Railroad and the Canadian Pacific Railway.

7. **Oakfield, Maine:** Oakfield (SE) was the northernmost crew change point on the Bangor & Aroostook Railroad. Today the station houses the Oakfield Railroad Museum.

8. **Houlton, Maine:** Both the Bangor & Aroostook Railroad and the Canadian Pacific Railway served Houlton, Maine (SW) but did not physically connect.

9. **Megantic, Quebec:** Megantic (SW) was a crew change point on the direct route of the Canadian Pacific Railway to the Maritime Provinces of eastern Canada; this line traversed the state of Maine from west to east.

10. **Jackman, Maine:** Canadian Pacific steam locomotives replenished their coal supply at Holeb, Maine (SW), 86 miles west of the division point at Brownville Jct., Maine.

11. **Somerset, Maine:** The northernmost branch line of the Maine Central Railroad crossed under the Canadian Pacific Railway five miles south of Kineo, Maine (SE), on the shores of beautiful Moosehead Lake.

12. **Millinocket, Maine:** In 1899, both the Great Northern Paper Company and the Great Northern Hotel were established in Millinocket (SE).

13. **Mattawamkeag, Maine:** The historic junction of the east-west Canadian Pacific and the north-south Maine Central, Mattawamkeag (SW) would later become the eastern end and zero milepost of the "freight main line" of the Guilford Transportation System. This line ended some 753 miles to the southwest at Sunbury, Pennsylvania.

14. **Danforth, Maine:** The Canadian Pacific Railway operated by trackage rights over the Maine Central for 55 miles, from Mattawamkeag to Vanceboro.

15. **Vanceboro, Maine:** At the eastern end of the Maine Central Railroad main line, 251 miles east of Portland, Maine (SW), Vanceboro was a railroad town.

16. **Connecticut Lakes, N.H.:** There was no railroad on the United States side, but the Canadian Pacific Railway formerly ran directly adjacent to the international boundary, in Canada.

17. **Kennebago, N.H.:** This Maine Central branch line ended at the fly-fishing center of Kennebago Lake (SW).

18. **Carrabassett, Maine:** The former end of the Maine Central Bigelow branch is now the focal point of the Bigelow Preserve.

19. Greenville, Maine: The westernmost point of the Bangor & Aroostook Railroad was located at Greenville, Maine (NE).

20. Brownville Jct., Maine: This community (NE) was the interchange point between the Bangor & Aroostook Railroad and the Canadian Pacific Railway and a crew change point on the latter railroad.

21. Milo, Maine: South Lagrange (SW) was the southerly junction of a by-pass on the Bangor & Aroostook Railroad, which permitted through trains to avoid Brownville Jct. and Milo (NW).

22. Passadumkeag Mountain, Maine: This mountain, visible from the Canadian Pacific and the Maine Central, rises abruptly some 960 feet above the surrounding landscape (SW).

23. Princeton, Maine: This Maine Central branch formerly reached Princeton (SE), which is now a small-mouth bass fishing center.

24. Calais, Maine: The Canadian Pacific Railway interchanged with the Maine Central via a short "belt line" between two stations, one in Canada and one in Maine, both named Milltown Jct. (SW). The president of the Canadian Pacific Railway, William Cornelius VanHorne, had his summer home on the island immediately northeast of St. Andrews, New Brunswick (SE).

25. Morristown, N.Y.: The New York Central's former St. Lawrence Division 134-mile main line north from Utica ended at Ogdensburg (SE).

26. Ogdensburg, N.Y.: At Ogdensburg (SW), the Rutland, which ran some 121 miles east to Alburgh, Vermont, did not interchange with either of the New York Central lines from both Syracuse and Utica.

27. Massena, N.Y.: The New York Central's branch to the Canadian capital at Ottawa originally began at Tupper Lake Jct., over 60 miles south of Helena (NE). The Massena Terminal Railroad, established in 1902, served a large copper plant immediately northeast of Massena (NW). The Hannawa Railroad, south of Potsdam (SW), was built in 1899 and served a pulp mill in Hannawa Falls.

28. Malone, N.Y.: The New York Central Railroad had trackage rights over the Rutland Railroad for 37 miles west from Malone (NW) to Norwood, the junction of its Syracuse-to-Massena main line.

29. Dannemora, N.Y.: Dannemora (SE), on the Delaware & Hudson branch west from Plattsburg, is the remote location of New York State's Clinton State Prison, built in 1845 with the objective of employing prisoners as miners.

30. St. Albans, Vt.: The Rutland Railroad main line across Grand Isle (SW) and Isle La Motte (NW) is now a uniquely beautiful "rail trail" for walkers and bicyclists.

31. Mount Mansfield, Vt.: The only railroad line on this map which survives today (albeit in an out-of-service status) is the Lamoille Valley Railroad, running westward from St. Johnsbury to Morrisville (SE).

32. Newport, Vt.: The Canadian Pacific Railway line through Newport (NE) was once a part of the Connecticut and Passumpsic Rivers Railroad, Boston & Maine Railroad subsidiary.

33. Island Pond, Vt.: Island Pond, Vermont (NW), was a lonely crew change point on the original Grand Trunk Railway, 148 miles northwest of Portland, Maine. In 1946, two trains to and from Montreal and two Montreal-Portland passenger through trains served Island Pond each day.

34. Coos, N.H.: This remote northern New Hampshire town was near the Nay-Fogg Mine at Copperville (SE), 6 miles east.

35. Rumford, Maine: Rumford (SE), the birthplace of presidential candidate and former Maine governor and senator, Edmund S. Muskie, was where the Maine Central branch from Livermore Falls ended in 1946.

36. Farmington, Maine: A network of narrow-gauge lines and branches once served northern Franklin County, Maine. An active railroad historical society preserves the history of the former Sandy River and Rangeley Lakes narrow-gauge system.

37. Waterville, Maine: Waterville (SE) was the location of the principal locomotive and car shops of the Maine Central Railroad.

38. Burnham, Maine: The Belfast & Moosehead Lake Railroad survives today as a tourist railroad, based at Belfast, at the head of Penobscot Bay.

39. Bangor, Maine: Northern Maine Jct. (NW) was a major interchange for the potato shipments via rail south over the Maine Central, Boston & Maine, and New Haven Railroads.

40. Ellsworth, Maine: The famous summer passenger train "Bar Harbor Express" originally ended its run from New York City at Mount Desert Ferry (SE), where vacationers boarded a ferry to Bar Harbor. By 1946, the rail trip ended at Ellsworth (SW), from which a bus continued to Bar Harbor.

41. Harrington, Maine: The Maine Central branch to Calais and Eastport passes along the less populous "Down East" portion of the Maine coast, long a favorite haunt of dedicated cruising yachtsmen.

42. Eastport, Maine: In 1921, Franklin D. Roosevelt, who would later be elected president, was stricken with polio at his summer home on Campobello Island, N.B., and traveled to his home in New York City by rail, starting from Eastport's Sea Street Station (NE).

43. Clayton, N.Y.: Cape Vincent, New York (SW), known as "the western gateway to the Thousand Islands," was served by a branch of the New York Central Railroad.

44. Philadelphia, N.Y.: Two of the three principal branch lines serving northern New York State crossed at Philadelphia (SE), a point almost equidistant from Syracuse and Utica, their respective origin points.

45. Gouverneur, N.Y.: The St. Joe Lead Company operated a branch to its zinc mine at Balmat (SW), east of Gouverneur (NW).

46. Newton Falls, N.Y.: The Grasse River Railroad formerly ran from Childwold (SE) on the New York Central some 16 miles west to the northern end of Cranberry Lake (SW).

47. Lake Clear, N.Y.: At Tupper Lake (SW), the New York Central formerly operated a branch line which ran 126 miles north to Ottawa, the capital of Canada.

48. Ausable Forks, N.Y.: Two Delaware & Hudson branches, both from Plattsburg, New York, ran to Lake Placid (NW) and Ausable Forks (NE).

49. Burlington, Vt.: The Central Vermont Railway entered the lakefront station at Burlington (NE) via a short tunnel just to the north.

50. Montpelier, Vt.: Montpelier (NE) is the smallest state capital city in the United States. Adjacent to it, Barre, Vermont (SE), is the granite capital of the world and was served by the Barre & Chelsea Railroad.

51. St. Johnsbury, Vt.: At White Mountains Transfer (SE), just east of Wells River, the Boston & Maine Railroad formerly maintained and operated its northernmost interlocking tower, which had a mechanical machine and lower quadrant semaphore signals.

52. Whitefield, N.H.: At the diamond crossing of the Boston & Maine and Maine Central Railroads, just east of Whitefield (NE), a ball signal was installed to protect train movements using this railroad crossing at grade. A similar signal also existed at Waumbek Jct. (NE), five miles east.

53. Mount Washington, N.H.: The peak of Mount Washington is the highest point in New England (6,288 ft.) and since 1869 it has been served by a cog railway. Both the Boston & Maine and Maine Central Railroads served the large paper mills of Berlin, New Hampshire (NE).

54. Norway, Maine: The two-foot narrow-gauge Bridgton & Harrison Railway crossed the north end of Long Lake on a causeway to serve its terminus at Harrison (SE).

55. Lewiston, Maine: The Canadian National (originally the Grand Trunk Railway) and the Maine Central crossed and recrossed each other three times in the nine miles between Danville Jct. (SW) and Mechanic Falls (SW).

56. Augusta, Maine: The capital of Maine, Augusta (NW) was served by the Maine Central main line or "Front Road." In more recent years, all through Maine Central (and later Springfield Terminal Rail System), trains were rerouted via Lewiston and Belgrade (NW), a route known as the "Back Road" between Royal Jct. and Waterville. This route is only three miles longer than the main line via Augusta.

57. Rockland, Maine: The Maine Central Rockland Branch brought summer tourists and vacationers bound for the many lovely Maine coast islands of Penobscot Bay, Isle au Haut Bay, Eggemoggin Reach, and Blue Hill Bay.

58. Searsport, Maine: The main line of the Bangor & Aroostook Railroad began on the northern shore of Penobscot Bay at Mack Point in Searsport (NW).

59. Pulaski, N.Y.: At Pulaski (SE), a New York Central predecessor railroad, the Rome, Watertown, and Ogdensburg ran along the cold and snowy shores of Lake Ontario some 175 miles westward to Suspension Bridge, just north of Niagara Falls.

60. Watertown, N.Y.: At Watertown (NW), the Syracuse-to-Massena main line of the New York Central was connected to four branch lines and was served by seventeen passenger trains daily in 1946.

61. Lowville, N.Y.: Two short lines, the Lowville & Beaver River Railroad (NW) and the Glenfield & Western Railroad (SW), connected with the Utica to Ogdensburg main line of the New York Central at Lowville and Glenfield, respectively. The latter railroad had been abandoned by 1946.

62. Raquette Lake, N.Y.: Deep in the heart of the Adirondack Forest Preserve, which had been established in 1885, the Raquette Lake Railway formerly ran from its junction with the New York Central's Herkimer-to-Malone line to the southwest corner of Raquette Lake (NE) from 1900 to 1933.

63. Indian Lake, N.Y.: To serve the war effort, in 1944 the National Lead Company, with the assistance of the U.S. Government, built a branch from the end of the Delaware & Hudson Railroad at North Creek to serve its iron and titatium mine and factory at Tahawus, just off the northeast corner of this map (Kudish, 1985, p. 58).

64. North Creek, N.Y.: In 1901, Vice President Theodore Roosevelt, vacationing on nearby Mt. Marcy, received the urgent message that President William McKinley had died from wounds inflicted by an assassin in Buffalo. Roosevelt traveled by special train from North Creek (SW) south to Albany, and then west to Buffalo.

65. Whitehall, N.Y.: Extensive marble deposits near Rutland and Pittsford, Vermont (SE), were served by the Clarendon & Pittsford Railroad.

66. Rutland, Vt.: The "heart" of the Rutland Railroad was its namesake city (SW). Shops, yards, roundhouse, and the junction of its two main lines from the south and its Bellows Falls branch were all located here.

67. White River Jct., Vt.: The most significant junction point in the northern Connecticut River Valley, where in 1946 passengers from New York City and Boston could connect with trains to and from Montreal and Quebec City, Canada.

68. Plymouth, N.H.: Plymouth (NE) is the gateway to the southern White Mountains, Pemigewasset wilderness camping, hiking, and hotel resort areas.

69. Laconia, N.H.: Beautiful Lake Winnipesaukee (SW) was served from both sides by the Boston & Maine Railroad.

70. Sebago Lake, Maine: When the Bridgton & Harrison Railroad (NW/NE) was abandoned in 1941, Ellis D. Atwood, who owned a cranberry plantation in South Carver, Massachusetts, purchased engines, cars, track, and equipment that would otherwise have been scrapped. At the end of World War II, he created the Edaville Railroad at his cranberry plantation, where it was enjoyed by thousands for over forty years.

71. Portland, Maine: A deepwater Atlantic Ocean port that attracted the Grand Trunk Railway from Canada and became the primary connection between the Boston & Maine and Maine Central

Railroads. Portland Union Station, a large, imposing stone structure, was built in 1888 and demolished in 1961. (See also detail map 71A, **Portland**.)

72. **Brunswick, Maine:** The Bath Iron Works (NW), established in 1884 and a major shipbuilding yard for the U.S. Navy, was served by the Maine Central.

73. **Niagara Falls, N.Y.:** The famous Suspension Bridge (SE), a railroad bridge originally constructed across the Niagara River Gorge in 1855 by John Roebling, and the extraordinary attraction of the falls themselves have combined to make Niagara Falls an important international rail interchange and tourist attraction. (See also detail map 73A, **Suspension Bridge**.)

74. **Tonawanda, N.Y.:** The western terminus of the New York State Barge Canal, Tonawanda (SW) was served by three Class I railroads—the Erie, the Lehigh Valley, and the New York Central. (See also detail map 74A, **Tonawanda**.)

75. **Medina, N.Y.:** Five east-west New York Central lines cross this map (from *top* to *bottom*): The former Rome, Watertown, and Ogdensburg from Pulaski to Suspension Bridge; the "Falls Road" from Rochester to Suspension Bridge; the "West Shore" from Weehawken, New Jersey, to Buffalo; the "Water Level Route" main line from New York City to Buffalo; and, finally, the Tonawanda Branch from Batavia to Tonawanda.

76. **Rochester, N.Y.:** Five Class I railroads served Rochester (SE) in 1946; four of them (Baltimore & Ohio, Erie, Lehigh Valley, and Pennsylvania) crossed the "West Shore" of the New York Central. This same West Shore line was later saved by Conrail as a through freight bypass route around Rochester. (See also detail map 76A, **Rochester**.)

77. **Wayneport, N.Y.:** The New York Central chose Wayneport (SW), where both its main line and West Shore closely paralleled each other, to construct both a major coaling station and an icing station. In addition, the New York State barge canal was located immediately to the south of these facilities.

78. **Oswego, N.Y.:** Oswego's (NE) dreams of becoming the major Lake Ontario commercial port, rivaling Buffalo, never fully materialized, but it did attract a major branch of the Erie Canal and three Class I railroads: the Delaware, Lackawanna & Western, the New York Ontario & Western, and the New York Central. (See also detail map 78A, **Oswego**.)

79. **Syracuse, N.Y.:** Dominated by the New York Central, with a fascinating history of complex junctions, relocated lines, and stations, Syracuse was able to attract only one other Class I railroad, the Delaware, Lackawanna & Western, which tunneled under the complex track work at Syracuse Jct. (SE) just west of the city center. (See also detail map 79A, **Syracuse Junction**.)

80. **Oneida, N.Y.:** Before the Lehigh Valley abandoned its line from Canastota (SE) north to Camden (NE) sometime between 1937 and 1942, it had been sufficiently important to have been grade-separated to run over the New York Central main line.

81. **Utica, N.Y.:** At Schuyler Junction (SE), the New York Central had an interesting two-level interconnection between its main line and West Shore routes. And on the West Shore, some five miles east, was a former, sizable yard and roundhouse at Frankfort (SE), as shown on the USGS Utica 15-foot quadrangle of 1898. (See also detail map 81A, **Utica**.)

82. **Little Falls, N.Y.:** Immediately east of Little Falls (SW), there was a sharp curve—called the "Gulf Curve"—in the New York Central main line. This was the site of a serious passenger train wreck in April 1940. Later, this curve was "eased" to safely allow for faster train speeds.

83. **Gloversville, N.Y.:** The Northville (SE) branch of the Fonda, Johnstown & Gloversville Railroad was abandoned in 1930 when the Sacandaga Reservoir (SE) was created.

84. **Saratoga Springs, N.Y.:** The Saratoga & Schuylerville Railroad (SE) was created in May 1945, after having been part of the Boston & Maine Railroad.

85. **Manchester, Vt.:** Even though the Delaware & Hudson Railroad branch line at Cambridge was only 35 miles from Albany geographically, its milepost mileage was 137 miles from Albany via the intervening D&H junction points at Whitehall, New York, and Castleton, Vermont.

86. **Chester, Vt.:** Chester Depot (NE) was later made famous as the northern terminus of the Green Mountain Railroad, part of the original "Steamtown" operation just north of Bellows Falls, Vermont.

87. **Claremont, N.H.:** The seven-mile-long Springfield Terminal Railway (NW) would, in 1995, have its corporate home partially applied to what had been the much larger railroad systems of the Boston & Maine and Maine Central Railroads—namely, the Springfield Terminal System.

88. **Concord, N.H.:** The capital of New Hampshire, Concord (SE), was for many years a historic locomotive and car repair shop of the Boston & Maine Railroad.

89. **Epping, N.H.:** The Suncook Valley Railroad (SW–NW) was, prior to 1924, part of the Boston & Maine Railroad. It was abandoned late in 1952. The original Portsmouth & Concord Railroad between Candia and Hooksett was abandoned in 1862, but its cuts and fills are still visible on USGS 7½-minute maps!

90. **Dover, N.H.:** North Berwick, Maine (NE), was one of three junctions of the eastern (via Portsmouth) and the western (via Dover) routes of the Boston and Maine Railroad between Boston and Portland. The Dover (SW) centralized train control installation was one of the earliest (1931) such main-line, double-track, signaling projects in the United States.

91. **Biddeford, Maine:** The former "AR" tower (NW) contained a similar (1931) installation to Dover.

92. **Silver Creek, N.Y.:** This map shows the multiple drawbridges over the Welland Canal (NW; later replaced in part by a rail tunnel under the canal). In the United States, the joint Nickel Plate Road/Pennsylvania double-track (SE) between Blasdell (map 93B) and Brockton (map 111), New York, is clearly shown.

93. **Buffalo, N.Y.:** Here is the most complex array of multiple railroad junctions east of Chicago! Buffalo in 1946 was served by no fewer than ten Class I railroad companies. (See also detail maps 93A, **Buffalo**; 93B, **Lackawanna**; 93C, **East Buffalo**; 93D, **Black Rock** and 93E, **Depew**.)

94. **Batavia, N.Y.:** Four Class I railroads closely paralleled each other eastward from Buffalo toward Batavia (NE): Delaware, Lackawanna & Western; Erie; Lehigh Valley; and the New York Central. The dramatic Portage Viaduct (SE), carrying Erie's Buffalo line over the Genessee River, and the Pennsylvania Railroad Rochester Branch 190' below are found at SE.

95. **Caledonia, N.Y.:** Wayland (SE) was the northeastern end of the Pittsburgh, Shawmut and Northern Railroad, which ran 147 miles southwestward to Brockway, Pennsylvania. By 1947, this railroad had been abandoned.

96. **Stanley, N.Y.:** The first Lehigh Valley crew change point east of Buffalo was located at Manchester (NE).

97. **Geneva, N.Y.:** No fewer than four parallel north-south Lehigh Valley main and branch lines ran through the "Finger Lakes Region" of New York State.

98. **Cortland, N.Y.:** Cortland Junction (SE) was once a passenger transfer point between the Delaware, Lackawanna & Western and Lehigh Valley Railroads. Both lines served numerous milk creameries before the arrival of the refrigerated milk truck.

99. **Cazenovia, N.Y.:** At one time, a branch of the Erie Canal, named the Chenango Canal (NE), was built southwestward from the Erie Canal at Utica toward Norwich, New York (SE). Initial plans were to continue to Binghamton.

100. **Bridgewater, N.Y.:** The Unadilla Valley Railway line from New Berlin Jct. (map 119) to New Berlin (SW) and the Edmeston branch were originally part of the New York, Ontario & Western railway system. John Hudson notes that the "Borden's Condensed Milk Company plant in New Berlin was the largest such facility in the world when it canned 'Eagle Brand' condensed milk in a former textile mill" (2002, p. 36).

101. **Cooperstown, N.Y.:** To overcome the steep eastward grade from Schenevus (SW) to Dante (SE), the Delaware & Hudson built a 12-mile-long "low-grade" line between these two points.

102. Rotterdam Jct., N.Y.: This New York Central junction (NE), between its "West Shore" line and the Stuyvesant-to-Hoffmans freight by-pass through Selkirk Yard, also marked the westernmost point on the Boston & Maine Railroad, 210 miles from Boston, Massachusetts.

103. Albany, N.Y.: In 1924, the New York Central Railroad began operating a new hump classification yard at Selkirk (SW), 8 miles south of Albany. This yard project included a 42-mile freight by-pass from Stuyvesant (map 122), south of Albany, to Hoffmans (map 102), west of Schenectady, using 21 miles of the "West Shore" line and crossing the Hudson River on a massive bridge, named for Alfred H. Smith, who was then president of the New York Central System. Selkirk Yard is now the primary hump classification yard for New York State and the New England region. (See also detail map 103A, **Albany-Troy** and detail map 103B, **Schenectady**.)

104. Hoosac, Mass.: The Boston & Maine Railroad maintained an electrified section through the Hoosac Tunnel (SE, built in 1874) to avoid dangerous steam locomotive exhaust through this nearly 4.75-mile-long tunnel bore. The portion of the Rutland Railroad from Chatham, New York (SW), to Bennington, Vermont (NE), was informally known as the "Corkscrew Division."

105. Greenfield, Mass.: The Boston & Maine Railroad maintained a freight classification yard at East Deerfield at the junction of its east-west freight main line and its north-south Connecticut River line. Also, in 1931, one of the earliest main line centralized control installations was installed between Montague (SE) and Soapstone (SW), with the control machine in the Greenfield dispatcher's office. (See also detail map 105A, **Greenfield**.)

106. Keene, N.H.: A "paired-track" train operation existed between the Boston & Maine Railroad and the Central Vermont Railway between East Northfield, Massachusetts (SW), and Brattleboro, Vermont (map 105, NE). The Boston & Maine line ran along the east bank of the Connecticut River, and the Central Vermont was located on the west bank.

107. Ayer, Mass.: Ayer, Massachusetts (SE), was another major junction on the Boston & Maine Railroad where the east-west Boston-to-Rotterdam-Jct. main line crossed the north-south Worcester-to-Portland route, including the remaining parts of what are still known as the "Worcester, Nashua and Portland Division."

108. Lowell, Mass.: This map clearly shows the tangle of Boston & Maine main lines and branches, many of which were already abandoned by 1946, which served the northern suburbs and the extensive water-powered mill complexes of the Merrimack River Valley. (See also detail map 108A, **Manchester** [with detail map 105A].)

109. Salem, Mass.: The so-called "Eastern Route" of the Boston & Maine crossed the Atlantic coastal fens and marshes on its direct route "down east" from Boston, Massachusetts, to Portland, Maine.

110. Westfield, N.Y.: Here is the junction between the New York Central Railroad's Buffalo-to-Chicago main line at Westfield and the Jamestown, Westfield, and Northwestern electric interurban railroad. This electric interurban line ran south along Chautauqua Lake, some 30 miles to the city of Jamestown, N.Y., where a connection could be made with the Erie Railroad main line.

111. Dunkirk, N.Y.: Dunkirk, on Lake Erie, was the western terminus of the original Erie Railroad, built to a 6-foot track gauge in 1851 westward from Piermont Dock on the Hudson River, just north of New York City. This 6-foot gauge, later converted to standard gauge, meant that bridge clearances on the Erie were higher and wider than those of other railroads.

112. Salamanca, N.Y.: Salamanca was the location of the zero milepost of one of the original railroad companies that were later merged to form the greater Erie. This railroad, the Atlantic and Great Western, was completed in 1864 and stretched from Salamanca, New York, to Dayton, Ohio, some 390 miles to the west.

113. Olean, N.Y.: Shown on this map is the "River Line" of the Erie Railroad. Built in 1910, it ran between River Junction and Cuba Junction (SW). Situated on this line was the 3,119-foot Belfast

Viaduct. This line permitted freight trains to by-pass two severe 1 percent grades, at Tip Top and Summit, on the main line between Hornell and Cuba Junction.

114. Hornell, N.Y.: Called the "Heart of the Erie," Hornell was a crew change point and the location of a large freight yard and the steam locomotive shops of the Erie Railroad.

115. Corning, N.Y.: Both the Erie Railroad main line and New York Central branch to Williamsport ran through the streets of downtown Corning (SE).

116. Elmira, N.Y.: Four Class I railroads ran through Elmira in 1946. Today, only the Norfolk Southern runs through Elmira, using the tracks of the former Erie.

117. Owego, N.Y.: The Lackawanna Railroad used a switchback (NW) to reach Elmira. Route 17 now runs in what was once the Lackawanna main line roadbed.

118. Binghamton, N.Y.: Here is the junction (SW) of two east-west major railroads, the Lackawanna and the Erie, and the north-south "bridge" railroad, the Delaware & Hudson. (See also detail map 118A, **Binghamton**.)

119. Sidney, N.Y.: Oneonta (NE) was a crew change point and a locomotive and car shop on the Delaware & Hudson Railroad.

120. Arkville, N.Y.: Creameries were so important along the New York Central Catskill Mountain branch (former Ulster & Delaware Railroad) that some were even listed as stations on certain maps.

121. Grand Hotel, N.Y.: This is the eastern part of the Catskill Mountain resort area, and it includes the abandoned Ulster & Delaware Railroad branch that ran from Phonecia (SW) to Kaaterskill (SE).

122. Chatham, N.Y.: The abandoned New Haven Railroad lines at the bottom of the map, especially the main route to "Campbell Hall" (SE), represent the initial rail routes between New England and the Poughkeepsie Railroad bridge, opened to traffic in 1889.

123. Pittsfield, Mass.: The summit or "top" of the Boston & Albany Division of the New York Central was (and is) just east of Washington, Massachusetts (NE). Between Chester (map 124 NW) and Washington (NE), the westward gradient of 1.65 percent is one of the steepest grades on the New York Central System.

124. Springfield, Mass.: Until abandoned by the New Haven Railroad in 1943, the line north of Williamsburg Jct. (NE) closely paralleled the Boston & Maine's Connecticut River line. Today, part of this right-of-way is used by Interstate 91. (See also detail map 124A, **Springfield**.)

125. Palmer, Mass.: During the period from 1910 to 1920, when the New Haven Railroad controlled the Boston & Maine, a track connection was cut, filled, and graded, and a single track was constructed from Athol Jct. on the Boston & Albany (map 124 SE) to a point near Forest Lake (SW) on the Boston & Maine's Central Massachusetts line from Boston to Northampton, Massachusetts. Built under the name of the Hampden Railroad, it was intended to be a direct connection between the two railroad systems. However, no regular trains ever operated over this connection, and both it and the portion of the Boston & Maine from Forest Lake (SW) and Berlin, Massachusetts (47 miles to the east [map 126, NE]), were abandoned some years prior to 1946.

126. Worcester, Mass.: At Worcester (NW) the famous "State of Maine Express," a summer vacation train from New York City to Portland, Maine, was transferred from the New Haven Railroad to the Boston & Maine Railroad. In 1930, this train left New York at 8:15 p.m. and Worcester at 2:00 a.m. and arrived in Portland at 6:45 a.m. the next morning. (See also detail map 126A, **Worcester** [with detail map 124A].)

127. Boston, Mass.: Using the most complex railroad network in New England, Boston's suburbs were extended outward in the first half of the twentieth century due to the suburban commuter service of the Boston & Albany, Boston & Maine, and New Haven Railroads. (See also detail maps 127A, **Boston**; 127B, **Lexington**; 127C, **Boston North**; 127D, **Newton**; 127E, **Boston South**; 127F, **Readville**; and 127G, **Framingham** [detail maps 127F and 127G are with detail map 124A].)

128. Boston Bay, Mass.: The original overhead electrical operation of the New Haven Railroad was a 600-volt DC installation from Nantasket Jct. (SW) to Pemberton (NW) in 1895. The second was a third-rail operation from Nantasket Jct. west to Weymouth (SW) in 1896. Later, this third rail (the first in the nation) was extended west to Braintree and east to Cohasset (SW). By 1902, both had been abandoned due to safety problems with the third rail and a change in management philosophy.

129. Provincetown, Mass.: This Cape Cod resort and fishing port was served by the farthest eastward extension (120 miles east of Boston) of the so-called "Old Colony Railroad Company," which had been operated under lease by the New Haven since 1893.

130. Corry, Pa.: Titusville (SE), the site of the first U.S. oil wells, was served by both the New York Central and Pennsylvania Railroads.

131. Warren, Pa.: Between 1906 and 1908, the Erie Railroad built a low-grade line through Lottsville (NW) to avoid the severe westbound grade via the original Brady Lake route (Crist, 1993, p. 39).

132. Kinzua, Pa.: In 1900 the Erie Railroad built a steel viaduct at Kinzua (NE) to replace a wrought iron bridge that was reportedly the longest (2,053 ft.) and highest (301 ft.) in the world (Crist, 1993, p. 56).

133. Emporium, Pa.: The summit of the Pennsylvania Railroad's Buffalo-to-Harrisburg route was located at Keating Summit (SE).

134. Galeton, Pa.: Galeton (SE) was the junction of two Baltimore & Ohio Railroad branch lines, which extended north into New York State to interchange with the Erie Railroad at Wellsville and Addison.

135. Wellsboro, Pa.: Both the Erie and New York Central Railroads shared their separate lines with each other.

136. Sayre, Pa.: The principal shops of the Lehigh Valley Railroad were located at Sayre (NE), which was also a crew change point.

137. Towanda, Pa.: A distinctive Lehigh Valley Railroad main line coaling station was located just west of Towanda station (NW).

138. Susquehanna, Pa.: Three famous major viaducts are shown on this map—Kingsley (NW) and Tunkhannock (SW) on the Lackawanna Railroad and Starrucca (NE) on the Erie Railroad.

139. Honesdale, Pa.: The principal junction of the New York, Ontario & Western was at Cadosia (NW).

140. Monticello, N.Y.: Route 17 (now Interstate 86) was built on or along the abandoned railroad grade of the New York, Ontario & Western Railway from Ferndale (NW) to Cook's Falls (NW) and westward to Hancock (map 139, NW).

141. Walkill, N.Y.: At Summitville (SW), two branches of the New York, Ontario & Western joined the main line—one southwest to Port Jervis and Monticello and the other northeast to Kingston (NE).

142. Poughkeepsie, N.Y.: The famous Poughkeepsie Railroad Bridge (SW) was opened in 1889 and rises 212 feet above the Hudson River. For most of its history, it was the primary rail freight route to points east of the Hudson. In 1969, when Penn Central was directed by the Interstate Commerce Commission to include the New Haven in its network, traffic east of Hudson was gradually diverted to the modern classification yard at Selkirk near Albany, across the A. H. Smith bridge, and into New England via the Boston & Albany. Service ended in 1974 as a result of a fire on the bridge.

143. Waterbury, Conn.: Waterbury (SE) was a major producer of brass products, which were a major source of regular freight revenue for the New Haven Railroad in 1946.

144. Hartford, Conn.: Central Connecticut was formerly served by an extensive network of rail lines and branches. Among these branches was a remnant of the also-extensive interurban electric trolley routes—namely, the former Connecticut Company line to South Glastonbury (SE) from a connection with the New Haven Railroad west of Burnside (NE). (See also detail map 144A, **Hartford**.)

145. Willimantic, Conn.: Bridge Street (SE) was the site of a singularly unique double-mounted "tilting board" signal that controlled movements to two New Haven routes: "Hartford" and "Air Line" and the "Central Vermont" route and also could display "stop all routes."

146. Putnam, Conn.: One of the New Haven's two track pans was formerly located near Klondike, north of Putnam (NW), along with an overhead main track coaling station.

147. Providence, R.I.: As late as 1937, a steamboat carried passengers between New York City and the Fall River Wharf (SE), where passengers boarded a train for the last 50 miles to Boston. (See also detail map 147A, **Providence**; detail map 147B, **Pawtucket**; and detail map 147C, **Taunton**.)

148. Middleboro, Mass.: At Woods Hole (SE) during summer months, direct trains from New York City connected with New Bedford, Martha's Vineyard, and Nantucket Steamship Line boats for service to both Martha's Vineyard and Nantucket.

149. Yarmouth, Mass.: During summer months, through trains operated from New York City to Hyannis (SW).

150. Port Jervis, N.Y.: Port Jervis (NE) was both a crew change point and a classification yard on the Erie Railroad, where cars destined for New England were separated from those headed for New York City.

151. Middletown, N.Y.: At Maybrook (NE), the New Haven Railroad's hump yard (with hand brakemen for each car!) classified cars at the principal freight gateway to and from New England.

152. Harmon, N.Y.: Piermont Dock (SW) was the original historic starting point of the Erie Railroad, which was built westward to Lake Erie at Dunkirk, New York.

153. Bridgeport, Conn.: One of the first New Haven Railroad lines to be abandoned in the twentieth century was part of the original Shepaug Railroad between Bethel and Hawleyville—abandoned in 1911 (NW). (See also detail map 153A, **Bridgeport**.)

154. New Haven, Conn.: The "heart" of the New Haven Railroad. The largest freight classification yard in New England was at Cedar Hill, north of New Haven. Built in 1918, it included two separate hump yards with electric retarders. (See also detail map 154A, **New Haven** [with detail map 153A].)

155. New London, Conn.: Prior to the construction of Cedar Hill Yard (map 154) in 1918, a smaller yard and crew change point was located at Midway (NE), five miles east of New London. Later, after Cedar Hill Yard was built, a massive concrete main track coaling station was built at Midway to service steam locomotives. (See also detail map 155A, **New London** [with detail map 153A].)

156. Kingston, R.I.: The original Stonington Railroad was constructed in 1837 between Providence and the steamboat wharf at Stonington (NW), from which steamships conveyed passengers and freight to New York City, thus avoiding an often rough water passage around Point Judith, Rhode Island (map 157 NW).

157. Narragansett, R.I.: Historians of the Narragansett Pier Railroad (NW) like to recall the legendary story of the president of the large and prosperous New Haven Railroad writing to the president of the little, eight-mile Narragansett Pier, asking how much money it would take to buy his railroad. The president of the Narragansett Pier promptly wrote back: "Mine not for sale. How much for yours?"

158. Newark, NJ: This is the most complex map in this atlas, with over 430 named places. Nine major railroads converge in northern New Jersey, the western gateway to New York City. However, only the Pennsylvania Railroad provided direct all-rail service to and through New York City. (See also detail map 158A, **Jersey City**.)

159. New York City, N.Y.: The complex rail network east of the Hudson River, included the

Long Island Railroad, controlled by the Pennsylvania Railroad, exclusively serving Long Island. The New York Connecting Railroad, jointly owned by the Pennsylvania and the New Haven Railroads, provided the key connecting link across the Hell Gate Bridge to New England. (See also detail maps 159A, **The Bronx**; 159B, **Sunnyside**; 159C, **Fresh Pond**; and 159D, **Jamaica**.)

160. Babylon, N.Y.: Despite the fact that Long Island was generally about 15 miles wide, the Long Island Railroad served central Long Island with three parallel east-west passenger lines.

161. Manorville, N.Y.: U. S. Army troops training for the Spanish-American and First World Wars assembled at Camp Upton, east of Yaphank, New York (NW).

162. East Hampton, N.Y.: During World War II in 1942, several German spies from a U-boat came ashore near Amagansett station (NE) and rode the train westward toward New York City.

References

In addition to the listed references, the following types of sources from a variety of railroad company publications were used: employee timetables; signal interlocking diagrams; station lists; system, division, and city/area maps; and track charts.

Atlases and Maps

Hammond's Modern Atlas of the World. 1929. New York: C. S. Hammond & Company, Inc.
Hammond's Universal World Atlas. 1946 (and earlier years). New York: C. S. Hammond & Company, Inc.
United States Geological Survey (USGS)
    United States 1:250,000—Scale Series (base map)
    United States 15-Minute Series
    United States 7.5-Minute Series

Books, Guides, and Manuals

ABC Pathfinder Shipping and Mailing Guide. 1907. Boston: New England Railway Publishing Co.
Crist, Edward J. 1993. *Erie Memories*. New York: Quadrant Press, Inc.
Edson, William D. 1999. *Railroad Names*. Potomac, Maryland: William D. Edson.
Hilton, George W. 1990. *American Narrow Gauge Railroads*. Stanford, California: Stanford University Press.
Hudson, II, John W. 2002. *Creameries of Upstate New York*. Ansonia, Connecticut: Bob's Photo.
Jowett, Alan. 1989. *Railway Atlas of Great Britain and Ireland from Pre-Grouping to the Present Day*. Wellingborough, Northamptonshire, England: Patrick Stephens, Ltd.
Karr, Ronald Dale. 1996. *Lost Railroads of New England*. 2d ed. Pepperell, Massachusetts: Branch Line Press.
Karr, Ronald Dale. 1995. *The Rail Lines of Southern New England*. Pepperell, Massachusetts: Branch Line Press.
Kudish, Michael. 1985. *Where Did the Tracks Go?* Saranac Lake, New York: The Chauncy Press.
Lindsell, Robert M. 1997. *The Rail Lines of Northern New England*. Pepperell, Massachusetts: Branch Line Press.
Moody, John. 1946. *Moody's Steam Railroads*. New York: Moody's Investors Service.
Official Guide of the Railways. 1946. New York: National Railway Publication Co. (and earlier years).
Poor's Manual of Railroads. 1923. New York: Poor's Publishing Co.
Preliminary System Plan. Vol. II. 1975. Washington, D.C.: United States Railway Association.
Railway Clearing House. [1915] 1969. *Railway Junction Diagrams*. New York: Augustus M. Kelley.
Saunders, Richard. 2001. *Merging Lines–American Railroads, 1900–1970*. 2001. DeKalb, Illinois: Northern Illinois University Press.
Ullman, Edward L. 1950. *U.S. Railroads (Map)*. New York: Simmons-Boardman Publishing Corp.
Walker, Mike. 1999. *Steam Powered Video's Comprehensive Railroad Atlas of North America: New England & Maritime Canada*. Dunkirk, Faversham, Kent, England: Ian Andrews.
Walker, Mike. 1997. *Steam Powered Video's Comprehensive Railroad Atlas of North America: North East*. Dunkirk, Faversham, Kent, England: Ian Andrews.
Wilner, Frank N. 1997. *Railroad Mergers: History, Analysis, Insight*. Omaha, Nebraska: Simmons-Boardman Books, Inc.

Indexes

INDEX OF COALING STATIONS

Steam locomotives normally were replenished with coal at engine terminals. However, to avoid delay on longer inter-city runs, some coaling stations were located directly on the main track, usually halfway between division or crew change points. The railroad owning the coaling station is signified by their reporting marks shown in parentheses. Coaling stations marked with an asterisk (*) were no longer in service in 1946.

Cadosia (NYO&W), NY, 139, NW
Chester (B&A), MA, 124, NW

East Deerfield (B&M), MA, 105, SE

Gorham (CN), NH, 53, NE

Holeb (CPR), ME, 10, SW
Hopewell Jct. (NH), NY, 142, SW
Hyde Park (NYC), NY, 142, SW *

Island Pond (CN), VT, 33, NW

Jefferson Jct. (ERIE), PA, 138, NE

Kingston (NYC), NY, 142, NW

Malone (NYC), NY, 28, NW
Midway (NH), CT, 155, NE

Northern Maine Jct. (BAR), ME, 39, NW

Pomfret (NKP), NY, 110, NE
Putnam (NH), CT, 146, NW

South Jct. (D&H), NY, 30, SW

South Plainfield (LV), NJ, 158, SW *

St. Johnsville (NYC), NY, 101, NE *

Star Brick (PRR), PA, 131, NE
Susquehanna (ERIE), PA, 138, NE

Sussex (LNE), NJ, 150, SE

Towanda (LV), PA, 137, NW

Wayneport (NYC), NY, 77, SW

INDEX OF INTERLOCKING STATIONS AND FORMER INTERLOCKING STATIONS

In 1946, the centralized traffic control technology, which is the universal railroad operational standard today, was only beginning to be used on American railroads, and then only to a limited degree in the seven states of the New York & New England volume. Then, interlocking stations (or "signal towers," as they are commonly known) were located every five to ten miles along primary main and branch lines and also at nearly every junction or railroad crossing at grade.

These interlocking stations are identified by their name and/or telegraphic call letters (in bold type). In a few instances, tower numbers are shown in bold type in place of call letters. Where needed for clarity, a geographic place name is given in parentheses. Finally the abbreviation of the railroad to whose building style the tower was built is also shown in parentheses, followed by the state, map number, and map quadrant.

The list of interlocking stations is followed by a separate list of former interlocking stations, identified from various historic collections and old railroad employee timetables. The list of former interlocking stations is not offered as a complete list of every interlocking station that ever existed. To the author's knowledge, no such list currently exists.

Interlocking Stations

106th St., **NK** (NYC), NY, 159A

Albany, East End Frt. Br., **D**, (NYC), 103A

Albany, East End Pass. Br., **101** (NYC), NY, 103, SE, 103A

Albany, East End Yard, **A** (NYC), NY, 103A

Albany, West End Yard, **B** (NYC), NY, 103A

Albany Viaduct, **1** (NYC), NY, 103, SW, 103A

Allens Point (R), VT, 30, SW

Allston, **10** (B&A), MA, 127D

Angola, **NA** (NYC), NY, 92, SE

Arlington (SIRT), NY, 158, SE

Arthur Kill Draw, **AK** (SIRT), NJ, 158, SE

Ashford, **AD** (B&O), NY, 112, NE

Ashland, **23** (B&A), MA, 127, NW

Athens, **RA** (LV), PA, 136, NE

Attleboro, Park St., **165** (NH), MA, 147, NW

Augusta, **AG** (L&NE), NJ, 150, SE

Ayer, **G** (B&M), MA, 107, SE

Babylon, **BJ** (LI), NY, 160, SW

Back Cove (CN), ME, 71, SE, 71A

Ballston, **JS** (D&H), NY, 103, NW

Bangor (MEC), ME, 39, NW

Bangor, **RD** (MEC), ME, 39, NW

Bank St. Jct., **202** (NH), CT, 143, SE

Barber, **BA** (B&A), MA, 126, NW

Barber Creek Draw, **WO** (Perth Amboy), (CNJ), NJ, 158, SW

Barrytown, **71** (NYC), NY, 142, NW

Batavia, Erie Crossing, **40** (NYC), NY, 94, NE

Batavia, Walnut St., **40A** (NYC), NY, 94, NE

Bay View, **BV** (NYC), NY, 93, NW, 93B

Beach, **WD** (LI), NY, 159, SW

Beacon, **50** (NYC), NY, 152, NW

Bedford, **MJ** (LI), NY, 159, SW

Bellows Falls, Ball House (B&M), VT, 87, SW

Bergen Jct., **BJ** (Rutherford), (Erie), NJ, 158, NE

Berkshire Jct., **199** (NH), CT, 153, NW

Bethpage, **B** (LI), NY, 160, SW

Beverly Draw (B&M), MA, 109, SW

Binghamton, **B** (D&H), NY, 118, SW, 118A

Binghamton, Court St., **BD** (Erie), NY, 118, SW, 118A

Binghamton, Liberty St., **LR** (Erie), NY, 118, SW, 118A

Black Rock, Erie Crossing, **H** (NYC), NY, 93D
Black Rock, **F** (NYC), NY, 93D
Black Rock, Military Road, **I** (NYC), NY, 93D
Blasdell, **GB** (PRR), NY, 93B
Blasdell Jct., **BJ** (NYC), NY, 93B
Bloomfield, **BF** (Erie), NJ, 158, NE
Boston, **A** (B&M), MA, 127A, 127E
Boston, Beacon Park, **9** (B&A), MA, 127A, 127E
Boston, Berkeley St., **5** (B&A), MA, 127A, 127E
Boston, Boylston St., **7** (B&A), MA, 127A, 127E
Boston, Broadway, **4** (B&A), MA, 127A, 127E
Boston, Chickering, **185** (NH), MA, 127A, 127E
Boston, East Somerville, **C** (B&M), MA, 127A, 127C
Boston, Huntington Ave., **6** (B&A), MA, 127A, 127E
Boston, Mystic Jct., **X** (B&M), MA, 127A, 127C
Boston, **ONE** (BTCo.), MA, 127A, 127E
Boston, Somerville, **H** (B&M), MA, 127A
Boston, South Bay Jct., **236** (NH), MA, 127A, 127E
Boston, South Boston, **237** (NH), MA, 127A, 127E
Botanical Garden, **BG** (NYC), NY, 159, NW
Bow Arrow Point (R), VT, 30, SW
Bradford, **125** (NH), RI, 156, NE
Branford, **87** (NH), CT, 154, NW
Brentwood, **SG** (LI), NY, 160, NW
Brewster, **B** (NYC), NY, 152, NE
Bridgeport, Bishop Ave., **63** (NH), CT, 153, SE
Bridgeport, Burr Road, **55** (NH), CT, 153, SE
Bridgeport, Central Ave., **62** (NH), CT, 153, SE, 153A
Bridgeport, Pequonnock River, **60** (NH), CT, 153, SE, 153A
Broadway, **RS** (NYS&W), NJ, 158, NE
Brockton, **BM** (NKP), NY, 111, NW

Brook, **FT** (LI), NY, 159, SW
Buffalo Creek, **BC** (NYC), NY, 93A
Buffalo River Draw (DL&W), NY, 93A
Buffalo, Bailey Ave., **DM** (DL&W), NY, 93C
Buffalo, Belt Line Jct., **T** (NYC), NY, 93C
Buffalo, Chicago St., **52** (NYC), NY, 93A
Buffalo, Chicago St., **CG** (LV), NY, 93A
Buffalo, Clinton St., **49A** (NYC), NY, 93A
Buffalo, Crosstown Jct., **B** (NYC), NY, 93A
Buffalo, East End, **48** (NYC), NY, 93C
Buffalo, Erie Railroad Crossing, **51** (NYC), NY, 93A
Buffalo, Erie Railroad Crossing, **EX** (NYC), NY, 93A
Buffalo, Michigan St. (DL&W), NY, 93A
Buffalo, PRR Crossing, **FW** (Erie), NY, 93A
Buffalo, River Bridge, **RB** (NYC), NY, 93A
Buffalo, Scott St. (LV), NY, 93A
Buffalo, Seneca St., **50** (NYC), NY, 93A
Buffalo, Tifft St., **FY** (NKP), NY, 93A
Buffalo, West End, **49** (NYC), NY, 93A
Buffalo Yard, West End, **FO** (NYC), NY, 93A
Burlington, **BJ** (CV), VT, 49, NE
Burnside, **RX** (NYO&W), NY, 151, NE
Buzzards Bay, **U-416** (NH), MA, 148, SE

Cadosia, **HD** (NYO&W), NY, 139, NW
Caledonia, G&W Jct. (GNW), NY, 95, NW
Callicoon, **CO** (Erie), NY, 139, NE
Campbell Hall, **CH** (NYO&W), NY, 151, NW
Canaan, **B-255** (NH), CT, 123, SW
Canastota, **41** (NYC), NY, 80, SW
Canastota, **CD** (NYC), NY, 80, SE

Canton Jct., **178** (NH), MA, 127, SE
Carbondale, **WC** (Erie), PA, 139, SW
Carman, **7** (NYC), NY, 103, NW
Central Square, **CQ** (NYO&W), NY, 79, NE
Charlotte, Genesee River (NYC), NY, 76, NE
Chatham, **65** (B&A), NY, 122, NE
Chatham, **66** (B&A), NY, 122, NE
Chelsea Creek, **E-2** (B&A), MA, 127C
Chester, **50** (B&A), MA, 124, NW
Chili Jct., **33** (NYC), NY, 76, SW
Clark Mills, **AK** (NYC), NY, 81, SW
Clark St., **GS** (NYC), NY, 79, SE
Clifton Jct., **JN** (SIRT), NY, 158, SE
Clinton, **95** (NH), CT, 154, NE
Communipaw, **C** (CNJ), NJ, 158A
Communipaw, **CF** (LV), NJ, 158A
Communipaw Ave. (LV), NJ, 158A
Concord, **CD** (B&M), NH, 88, SE
Connecticut River, **104** (NH), CT, 155, NW
Corfu, **42** (NYC), NY, 94, NW
Corning, **AQ** (Erie), NY, 115, SE
Cornwall, **CN** (NYC), NY, 151, NE
Corry, **EYE** (PRR), PA, 130, NE
Corry, **MS** (Erie), PA, 130, NE
Cortland Jct., **JA** (LV), NY, 98, SE
Cotley Jct., **O-347** (NH), MA, 147, NE
Cranford Jct., **XC** (CNJ), NJ, 158, SW
Cranston, **141** (NH), RI, 147, NW, 147A
Crawbuckie, **PF** (NYC), NY, 152, SW
Crescent, **QS** (D&H), NY, 103, NW
Croton River, **CR** (NYC), NY, 152, SW
Croton-on-Hudson, **CD** (NYC), NY, 152, SW
Croxton, **BR** (Erie), NJ, 158A
Cuba Jct., **CB** (Erie), NY, 113, SW

Danbury, **198** (NH), CT, 153, NW
Danville Jct., **J** (MEC), ME, 55, SW

Davisville, **136** (NH), RI, 147, SW
Deering Jct., **DJ** (PTM), ME, 71, SW, 71A
Denville, **V** (DL&W), NJ, 158, NW
Depew, **46** (NYC), NY, 93, NE, 93E
Derby Jct., **B-253** (NH), CT, 153, NE
Devon, **71** (NH), CT, 153, SE
Divide, **HN** (LI), NY, 159, NE
Dover, **BM** (Disp.) (B&M), NH, 90, SW
Dover, **DT** (B&M), NH, 90, SW
Draw, **HJ** (LI), NY, 159, SW
Drawbridge (MC), ON, 92, NW
Dumont, **DU** (NYC), NJ, 159, NW
Dunellen, **DN** (CNJ), NJ, 158, SW
Dunkirk, **AK** (NKP), NY, 111, NW
Dunkirk, Plate (NKP), NY, 111, NW
Dunkirk, **X** (NYC), NY, 111, NW
Dunton, **MP** (LI), NY, 159D
Durham, **SN** (RDG), NJ, 158, SW
Dutch Kills Draw, **DB** (LI), NY, 159B
Dutch Kills Draw, **M** (LI), NY, 159B

East Alburgh (CV), VT, 30, NE
East Binghamton Yard, **BY** (DL&W), NY, 118, SW
East Buffalo (DL&W), NY, 93C
East Buffalo, **47** (NYC), NY, 93C
East Buffalo, **IQ** (Erie), NY, 93C
East Deerfield, **WX** (B&M), MA, 105, SE
East Greenwich, **137** (NH), RI, 147, SW
East Hartford, **217** (NH), CT, 144, NE, 144A
East Northfield, **SV** (B&M), MA, 106, SW
East Port Chester, **27** (NH), CT, 152, SE
East Rochester, **22** (NYC), NY, 77, SW
East Syracuse, **48** (NYC), NY, 79, SE
Edison, **HU** (PRR), NJ, 158, SW
Eldred, **CR** (PRR), PA, 133, NW
Elizabeth River Draw, **WY** (Elizabeth), (CNJ), NJ, 158, SE

Index of Interlocking Stations / 215

Elmira, Fifth St., **FS** (Erie), NY, 116, SW
Elmira, **MI** (DL&W), NY, 116, SW
Elmira, **MS** (DL&W), NY, 116, SW
Elmora, **SA** (PRR), NJ, 158, SE
Emporium, **JN** (PRR), PA, 133, SE
Everett, **E-5** (B&A), MA, 127A, 127C
Fair (LI), NY, 159, SW
Fairfield, **F** (MEC), ME, 37, SE
Falconer Jct., **DV** (Erie), NY, 111, SE
Fassett, **F** (PRR), PA, 136, NW
Fitchburg, **BX** (B&M), MA, 107, SW
Five, **BR** (PTM), ME, 71, SW, 71A
Fonda, **16** (NYC), NY, 102, NW
Forest Hill, **OJ** (Erie), NJ, 158, NE
Forest Hills, **FH** (NH), MA, 127E
Fort Yard, **112** (NH), CT, 155, NE
Four, **GD** (PTM), ME, 71, SW, 71A
Framingham, **20** (B&A), MA, 127, NW, 127G
Framingham, **21** (B&A), MA, 127, NW, 127G
Framingham Centre, **O-364** (NH), MA, 127, NW
Framingham Yard, **22** (B&A), MA, 127, NW, 127G
Franklin Jct., **FJ** (L&HR), NJ, 150, SE
Fremont, **FN** (LI), NY, 159C

Garden, **HC** (LI), NY, 159, SE
Gardenville Jct., **GJ** (NYC), NY, 93, NW
Gardner, **D** (B&M), MA, 107, SW
Garrison, **43** (NYC), NY, 152, NW
Genessee Jct., **G** (NYC), NY, 76, SE, 76A
Geneva, **GY** (NYC), NY, 97, NW
Geneva Jct., **VJ** (LV), NY, 97, NW
Glens Bridge, **GB** (D&H), NY, 119, NE
Glenwood, **GD** (NYC), NY, 159, NW
Gloucester Draw (B&M), MA, 109, SE
Goshen, **GP** (Erie), NY, 151, NW
Graham, **FX** (Erie), NY, 150, NE
Granton Jct., **GR** (NYS&W), NJ, 158, NE
Great Neck, **G** (LI), NY, 159, NE
Great Notch, **GA** (Erie), NJ, 158, NE
Greenfield Disp., **DS** (B&M), MA, 105, SE
Greens Farms, **53** (NH), CT, 153, SW
Greenwich, **28** (NH), CT, 152, SE
Groton, **119** (NH), CT, 155, NE
Grove St., **GS** (Erie), NJ, 158A
Grove St., **Z** (DL&W), NJ, 158A
Groveland (DL&W), NY, 95, SW
Guilford, **92** (NH), CT, 154, NE

Hack (PRR), NJ, 158A
Hackensack Draw, **HA** (CNJ), NJ, 158A
Hackensack Draw, **HD** (DL&W), NJ, 158A
Hackensack Draw, **HX** (Erie), NJ, 158A
Hall, **JE** (LI), NY, 159D
Harlem River Draw, **DB** (NYC), NY, 159A
Harmon, **HM** (NYC), NY, 152, SW
Harold, **H** (LI), NY, 159B
Hartford, Avon St., **214** (NH), CT, 144, NE, 144A
Hastings, **HS** (NYC), NY, 159, NW
Haven, **WT** (LI), NY, 159, SW
Herkimer, **26** (NYC), NY, 82, SW
Hills Grove, **139** (NH), RI, 147, SW
Hinsdale, **53** (B&A), MA, 123, NE
Hoboken Terminal, **PY** (DL&W), NJ, 158A
Hoffmans, **11** (NYC), NY, 102, NE
Hollis, **IS** (LI), NY, 159D
Hoosac Tunnel, East Portal, **NY** (B&M), MA, 105, SW
Hornell, Cass St. (Erie), NY, 114, NE
Hornell Yard, **ZY** (Erie), NY, 114, NE
Horseheads, **HO** (Erie), NY, 116, SW
Hudson, **84** (NYC), NY, 122, NW
Hudson Siding, **82** (NYC), NY, 122, SW

India Point Draw, **L-321** (NH), RI, 147A
Irv, **VA** (Irvineton), (PRR), PA, 131, NW

Jacy, **Z** (Jersey City), (PRR), NJ, 158A
Jay, **J** (LI), NY, 159D
Jefferson Jct., **JN** (Erie), PA, 138, NE
Jersey City, **A** (CNJ), NJ, 158A
Jersey City, **JC** (Erie), NJ, 158A
Jersey City, **JR** (LV), NJ, 158A
Jersey City Eng. Term., **B** (CNJ), NJ, 158A
Johnsonville, **JV** (B&M), NY, 103, NE
Jones Crossing, **3** (NYC), NY, 103, SW, 103A
Jordan, **6** (NYC), NY, 79, SW
Journal Square, **WR** (H&M), NJ, 158A

Kendall, **SJ** (PRR), NY, 116, SW
Kennebec River Draw (MEC), ME, 72, NW
Kenwood Jct., **KN** (D&H), NY, 103, SW, 103A
Kilmer, **GR** (PRR), NJ, 158, SW
Kingston, **133** (NH), RI, 156, NE
Kingston, **KI** (NYC), NY, 141, NE
Kingston Yard, **KY** (NYC), NY, 142, NW
Kirkville, **44** (NYC), NY, 80, SW

Lackawaxen, **BQ** (Erie), PA, 150, NW
Lake View, **KN** (PRR), NY, 93, SW
Lanesboro, **JA** (Erie), PA, 138, NE
Lawrence, **FA** (B&M), MA, 108, SE
Lead, **WL** (LI), NY, 159, SE
Little Falls, **24** (NYC), NY, 82, SW
Little Ferry, **FY** (NYC), NJ, 158, NE
Little Ferry Draw, **FY** (NYS&W), NJ, 158, NE
Lockport, **B** (NYC), NY, 74, SE
Locust, **OY** (LI), NY, 159, NE
Long Island City, **F** (PRR), NY, 159B

Lorraine, **QR** (CNJ), NJ, 158, SW
Lowell, Bleachery, **BO** (B&M), MA, 108, SW
Lowell, Hale St., **HS** (B&M), MA, 108, SW
Lowell Jct., **JA** (B&M), MA, 108, SE
Lynn, Green St., **GS** (B&M), MA, 128, NW
Lyons, **15** (NYC), NY, 78, SW
Lyons, **WD** (NYC), NY, 78, SW
Lyons Jct., **ON** (NYC), NY, 78, SW

Machias, **CH** (PRR), NY, 113, NW
Magnolia St. (DL&W), NY, 79, SE, 79A
Manchester, **CH** (LV), NY, 96, NE
Manchester, **MA** (B&M), NH, 108, NW
Manchester Draw (B&M), MA, 109, SW
Mansfield, **169** (NH), MA, 127, SE
Marble Hill, **FN** (NYC), NY, 159, NW
Marsh River Draw (MEC), ME, 56, SE
Mechanicville, **XO** (D&H), NY, 103, NE
Medfield Jct., **O-361** (NH), MA, 127, SW
Mianus River Draw, **29** (NH), CT, 152, SE
Middletown-Conn. River Draw, **F-280** (NH), CT, 144, SE
Millburn, **MN** (DL&W), NJ, 158, SW
Mohawk, **GE** (D&H), NY, 103, NW
Monmouth St., **OS** (Erie), NJ, 158A
Montclair, **MO** (DL&W), NJ, 158, NE
Montgomery Crossing, **MQ** (Erie), NY, 151, NW
Moodna Viaduct, **BS** (Erie), NY, 151, NE
Morristown, **OW** (DL&W), NJ, 158, NW
Mortimer, **RQ** (NYC), NY, 76, SE, 76A
Mott Haven Jct., **MO** (NYC), NY, 159A

Mount Morris-PRR, **BY** (DL&W), NY, 95, SW
Mount Vernon, **VO** (NYC), NY, 159, NW
Mountain View, **MV** (DL&W), NJ, 159, NW
Myricks, **O-346** (NH), MA, 147, NE
Mystic River, **122** (NH), CT, 156, NW
Mystic River, **Draw 7** (B&M), MA, 127A, 127C
Mystic River, **Draw 8** (B&M), MA, 127A, 127C

Nashua, **S** (B&M), NH, 108, NW
Nassau, **MT** (LI), NY, 159, SE
National Jct., **NJ** (NYC), NJ, 158A
Neponset Draw, **U-436** (NH), MA, 127E
New Hamburg, **54** (NYC), NY, 142, SW
New Haven, Air Line Jct., **80** (NH), CT, 154, NW, 154A
New Haven, Cedar Street, **75** (NH), CT, 154, NW, 154A
New Haven, Fair Street, **78** (NH), CT, 154, NW, 154A
New Haven, Mill River Jct., **79** (NH), CT, 154, NW, 154A
New Haven, Water Street, **77** (NH), CT, 154, NW, 154A
New Rochelle Jct., **22** (NH), NY, 159, NW
New Rochelle Yard, **23** (NH), NY, 159, NW
New York Mills, **NF** (NYC), NY, 81, SW, 81A
Newberryport Draw (B&M), MA, 109, NW
Newburgh, **GY** (NYC), NY, 151, NE
Newburgh Jct., **NJ** (Erie), NY, 151, NE
Newton Jct., **QM** (B&M), NH, 108, NE
Niagara Falls, **63** (NYC), NY, 73, SE, 73A
Niagara Jct., **GJ** (LV), NY, 93, NE, 93E
Niantic River, **108** (NH), CT, 155, NE
Nineveh Jct., **SW** (D&H), NY, 118, SE

Niobe Jct., **NE** (Erie), NY, 111, SW
Niverville, **70** (B&A), NY, 122, NE
North Adams (B&M), MA, 104, SE
North Adams Jct., **55** (B&A), MA, 123, NE
North Buffalo Jct., **55** (NYC), NY, 93D
North Chelmsford, **NS** (B&M), MA, 108, SW
North East, **N** (NYC), PA, 110, SW
North Haven Cabin (NH), CT, 154, NW
North Hawthorne, **NH** (NYS&W), NJ, 158, NE
North Port Byron, **8** (NYC), NY, 78, SE
North Tonawanda, **Erie 2** (NYC), NY, 74, SW, 74A
North Tonawanda, **Erie 3** (NYC), NY, 74, SW, 74A
North Tonawanda, Erie Crossing, **59** (NYC), NY, 74, SW, 74A
North Tonawanda, **LL** (NYC), NY, 74, SW, 74A
Northampton, **NO** (B&M), MA, 124, NE
Northern Maine Jct., **ONE** (BAR), ME, 39, NW
Norwalk River, **45** (NH), CT, 153, SW
Norwood, **J** (NYC), NY, 27, SW
Nutt Street (CV), VT, 67, SW

Oak Point, **4** (NH), NY, 159, NW
Olean, **X** (Erie), NY, 113, SW
One (PTM), ME, 71, SW, 71A
Oneida, **39** (NYC), NY, 80, SE
Oneonta, Fonda Ave., **FA** (D&H), NY, 119, NE
Orange, **OR** (DL&W), NJ, 158, NE
Otts, **DW** (Warren), (PRR), PA, 131, NE
Owego, **OG** (Erie), NY, 117, SW
Ozone, **RK** (LI), NY, 159, SW

Palatine Bridge, **19** (NYC), NY, 101, NE
Palmer, **35** (B&A), MA, 125, SW
Palmyra, **18** (NYC), NY, 77, SE

Park, **FK** (LI), NY, 159, SE
Passaic Draw, **PW** (Lyndhurst), (DL&W), NJ, 158, NE
Passaic Jct., **BT** (Erie), NJ, 158, NE
Passaic River Draw, **BE** (Erie), NJ, 158, NE
Passaic River Draw, **WR** (Erie), NJ, 158, NE
Patchogue, **PD** (LI), NY, 160, NE
Patterson Jct., **JN** (DL&W), NJ, 158, NE
Patterson Jct., **XW** (Erie), NJ, 158, NE
Peekskill, **37** (NYC), NY, 152, NW
Pelham Bay Draw, **14** (NH), NY, 159, NW
Pelots Point (R), VT, 30, NW
Penn (East New York), (LI), NY, 159, SW
Petersburg Jct., **PI** (B&M), NY, 104, NW
Phillips St., **F** (CNJ), NJ, 158A
Pittsburgh & Lehigh Jct., **JN** (LV), NY, 95, NW
Pittsfield, **57** (B&A), MA, 123, NW
Plainfield, **JA** (CNJ), NJ, 158, SW
Plainville, **212** (NH), CT, 144, SW
Pond, **DF** (LI), NY, 159C
Port, **FY** (LI), NY, 159, SE
Port Dickenson, **YO** (D&H), NY, 118, SW
Port Morris, **3** (NH), NY, 159, NW
Port Reading Jct., **PD** (CNJ), NJ, 158, SW
Portal, **W** (PRR), NJ, 158A
Portsmouth, Kittery Draw (B&M), NH, 90, SW
Poughkeepsie, **58** (NYC), NY, 142, SW
Poughkeepsie, **60** (NYC), NY, 142, SW
Poughkeepsie Bridge, **194** (NH), NY, 142, SW
Providence, Brayton Ave., **149** (NH), RI, 147, NW, 147A
Providence, Promenade St., **151** (NH), RI, 147, NW, 147A
Putnam, **227** (NH), CT, 146, NW

Queens, **QU** (LI), NY, 159, SE

Rahway River Draw, **RH** (West Carteret), (CNJ), NJ, 158, SE
Readville, **181** (NH), MA, 127F
Rensselaer, **100** (NYC), NY, 103, SE, 103A
Rensselaer, **98** (NYC), NY, 103, SE, 103A
Rensselaer, **99** (NYC), NY, 103, SE, 103A
Ridgewood Jct., **WJ** (Erie), NJ, 158, NE
River Jct., **NT** (Erie), NY, 94, SE
Riverside, **15** (B&A), MA, 127, SW
Riverside Jct., **RJ** (PRR), NY, 112, SE
Rochester, Ames Street Jct., **29** (NYC), NY, 76, SE, 76A
Rochester, B&O Jct., **30** (NYC), NY, 76, SE, 76A
Rochester, Clinton Ave., **27** (NYC), NY, 76, SE, 76A
Rochester, Goodman Street, **25A** (NYC), NY, 76, SE
Rochester, Jay St. Jct., **29A** (NYC), NY, 76, SE, 76A
Rochester Jct., **RJ** (LV), NY, 95, NE
Rochester Yard, **25** (NYC), NY, 76, SE
Rock, **FW** (LI), NY, 159, SE
Rockingham, **SK** (B&M), NH, 90, SW
Rome, **34** (NYC), NY, 81, SW
Rotterdam Jct., **RJ** (NYC), NY, 102, NE
Rouses Point (R), NY, 30, NW
Rouses Point, **RO** (D&H), NY, 30, NW
Royal Jct., **J** (MEC), ME, 71, NE
Rye, **26** (NH), NY, 159, NE

Sail Loft, **113** (NH), CT, 155, NE
Salamanca, **WC** (Erie), NY, 112, SE
Salem, **SA** (B&M), MA, 109, SW
Saugatuck River, **52** (NH), CT, 153, SW
Saugus River Draw (B&M), MA, 128, NW
Saybrook Jct., **102** (NH), CT, 155, NW
Schenectady, **8** (NYC), NY, 103, NW

Schodack Jct., **SM** (NYC), NY, 103, SW
Schuyler Jct., **29** (NYC), NY, 81, SE, 81A
Sedgwick Ave., **SK** (NYC), NY, 159, NW
Seekonk River, **K-315** (NH), RI, 147A
Selkirk Jct., **SK** (NYC), NY, 103, SW
Seneca, **D** (NYC), NY, 93B
Sharon Heights, **171** (NH), MA, 127, SE
Sheepscot Draw (MEC), ME, 56, SE
Shore Line Jct., **81** (NH), CT, 154, NW, 154A
Sidney, **GX** (D&H), NY, 119, NW
Silver Creek, **CK** (NKP), NY, 92, SE
Sound View, **106** (NH), CT, 155, NW
South Jct., **JC** (D&H), NY, 30, SW
South Mount Vernon, **20** (NH), NY, 159, NW
South Norwalk, **44** (NH), CT, 153, SW
South Orange, **J** (DL&W), NJ, 158, SW
South Plainfield, **SP** (LV), NJ, 158, SW
South Sudbury, **O-365** (NH), MA, 127, NW
Sparrowbush, **WX** (Erie), NY, 150, NE
Spring St., **GW** (Elizabeth), (CNJ), NJ, 158, SE
Springfield, **40** (B&A), MA, 124, SE, 124A
Springfield, **E-274** (NH), MA, 124, SE, 124A
Springfield, **WA** (B&M), MA, 124, SE, 124A
Spuyten Duyvil, **DV** (NYC), NY, 159, NW
St. George, **TA** (SIRT), NY, 158, SE
St. George, **TB** (SIRT), NY, 158, SE

St. Johnsville, **22** (NYC), NY, 101, NE
Stamford, **38** (NH), CT, 152, SE
Stanley, **YN** (LV), NY, 96, NE
State Line, **60** (B&A), MA, 123, NW
State Line, **AN** (LV), NY, 116, SE
Steamburg, **RH** (Erie), NY, 112, SW
Stuyvesant, **90** (NYC), NY, 122, NW
Suffern, **SF** (Erie), NY, 151, SE
Summit, **ST** (DL&W), NJ, 158, SW
Summitville, **SV** (NYO&W), NY, 141, SW
Sunnyside Yard, **Q** (PRR), NY, 159B
Sunnyside Yard, **R** (PRR), NY, 159B
Suspension Bridge, **65** (NYC), NY, 73, SE, 73A
Susquehanna, **SR** (Erie), PA, 138, NE
Syosset, **S** (LI), NY, 160, NW
Syracuse, **1** (NYC), NY, 79, SE, 79A
Syracuse, Magnolia St., (DL&W), NY, 79, SW, 79A
Syracuse, Salina, **JG** (NYC), NY, 79, SE, 79A
Syracuse Jct., **2** (NYC), NY, 79, SE, 79A

Tarrytown, **OW** (NYC), NY, 152, SW
Taunton, Oak St., **O-353** (NH), MA, 147, NE, 147C
Taunton, Wales St., **O-354** (NH), MA, 147, NE, 147C
Taunton Yard, Danforth St., **O-350** (NH), MA, 147, NE, 147C
Taunton Yard, Tremont St., **O-349** (NH), MA, 147, NE, 147C
Thomaston Draw (MEC), ME, 57, SE
Three, **DF** (PTM), ME, 71, SW, 71A
Tiverton Draw, **Q-382** (NH), RI, 147, SE

Tottenville (SIRT), NY, 158, SW
Troy, Fifth Ave., **4** (TU), NY, 103, SE, 103A
Troy, Grand St., **2** (TU), NY, 103, SE, 103A
Troy, State St., **1** (TU), NY, 103, SE, 103A
Two, **SH** (PTM), ME, 71, SW, 71A

Union, **DK** (PRR), NJ, 158, SW
University Heights, **BN** (NYC), NY, 159, NW
Utica, **30** (NYC), NY, 81, SE, 81A
Utica, **31** (NYC), NY, 81, SE, 81A

Valley (LV), NJ, 158, SW
Valley, **VA** (LI), NY, 159, SE
Van, **VD** (LI), NY, 159, SW
Van Cortlandt Park Jct., **JS** (NYC), NY, 159, NW
Van Nostrand Ave., **HY** (CNJ), NJ, 158A
Voorheesville, **NS** (NYC), NY, 103, SW

Wakefield Jct. (B&M), MA, 127C
Waldo, **SC** (PRR), NJ, 158A
Waldwick, **WC** (Erie), NJ, 151, SE
Walpole, **232** (NH), MA, 127, SW
Waltham, **OD** (B&M), MA, 127D
Washington, **52** (B&A), MA, 123, NE
Waterboro, **WO** (Erie), NY, 111, SE
Waterford, **110** (NH), CT, 155, NE
Waterville, East Tower, **B** (MEC), ME, 37, SE
Waterville, West Tower, **A** (MEC), ME, 37, SE
Wayneport, **20** (NYC), NY, 77, SW
Webster Jct., **31** (B&A), MA, 126, SW
Weehawken, **TU** (NYC), NJ, 158A
Welland Canal (CN), ON, 92, NW
Welland Jct., **WX** (MC), ON, 92, NE
Welland Ship Canal (CN), ON, 73, SE
West Bergen, **36** (NYC), NY, 75, SE

West End, **WD** (DL&W), NJ, 158A
West End, **WE** (North Bergen), (NYC), NJ, 158A
West Haverstraw, **HN** (NYC), NY, 152, SW
West Secaucus, **SY** (DL&W), NJ, 158A
West Springfield, **42** (B&A), MA, 124, SE, 124A
West Springfield, **43** (B&A), MA, 124, SE
Westerly, **124** (NH), RI, 156, NW
Westfield, **45** (B&A), MA, 124, SE
Westfield, **WF** (CNJ), NJ, 158, SW
Westfield, **WX** (NYC), NY, 110, NE
White House, **WH** (PRR), NY, 113, SW
White Plains North, **NW** (NYC), NY, 152, SW
Whittenton Jct., **O-351** (NH), MA, 147, NE, 147C
Willow Ave., **AV** (NYC), NJ, 158A
Willowglen Wye, **WY** (D&H), NY, 103, NE
Win, **WJ** (LI), NY, 159, SW
Winchester, **Q** (B&M), MA, 127B
Woodbridge Jct., **WC** (Perth Amboy), (CNJ), NJ, 158, SW
Woodlawn, **154** (NH), RI, 147, NW, 147B
Woodlawn, **JO** (NYC), NY, 159, NW
Woodmont, **73** (NH), CT, 153, SE
Worcester, **28** (B&A), MA, 126, NW, 126A
Worcester, Franklin St., **M-334** (NH), MA, 126, NW, 126A
Worcester, Jackson St., **M-333** (NH), MA, 126, NW, 126A
Worcester Yard, **26** (B&A), MA, 126, NW, 126A
Wrights, **NR** (PRR), PA, 133, SE

Yarmouth Jct., **YJ** (MEC), ME, 71, NE

Former Interlocking Stations

156th St., **KY** (Melrose), (NYC), NY, 159A
180th St., **UP** (NYW&B), NY, 159, NW

Adamsdale Jct., **N-340** (NH), MA, 147, NW
Agawam Jct., **44** (B&A), MA, 124, SE
Albany, Columbia St., **CS** (D&H), NY, 103A
Albany, Livingston Ave., **LA** (D&H), NY, 103A
Albany, West End Frt. Bridge, **C** (NYC), NY, 103, SE, 103A
Albany Grade, **2** (NYC), NY, 103, SW
Alford (DL&W), PA, 138, NW
Amsterdam, **13** (NYC), NY, 102, NE
Apponaug, **138** (NH), RI, 147, SW
Athol Jct., **38** (B&A), MA, 124, SE
Atlantic, **U-434** (NH), MA, 127E
Attleboro, Olive St., **164** (NH), MA, 147, NW
Attleboro Jct., **O-352** (NH), MA, 147, NE, 147C
Auburn-Monroe St. (LV), NY, 97, NE
Ayer, Hill Yard, **AY** (B&M), MA, 107, SE

Back Bay, **186** (NH), MA, 127A
Baldwinville, **BV** (B&M), MA, 106, SE
Barrett's Jct., **BR** (B&A), MA, 125, SW
Battery Jct., **BT** (BRB&L), MA, 128, NW
Baychester Ave., **BJ** (NYW&B), NY, 159, NW
Bergen, **35** (NYC), NY, 76, SW
Berkshire Jct. (Old), **199** (NH), CT, 153, NW
Beverly Draw, **BN** (B&M), MA, 109, SW
Beverly Draw, **BS** (B&M), MA, 109, SW
Beverly Jct., **BI** (B&M), MA, 109, SW

Biddeford, Alfred Road, **AR** (B&M), ME, 91, NW
Black Rock, Amherst St., **G** (NYC), NY, 93D
Black Rock, Erie Crossing, **H** (NYC), NY, 93D
Black Rock, **F** (NYC), NY, 93D
Bluff Point, **BX** (D&H), NY, 30, SW
Boonton, **BN** (DL&W), NJ, 158, NW
Boonton, **BO** (DL&W), NJ, 158, NW
Botsford, **B-254** (NH), CT, 153, NW
Boylston St., **184** (NH), MA, 127, NE, 127E
Braintree Highlands, **U-430** (NH), MA, 127, SE
Brayton Draw, **K-308** (NH), MA, 147, SE
Bridgeport, Fairfield Ave., **59** (NH), CT, 153A
Bridgeport, Lower Yard, **57** (NH), CT, 153A
Bridgeport, Main St., **56** (NH), CT, 153A
Bridgeport, Pequonnock River, **61** (NH), CT, 153A
Bridgeport, State St., **58** (NH), CT, 153A
Brighton, **24** (NYC), NY, 76, SE
Broadway, **188** (NH), MA, 127A
Brockton, **U-423** (NH), MA, 127, SE
Bronx River Draw, **7** (NH), NY, 159A
Brookline Jct., **8** (B&A), MA, 127A
Buffalo, Exchange St., **53** (NYC), NY, 93A
Buffalo, Exchange St., **54** (NYC), NY, 93A
Buffalo, Porter Ave., **C** (NYC), NY, 93NW, 93D
Buffalo, Tifft St., **TS** (NYC), NY, 93B
Buffalo, William St., **48** (NYC), NY, 93A

Buffalo St., **BS** (Olean), (PRR), NY, 113, SW
Burnside, **218** (NH), CT, 144, NE
Burnwood, **BS** (Erie), PA, 138, NE

Cadosia, Wheeler, **HX** (NYO&W), NY, 139, NW
Cameron, **CN** (Erie), NY, 115, SW
Campbell Hall, **CP** (NYO&W), NY, 151, NW
Campello, **U-422** (NH), MA, 127, SE
Canton, **C** (PRR), PA, 136, SW
Carbondale, Lookout Jct., **OU** (D&H), PA, 138, SE
Carrollton, **CT** (Erie), NY, 112, SE
Castle Square, **187** (NH), MA, 127A
Castleton-on-Hudson, **95** (NYC), NY, 103, SW
Cedar, **O-360** (NH), MA, 127, SW
Cedar Hill Yard, Scissors Jct., **G-290** (NH), CT, 154, NW
Charles River, **E-8** (B&M), MA, 127A
Charlestown, **B** (B&M), MA, 127A
Charlton, **32** (B&A), MA, 126, SW
Chase's, **C-265** (NH), CT, 143, SE
Chatsworth Ave., **AT** (Larchmont), (NYW&B), NY, 159, NW
Chelsea, **53** (NYC), NY, 142, SW
Chenango Forks, **CG** (DL&W), NY, 118, SW
Cherry Valley Jct., **KF** (D&H), NY, 101, SE
Chittenango, **CT** (NYC), NY, 80, SW
Churchville, **CV** (NYC), NY, 76, SW
Churchville, **X** (NYC), NY, 76, SW
Churchville Jct., **CJ** (NYC), NY, 76, SW
Clarendon Hills, **182** (NH), MA, 127, NE, 127E
Clinton, **P-374** (NH), MA, 126, NE
Clinton Jct., **CN** (B&M), MA, 126, NE

Clinton Point, **55** (NYC), NY, 142, SW
Clyde, **12** (NYC), NY, 78, SW
Cohasset Narrows, **U-415** (NH), MA, 148, SE
Cold Spring, **45** (NYC), NY, 152, NW
Cold Spring, **46** (NYC), NY, 152, NW
Colonie, **NG** (D&H), NY, 103, SE, 103A
Colonie, **SG** (D&H), NY, 103, SE
Columbus Ave., **CA** (NYW&B), NY, 159, NW
Columbus Jct., **CM** (Erie), PA, 130, NE
Conklin, **RU** (DL&W), NY, 118, SW
Constable Jct., **CB** (Greenville), (LV), NJ, 158A
Cook St., **H-9** (B&M), MA, 127D
Cooperstown Jct., **N** (D&H), NY, 120, NW
Crawford Jct., **RF** (NYO&W), NY, 151, NW
Cromwell, **F-282** (NH), CT, 144, SE
Croxton, **JR** (NNJ), NJ, 158A

Dante, **DA** (D&H), NY, 101, SE
Darien, **40** (NH), CT, 153, SW
Dedham, **R-397** (NH), MA, 127F
Dedham, **R-398** (NH), MA, 127F
Dedham Road, **179** (NH), MA, 127, SE
Delanson, **DJ** (D&H), NY, 102, NE
Dellwood, **44** (NYC), NY, 93, NE
Denville (Old), **V** (DL&W), NJ, 158, NW
Depew, W.S. Conn., **46A** (NYC), NY, 93, NE, 93E
Depew Jct., **45** (NYC), NY, 93, NE, 93E
Dorchester, **235** (NH), MA, 127E
Douglas Jct., **229** (NH), MA, 126, SW
Dykemans, **X** (NYC), NY, 152, SE

East Binghamton Yard, West End, **RD** (DL&W), NY, 118A

Index of Interlocking Stations / 219

East Boston, **T** (BRB&L), MA, 127, NE, 127E
East Brookfield, **33** (B&A), MA, 125, SE
East Buffalo, **47** (NYC), NY, 93C
East Foxboro, **170** (NH), MA, 127, SE
East Haven, **85** (NH), CT, 154, NW
East Jct., **163** (NH), MA, 147, NW
East Karner, **5** (NYC), NY, 103, SW
East Lyme, **107** (NH), CT, 155, NE
East Providence, **K-313** (NH), RI, 147A
East Providence, Red Bridge, **L-323** (NH), RI, 147A
East Providence, Water St., **K-314** (NH), RI, 147A
East River, **93** (NH), CT, 154, NE
East Somerville, **C** (B&M), MA, 127A, 127C
East Somerville, **D** (B&M), MA, 127A, 127C
East St. Johnsville, **21** (NYC), NY, 101, NE
East Syracuse, **47** (NYC), NY, 79, SW
East Thompson, **228** (NH), CT, 126, SW
East Weymouth, **X-459** (NH), MA, 128, SW
Easton, **Q-389** (NH), MA, 127, SE
Eastwood, **WX** (NYC), NY, 79, SW
Elmira Heights, **VO** (Erie), NY, 116, SW
Erwin's Crossing (DL&W), NY, 115, SE
Everett Jct., **NA** (B&M), MA, 127A, 127C

Factoryville, **FO** (DL&W), PA, 138, SW
Fairfield, **54** (NH), CT, 153, SW
Fall River, **Q-383** (NH), MA, 147, SE
First St., **WA** (Olean), (PRR), NY, 113, SW
First St., **Y-465** (NH), MA, 127A
Fitchburg, **P-376** (NH), MA, 107, SW

Forest Hills, **183** (NH), MA, 127, NE, 127E
Framingham Yard, **O-363** (NH), MA, 127, NW, 127G
Franklin Jct., **231** (NH), MA, 127, SW
Franklin Jct., **FR** (DL&W), NJ, 150, SE

Garrison, **44** (NYC), NY, 152, NW
Germantown, **77** (NYC), NY, 122, SW
Ghent, **BA** (NYC), NY, 122, NE
Glenbrook, **39** (NH), CT, 152, SE
Glenville Jct., **GV** (D&H), NY, 103, NW
Goldens Bridge, **GN** (NYC), NY, 152, NE
Grand Crossing, **JU** (ERIE), NY, 93A
Granton, **GO** (NYC), NJ, 158, NE
Gratwick, **60** (NYC), NY, 74, SW
Greendale, **81** (NYC), NY, 122, SW
Greens Farms (Old) **53** (NH), CT, 153, SW
Greenway, **37** (NYC), NY, 80, SE
Groton (Old), **119** (NH), CT, 155A
Gulf Summit, **GF** (Erie), NY, 118, SE

Hackensack Draw, **HN** (DL&W), NJ, 158A
Hallstead, **HY** (DL&W), PA, 138, NW
Harbor, **WF** (NYC), NY, 81, SE, 81A
Harlem River, **1** (NH), NY, 159A
Harrison, **HA** (NYW&B), NY, 159, NE
Harrison Square, **U-437** (NH), MA, 127E
Harsimus Cove, **JH** (PRR), NJ, 158A
Hartford, Branch Switch Jct., **215** (NH), CT, 144, NE, 144A
Hawleyville Jct., **200** (NH), CT, 153, NW
Heathcote, **HC** (NYW&B), NY, 159, NW
Herkimer, **26** (NYC), NY, 82, SW
High View, **BX** (NYO&W), NY, 141, SW

Highland Jct., **204** (NH), CT, 143, SE
Hingham, **X-458** (NH), MA, 128, SW
Hinsdale, **HD** (PRR), NY, 113, SW
Holbrook, **U-424** (NH), MA, 127, SE
Hoosac Tunnel, West Portal, **WO** (B&M), MA, 104, SE
Hopewell Jct., **196** (NH), NY, 142, SW
Hudson, **HU** (B&A), NY, 122, NW
Hudson River, **TR** (D&H), NY, 103, SE, 103A

Iona Island, **MX** (NYC), NY, 152, NW

Jersey City, **RU** (PRR), NJ, 158A
Jordan, **JR** (NYC), NY, 79, SW

Kingsley, **KS** (DL&W), PA, 138, NW
Kingston (Old), **133** (NH), RI, 156, NE
Kingston, **KI** (NYC), NY, 142, NW
Kirkville Jct., **KR** (NYC), NY, 80, SW
Klondike, **228** (NH), CT, 146, NW

Lake Mahopac, **XC** (NYC), NY, 152, NE
Lawrence, Engine House, **JK** (B&M), MA, 108, SE
Leetes Island, **91** (NH), CT, 154, NE
Lincoln Park, **KN** (DL&W), NJ, 158, NW
Linlithgo, **79** (NYC), NY, 122, SW
Little Falls, **FA** (Erie), NJ, 158, NE
Lockport Jct., **JC** (NYC), NY, 74, SW
Lyme, **105** (NH), CT, 155, NW
Lynn, **L** (BRB&L), MA, 128, NW
Lyons, **YO** (NYC), NY, 78, SW

Madison, **94** (NH), CT, 154, NE
Mamakating, **WX** (NYO&W), NY, 141, SW
Mamaroneck, **24** (NH), NY, 159, NE

Mamaroneck, **MQ** (NYW&B), NY, 159, NE
Manchester, **MR** (B&M), NH, 108, NW, 108A
Manchester Diamond, **MD** (B&M), NH, 108, NW, 108A
Manchester Yard, **MA** (B&M), NH, 108, NW, 108A
Mansfield, **168** (NH), MA, 127, SE
Mansfield, **O-358** (NH), MA, 127, SE
Mansfield Yard, **O-359** (NH), MA, 127, SE
Marlboro Jct., **P-373** (NH), MA, 126, NE
Matfield, **U-421** (NH), MA, 128, SW
Mayville Jct., **N** (JW&NW), NY, 111, NW
Mechanic Falls, **MF** (MEC), ME, 55, SW
Medford Jct., **JC** (B&M), MA, 127C
Melrose, **MX** (NYC), NY, 159A
Middleboro, **U-418** (NH), MA, 148, NW
Middletown, Bridge St., **F-281** (NH), CT, 144, SE
Middletown, **F-279** (NH), CT, 144, SE
Middletown, Main St., **MS** (NYO&W), NY, 151, NW
Midway, **120** (NH), CT, 155, NE
Midway Jct., **121** (NH), CT, 155, NE
Milford, **72** (NH), CT, 153, SE
Millers Crossing, **4** (NYC), NY, 103, SW
Milton Jct., **V-445** (NH), MA, 127E
Minoa, **45** (NYC), NY, 79, SW
Morris Park, **R** (LI), NY, 159D
Mott Haven Jct. (Old), **MO** (NYC), NY, 159A
Mott Haven Jct., **MJ** (NYC), NY, 159A
Mount Morris-Erie (DL&W), NY, 95, SW
Mount St. Vincent, **MV** (NYC), NY, 159, NW

NJ & NY Jct., **HB** (Erie), NJ, 158, NE

Nantasket Jct., **X-460** (NH), MA, 128, SW
Nash Road, **O-348** (NH), MA, 148, SW
Natick, **19** (B&A), MA, 127, NW
Needham Jct., West St., **S-403** (NH), MA, 127, NE, 127D
Neversink, **GX** (NYO&W), NY, 140, SE
New Milford, **FS** (DL&W), PA, 138, NW
Newburyport, **NY** (B&M), MA, 109, NW
Newport Road, **157** (NH), MA, 147, NW
Newton Hook, **88** (NYC), NY, 122, NW
Newton Hook, **89** (NYC), NY, 122, NW
Newtonville, **12** (B&M), MA, 127D
North Ave., **NA** (NYW&B), NY, 159, NW
North Bergen, **WE** (NYC), NJ, 158A
North Berwick, **BT** (B&M), ME, 90, NE
North East, **NH** (NKP), PA, 110, SW
North River, **X-463** (NH), MA, 128, SE
North Woburn Jct., **WF** (B&M), MA, 108, SE
Northeys Point, **NP** (B&M), MA, 109, SW
Northrup Ave., **153** (NH), RI, 147, NW
Norwood Central, **233** (NH), MA, 127, SE

Oakley (Erie), NY, 114, NE
Ocean Ave., **111** (NH), CT, 155, NE
Oneida Castle, **YO** (NYC), NY 80, SE
Oneonta, **MX** (D&H), NY, 119, NE
Orient Heights, **OH** (BRB&L), MA, 127C
Oriskany, **33** (NYC), NY, 81, SW

Passaic, **WA** (DL&W), NJ, 158, NE

Pawtucket, Boston Switch, **156** (NH), RI, 147, NW, 147B
Pawtucket, Dexter St., **155** (NH), RI, 147, NW, 147B
Pecks Bridge, **56** (B&A), MA, 123, NE
Peekskill Draw, **39** (NYC), NY, 152, NW
Perth Amboy, **VG** (CNJ), NJ, 158, SW
Phelps Jct., **JC** (PRR), NY, 96, NE
Plainfield, **J-301** (NH), CT, 146, SW
Plymouth Cordage, **W-451** (NH), MA, 148, NE
Pompton Jct., **PJ** (NY&GL), NJ, 151, SW
Port Chester, **PC** (NYW&B), NY, 152, SE
Port Jervis, **BC** (Erie), PA, 150, NE
Poughkeepsie (East of), **57** (NYC), NY, 142, SW
Poughkeepsie Bridge-West End, **193** (NH), NY 142, SW
Poughkeepsie Jct., **195** (NH) NY 142, SW
Pratts Jct., **P-375** (NH), MA, 126, NE
Providence, Dike St., **148** (NH), RI, 147, NW, 147A
Providence, Elmwood, **142** (NH), RI, 147, NW, 147A
Providence, Gaspee St., **150** (NH), RI, 147A
Providence, India Point, **L-321** (NH), RI, 147A
Providence, Orms St., **152** (NH), RI, 147, NW, 147A
Putnam Jct., **XN** (NYC), NY, 152, NE

Ramsey, **R** (Erie), NJ, 151, SE
Randolph, **Q-391** (NH), MA, 127, SE
Ravena, **Q** (NYC), NY, 122, NW
Ravena, **QA** (NYC), NY, 122, NW
Raynham, **Q-388** (NH), MA, 147, NE
Readville Transfer, **180** (NH), MA, 127, SW, 127F
Readville-Midland, **234** (NH), MA, 127, SW, 127F

Rensselaer Yard, **72** (B&A), NY, 103, SE
Rhinecliff, **68** (NYC), NY, 142, NW
River Road, **70** (NH), CT, 153, SE
Riverside, **30** (NH), CT, 152, SE
Rochester, Center Park, **28** (NYC), NY, 76, SE, 76A
Rochester, Hague St. (NYC), NY, 76, SE, 76A
Rosas, **RJ** (Erie), PA, 150, NW
Roundhouse, **R** (Hoboken), (DL&W), NJ, 158A
Rowayton, **41** (NH), CT, 153, SW
Rye, **RY** (NYW&B), NY, 159, NE

Saratoga Springs, **SR** (D&H), NY, 84, SW
Schenectady, Center St., **X** (D&H), NY, 103, NW, 103B
Schenevus, **WN** (D&H), NY, 101, SW
Schoharie Jct., **JX** (D&H), NY, 102, SW
Scotia, Sand Bank, **9** (NYC), NY, 103, NW
Secaucus, **CU** (DL&W), NJ, 158A
Secaucus, **SJ** (Kingsland), (DL&W), NJ, 158, NE
Seneca River, **SR** (NYC), NY, 78, SE
Shannock, **127** (NH), RI, 156, NE
Sharon, **172** (NH), MA, 127, SE
Simsbury, **D-267** (NH), CT, 144, NW
Slocums, **134** (NH), RI, 146, SE
Somerset Draw, **Q-386** (NH), MA, 147, NE
Somerset Jct., **Q-384** (NH), MA, 147, NE
South Auburn, **140** (NH), RI, 147, NW
South Braintree, **U-431** (NH), MA, 127, SE
South Norwalk, **43** (NH), CT, 153, SW
South Schenectady, **NE** (NYC), NY, 103, NW
Southboro, **P-372** (NH), MA, 126, NE
Southhampton St., **Y-467** (NH), MA, 127A
Southport Jct., **JF** (Erie), NY, 116, SW

Spragues Cabin (B&M), MA, 104, SE
Spring Valley, **ZY** (Erie), NY, 151, SE
Springfield, **40** (B&A), MA, 124, SE, 124A
Springfield, **41** (B&A), MA, 124, SE, 124A
Staatsburg, **65** (NYC), NY, 142, NW
Stamford, **37** (NH), CT, 152, SE
Stanley, **U-419** (NH), MA, 128, SW
Sterlington, **SJ** (Erie), NY, 151, SE
Stonington, **123** (NH), CT, 156, NW
Stony Creek, **90** (NH), CT, 154, NE
Storm King, **47** (NYC), NY, 152, NW
Stoughton Jct., **O-390** (NH), MA, 127, SE
Sullivan, **42** (NYC), NY, 80, SW
Suspension Bridge (Old), **65** (NYC), NY, 73, SE, 73A
Susquehanna (Old), **SR** (Erie), NY, 138, NE
Swampscott, **SR** (B&M), MA, 128, NW
Swett St., **U-439** (NH), MA, 127E
Sylvan Jct., **SX** (NYO&W), NY, 79, SE
Syracuse, **1** (NYC), NY, 79A
Syracuse, Geddes St., **1A** (NYC), NY, 79A
Syracuse, Salina, **SX** (NYC), NY, 79A
Syracuse, W.S., **RW** (NYC), NY, 79A
Syracuse, West St., **99** (NYC), NY, 79A
Syracuse Jct., Willis Ave., **2B** (NYC), NY, 79A
Thames River, **118** (NH), CT, 155A
Tivoli, **74** (NYC), NY, 122, SW
Tower No. 1 (BAR), ME, 39, NW
Tremont, **U-417** (NH), MA, 148, NW
Troy, Jacob St., **3** (TV), NY, 103, SW

Tunnel Jct., **201** (NH), CT, 153, NW
Tusten, **TU** (Erie), PA, 139, SE

Vails Gate Jct., **VG** (Erie), NY, 151, NE
Valley Falls, **M-329** (NH), RI, 147, NW, 147B
Van Etten, **VW** (LV), NY, 116, SE
Van Etten Jct., **VC** (LV), NY, 116, SE
Vernon, **219** (NH), CT, 145, NW
Vernon, **VO** (NYC), NY, 80, SE
Vernon Grade, **VG** (NYC), NY, 81, SW

Wakefield Jct., **WQ** (B&M), MA, 127C
Warbasse, Hyper Humus, **WA** (NYS&W), NJ, 150, SE
Warbasse Siding, **WX** (NYS&W), NJ, 150, SE
Warners, **4** (NYC), NY, 79, SW

Warren, **K-309** (NH), RI, 147, SW
Warren, **K-310** (NH), RI, 147, SW
Warren, **K-311** (NH), RI, 147, SW
Warren, **K-312** (NH), RI, 147, SW
Watertown Jct., **WJ** (NYC), NY, 60, NW
Watervliet, **WV** (D&H), NY, 103, SE, 103A
Watervliet Jct., **WX** (D&H), NY, 103, SE, 103A
Webster, **J-302** (NH), MA, 126, SW
Weir Jct., **O-348** (NH), MA, 147, NE, 147C
West Bowmansville, **A** (NYC), NY, 93, NE, 93E
West Cambridge, (B&M), MA, 127B
West Concord, **O-366** (NH), MA, 127, NW

West Englewood, **WG** (NYC), NJ, 158, NE
West Haven, **74** (NH), CT, 154, NW
West Karner, **6** (NYC), NY, 103, SW
West Lynn, **GS** (B&M), MA, 128, NW
West Mansfield, **167** (NH), MA, 147, NE
West Roxbury Jct., **S-404** (NH), MA, 127, NE, 127D
West Warren, **34** (B&A), MA, 125, SE
Westbrook, **96** (NH), CT, 155, NW
Westchester Ave., **WA** (NYW&B), NY, 152, SW
Westchester Yard, **12** (NH), NY, 159, NW
Westdale, **U-420** (NH), MA, 128, SW
Westfield, **FD** (NKP), PA, 110, NE
Westfield Yard, **D-269** (NH), MA, 124, SE

Westport, **51** (NH), CT, 153, SW
Whittenton Wye, **T-410** (NH), MA, 147, NE, 147C
Wickford Jct., **135** (NH), RI, 147, SW
Wilmington Jct., **U** (B&M), MA, 108, SE
Wood River Jct., **126** (NH), RI, 156, NE
Woodsville, White Mountains Transfer, **XO** (B&M), NH, 51, SE
Woonsocket, **M-330** (NH), RI, 126, SE
Woonsocket Jct., **230** (NH), RI, 126, SE
Worcester, Millbrook St., **MB** (B&M), MA, 126, NW
Worcester, Southbridge St., **M-331** (NH), MA, 126A
Wykagyl, **WY** (NYW&B), NY, 159, NW

Yosts, **17** (NYC), NY, 102, NW

INDEX OF PASSENGER AND NON-PASSENGER STATIONS

This is an alphabetical list of all the stations shown on the maps of this atlas, with the exception of interlocking stations, former interlocking stations, coaling stations, track pans, tunnels, and viaducts. These are shown in separate lists.

In those locations where two or more railroads serve the same place, the station name of that place is followed, in brackets, by the "reporting marks," or standard abbreviations, of the railroads concerned. In a few instances, adjacent places are given in brackets.

No distinction is made in this index as to the type of station (e.g., passenger stations, remote control interlockings, block stations, or freight-only stations). For that information, the reader may refer to "Map Symbols and Abbreviations" and to the atlas map concerned. Every index entry shows the postal abbreviation of the state within which the station is located.

The last items in each entry are the atlas map number and one of four possible map quadrant designations—NW, NE, SW, or SE—each of which covers one of the four areas of the 15 x 15 minutes each, which, taken together, constitute each atlas sheet. In the case of detail maps, there are no map quadrant designations; instead, the map number is followed by the letter A, B, C, etc., which signifies a particular detail map.

8th Ave., NY, 158, SE
33rd St., NY, 158, NE
60th St., NY, 159, NW
125th St., NY, 159A
130th St., NY, 159, NW
138th St., NY, 159A
152nd St., NY, 159, NW
183rd St., NY, 159, NW

Abbot Road (DL&W), NY, 93A
Abbot Village, ME, 20, SW
Abbotts Run, RI, 147, NW
Abington, CT, 145, NE
Abington, MA, 128, SW
Abnaki, VT, 30, NW
Academy Corners, PA, 135, NW
Accord, NY, 141, NE
Ackerson, NJ, 150, SE
Acton, MA, 127, NW
Acushnet, MA, 148, SW
Adaline, ME, 5, NE
Adams, MA, 104, SE
Adams, ME, 21, NW
Adams, NY, 59, NE

Adams Basin, NY, 76, SW
Adams Center, NY, 59, NE
Adams St., NY, 103, SE, 103A
Adamsdale, MA, 147, NW
Adamsdale Jct., MA, 147, NW
Addison (B&O; Erie), NY, 115, SE
Addison, ME, 41, SW
Addison Jct. (D&H), NY, 65, NW
Adrian, NY, 114, NE
Afton, NY, 118, SE
Agamenticus, ME, 90, SW
Agawam Jct., MA, 124, SE
Air Base, ME, 5, SE
Airport Switch, CT, 144, NE
Akeley, PA, 131, NE
Akron, NY, 75, SW
Akron Jct., NY, 74, SE
Alabama, NY, 75, SW
Alba, NY, 151, NW
Alba, PA, 136, SW
Albertine, NB, 2, NW
Albertson, NY, 159, NE
Albion, ME, 38, SW
Albion, NY, 75, SE

Albion, RI, 147, NW
Alburgh, VT, 30, NW
Alburgh Springs, VT, 30, NE
Alden, MA, 148, NW
Alden, NY, 94, NW
Alder Creek, NY, 81, NE
Alder Stream, ME, 36, NE
Alder Stream Jct., ME, 36, NE
Aldrich, NY, 45, SE
Alexander, NY, 94, NE
Alford, PA, 138, NW
Alfrecha, VT, 66, SW
Alfred, ME, 90, NE
Alfred, NY, 114, NW
Allandale, NB, 15, NW
Allegany (Erie), NY, 113, SW
Allegany (PRR), NY, 113, SW
Allendale, NJ, 151, SE
Allens, ME, 54, NW
Allens, ME, 57, SW
Allenstown, NH, 89, SW
Allerton, MA, 128, NW
Allerton Farms, CT, 143, SE
Allston, MA, 127D

Allwood, NJ, 158, NE
Allyns Point, CT, 155, NE
Almond, NY, 114, NE
Alna Center, ME, 56, SE
Alphaus, NY, 103, NW
Alsen, NY, 122, SW
Altamont, NY, 102, SE
Althom, PA, 131, NW
Altmar, NY, 59, SE
Alton, ME, 21, SE
Alton, NH, 89, NE
Alton, NY, 78, SW
Altona, NY, 29, NE
Alton Bay, NH, 89, NE
Aluminum Plant, NY, 27, NW
Alversons, NY, 59, NE
Amagansett, NY, 162, NE
Amawalk, NY, 152, NW
Amboy, NY, 79, SW
Ambulco, NY, 94, SE
Amenia, NY, 142, NE
Ames, NH, 69, SW
Amesbury, MA, 109, NW
Amherst, MA, 124, NE
Amherst, NH, 107, NE
Amityville, NY, 160, SW
Amoskeag, NH, 89, SW
Ampere, NJ, 158, NE
Amsterdam, NY, 102, NE
Amston, CT, 145, NW
Anburn, MA, 126, SW
Ancram, NY, 122, SE
Ancram Lead Mines, NY, 122, SE
Anderson, NH, 108, SW
Anderson St., NJ, 158, NE
Andes, NY, 120, SW
Andover, CT, 145, SE
Andover, MA, 108, SE
Andover, NB, 6, NE
Andover, NH, 88, NW
Andover, NY, 114, SW
Andrews Settlement, PA, 134, NW
Angelica, NY, 113, NE
Angola, NY, 92, SE
Annadale, NY, 158, SE
Anos Siding, NY, 81, NE
Ansonia (B&O; NYC), PA, 135, SW
Ansonia, CT, 153, NE
Anson's Crossing, NY, 142, NE
Anthony, RI, 146, SE
Anthony, VT, 104, NW

Antrim, NH, 88, SW
Antrim, PA, 135, SW
Antwerp, NY, 44, SE
Apalachin, NY, 117, SE
Apex, NY, 119, SW
Appalachia, NH, 53, NW
Appleton, NY, 74, NE
Apponaug, RI, 147, SW
Apulia, NY, 98, NE
Aquebogue, NY, 161, NE
Aqueduct, NY, 103, NW
Aqueduct, NY, 159, SW
AR, ME, 91, NW
Ararat, PA, 138, NE
Arcade, NY, 94, SW
Arcade Center, NY, 94, SW
Arcade Jct., NY, 94, SW
Archdale, NY, 85, SW
Arctic, RI, 146, SE
Arden, NY, 151, NE
Ardsley, NY, 152, SW
Ardsley-on-Hudson, NY, 152, SW
Arena, NY, 120, SE
Arey, ME, 39, SW
Argosy, NB, 6, NE
Arkport (Erie; PS&N), NY, 114, NE
Arkville, NY, 120, SE
Arkwright, RI, 146, SW
Arlington, MA, 127B
Arlington, NJ, 158, NE
Arlington, NY, 158, SE
Arlington, RI, 147A
Arlington, VT, 85, SE
Arlington Ave., NJ, 158A
Arlington Heights, MA, 127B
Armory, MA, 124, SE
Armstrong, NJ, 150, SE
Arnolds Mills, RI, 147, NW
Arnot, PA, 135, SE
Aroostook, NB, 6, NE
Aroostook Farm, ME, 5, SE
Arrowchar, NY, 158, SE
Arrowhead, NY, 79, NW
Arthursburg, NY, 142, SW
Arverne, NY, 159, SW
Ash Jct., NY, 122, SW
Ashford, NY, 112, NE
Ashland, MA, 127, NW
Ashland, ME, 5, SW
Ashland, NH, 68, SE
Ashland Jct., ME, 7, SE
Ashley Falls, MA, 123, SW

Ashley Pond, MA, 124, SE
Ashmont, MA, 127E
Ashokan, NY, 141, NE
Ashton, RI, 147, NW
Ashuelot, NH, 106, NW
Ashville, NY, 111, SW
Ashwood, NY, 75, NW
Assonet, MA, 147, NE
Atco, ME, 12, SE
Athenia (DL&W; Erie), NJ, 158, NE
Athens, ON, 25, SW
Athol, MA, 106, SE
Athol Springs, NY, 93, NW
Atkinson, NH, 108, NE
Atlanta, NY, 96, SW
Atlantic, MA, 127E
Atlantic, NH, 109, NW
Atlantic, NY, 158, SW
Atlantic Ave., NY, 159, SE
Attean, ME, 10, SW
Attica, NY, 94, NW
Attleboro, MA, 147, NW
Attleboro Jct., MA, 147, NE, 147C
Attlebury, NY, 142, NE
Atwell Crossover, ME, 38, NW
Atwood, CT, 146, SW
Auburn, ME, 55, SE
Auburn, NH, 89, SW
Auburn, NY, 97, NE
Auburndale, MA, 127D
Auburndale, NY, 159, NW
Augusta, ME, 56, NW
Auguston, PA, 131, SE
Aukland, QC, 16, SW
Aultsville, ON, 26, NE
Auriesville, NY, 102, NW
Aurora, NY, 97, NE
Ausable Chasm, NY, 30, SW
Ausable Forks, NY, 48, NE
Austin, PA, 133, SE
Austin Jct., ME, 19, SW
Avenel, NJ, 158, SW
Avery (Needham Heights), MA, 127D
Avery, PA, 138, SW
Avoca (DL&W; Erie), NY, 115, NW
Avon, CT, 144, NW
Avon, MA, 127, SE
Avon, ME, 36, NW
Avon, NY, 95, NE
Awosting, NJ, 151, SW

Axel, ME, 2, SE
Ayer, MA, 107, SE
Ayers Jct., ME, 42, NE

B&O Jct., NY, 94, NE
Baboosic Lake, NH, 107, NE
Babylon, NY, 160, SW
Back Bay (NH), MA, 127A
Backus, PA, 132, NE
Bagley, NH, 88, NW
Baileyville, ME, 23, SE
Bainbridge, NY, 119, NW
Baker, NY, 74, NE
Baker Brook, NB, 1, NE
Bakers, ME, 19, SW
Baldpate, MA, 109, SW
Baldwin, NY, 159, SE
Baldwin, VT, 65, NW
Baldwin Place, NY, 152, NW
Baldwinsville, NY, 79, SW
Baldwinville, MA, 106, SE
Baldwinville Crossover, MA, 106, SE
Baldwinville East, MA, 106, SE
Baleville, NJ, 150, SW
Ballardvale, MA, 108, SE
Balloch, NH, 87, NW
Ballston, NY, 84, SW
Ballston Lake, NY, 103, NW
Balmat, NY, 45, NW
Balmat Siding, NY, 45, NW
Baltic, CT, 145, SE
Baltusrol, NJ, 158, SW
Bancroft, ME, 13, SE
Bangall, NY, 142, NE
Bangor (BO&M; MEC), ME, 39, NW
Bangor, NY, 28, NW
Bangor Freight Yard, ME, 39, NW
Bangs, ME, 37, SW
Banta, NJ, 158, SW
Bantam, CT, 143, SE
Bar Mills, ME, 70, SE
Bar Road, NB, 24, SE
Barber, MA, 126, NW
Barber Dam, NB, 24, NE
Barclay, PA, 136, SE
Bardonia, NY, 151, SE
Bardwell, MA, 105, SE
Baring, ME, 24, SW
Barnard, ME, 20, NE
Barnard, NY, 76, SE
Barnes, NY, 116, NW

Barnes (S&T; TIV), PA, 131, SE
Barnet, VT, 51, NE
Barnjum, ME, 36, NW
Barnstable, MA, 149, SW
Barnstead, NH, 89, NW
Barnveld, NY, 81, NE
Baron, NB, 15, NE
Barre, MA, 125, NE
Barre, VT, 50, SE
Barre Jct., VT, 50, SE
Barre Plains, MA, 125, NE
Barre Transfer, VT, 50, SE
Barrett, ME, 5, NE
Barrett, NH, 52, NW
Barretts Jct., MA, 125, SW
Barrington, RI, 147, SW
Barrowsville, MA, 147, NE
Barrytown, NY, 142, NW
Bartlett, NB, 24, SE
Bartlett, NH, 53, SW
Barton, MA, 126, SW
Barton, NY, 117, SW
Barton, VT, 32, SE
Bartonsville, VT, 86, SE
Bartow, NY, 159, NW
Base, NH, 53, NW
Base Station, NH, 53, NW
Basin Mills, ME, 39, NE
Bass River, MA, 149, SE
Batavia, NY, 94, NE
Bates, ME, 54, NE
Bath (B&H; DL&W; Erie), NY, 115, NW
Bath, ME, 72, NW
Bath, NB, 6, SE
Bath, NH, 52, SW
Bath House, MA, 128, NW
Battenville, NY, 85, SW
Battery, MA, 128, NW
Bay Pond, NY, 47, NW
Bay Ridge, NY, 158, SE
Bay Terrace, NY, 158, SE
Bay View House, ME, 91, NW
Baychester, NY, 159, NW
Baychester Ave., NY, 159, NW
Bayport, NY, 160, SE
Bayshore, NY, 160, SW
Bayside, MA, 128, NW
Bayside, NH, 90, SW
Bayside, NY, 159, NW
Bayview, ME, 72, NW
Bayway (CNJ; SIRT), NJ, 158, SE
Beach Bluff, MA, 128, NW

Beach Glen, NJ, 158, NW
Beach Jct. (BCK), NY, 93A
Beach Ridge, NY, 74, SW
Beachmont, MA, 128, NW
Beacon (NH; NYC), NY, 152, NW
Beacon Falls, CT, 153, NE
Beacon Park Yard, MA, 127E
Beaconsfield, MA, 127D
Beamsville, ON, 73, SW
Beans, NY, 115, NW
Bear Lake, PA, 130, NE
Bear Mountain, NY, 152, NW
Beatties, NH, 33, SE
Beaufort, NJ, 158, NW
Beaver Brook, MA, 127B
Beaver Dam, NY, 116, NW
Beaver Falls, NY, 61, NW
Beaver Lake, NJ, 150, SE
Beaver River, NY, 62, NW
Becket, MA, 123, NE
Becket Quarry, MA, 123, NE
Beckley, CT, 144, SE
Bedell, ME, 90, SE
Bedell, ON, 25, NE
Bedells, NY, 60, NE
Bedford, MA, 127, NW
Bedford, NH, 107, NE
Bedford Hills, NY, 152, SE
Bedford Springs, MA, 108, SW
Bee Hive Crossing, NY, 104, NW
Beebe River, NH, 68, NE
Beecher Falls (CPR; MEC), VT, 33, NE
Beechwood, NB, 6, SE
Beekmans, NY, 142, SE
Beekmantown, NY, 30, NW
Beerston, NY, 119, SE
Belair Road, NY, 158, SE
Belchertown, MA, 125, NW
Belden, NY, 118, SE
Beldens, VT, 49, SE
Belfast (Erie), NY, 113, NE
Belfast, ME, 57, NE
Belfast (W&B), NY, 113, NE
Belgrade, ME, 56, NW
Belknap Point, NH, 69, SW
Bell, VT, 51, SE
Bell Rock, MA, 127C
Bellaire, NY, 159, SE
Bellamy, ON, 25, SW
Belle Isle, NY, 79, SW
Bellefleur, NB, 3, SW
Bellerose, NY, 159, SE

Belleville, NB, 8, SE
Belleville, NJ, 158, NE
Belleville, RI, 147, SW
Bellevue, MA, 127D
Bellmore, NY, 159, SE
Bellona, NY, 96, SE
Bellows Falls, VT, 87, SW
Bellport, NY, 161, NW
Bellwood Park, NJ, 158, NE
Belmont (Erie), NY, 113, SE
Belmont, MA, 127B
Belmont, NH, 89, NW
Belmont (W&B), NY, 113, SE
Belmont Jct., NH, 88, NE
Belmont Jct., NY, 160, SW
Belmont Park, NY, 159, SE
Belsprings, CT, 143, NW
Belvedere, ME, 7, SW
Belvidere (Erie), NY, 113, SE
Belvidere (PS&N), NY, 113, NE
Bemis, MA, 127D
Bemis, ME, 35, NE
Bemus Point, NY, 111, SW
Ben Thomas, ME, 5, NE
Bennett, NY, 97, NE
Bennett Hall, MA, 108, SW
Bennetts, ME, 7, SE
Bennetts (NY&P), NY, 114, SE
Bennetts (PS&N), NY, 114, NW
Bennington, NH, 88, SW
Bennington, VT, 104, NE
Benson, ME, 20, NW
Benson Mines, NY, 45, SE
Benton, ME, 14, NE
Benton, NY, 96, SE
Berea, NY, 141, SE
Bergen, NY, 76, SW
Bergenfield, NJ, 159, NW
Berkeley, RI, 147, NW
Berkeley Heights, NJ, 158, SW
Berkshire, MA, 104, SE
Berkshire, NY, 117, NE
Berlin, CT, 144, SW
Berlin, MA, 126, NE
Berlin, NH, 53, NE
Berlin, NY, 104, SW
Berlin Labor Camp, CT, 144, SW
Bernardston, MA, 105, SE
Bernhards, NY, 80, SW
Besemer, NY, 117, NW
Bethel, CT, 153, NW
Bethel (CV; WRR), VT, 66, NE
Bethel, ME, 54, NW

Bcthcl, NY, 142, NE
Bethel Lower Siding, CT, 153, NW
Bethlehem, NH, 52, NE
Bethlehem Jct., NH, 52, NE
Bethpage, NY, 160, SW
Beverly, MA, 109, SW
Beverly Farms, MA, 109, SW
Biddeford (E), ME, 71, SW
Biddeford, ME, 91, NW
Big Bend, PA, 131, NE
Big Fill, PA, 132, SW
Big Flats (DL&W; Erie), NY, 116, SW
Big Indian, NY, 121, SW
Big Moose, NY, 62, NW
Big Shanty, PA, 132, NE
Bigelow, ME, 18, SW
Billerica, MA, 108, SW
Billings, NY, 142, SW
Bilsborrow, NY, 97, NW
Bingham, ME, 19, SW
Bingham, PA, 132, NE
Bingham, PA, 134, NE
Binghamton (D&H; DL&W; Erie), NY, 118, SW, 118A
Binnewater, NY, 141, NE
Binney, ME, 21, SW
Bird Mills, MA, 127, SE
Bird St., MA, 127E
Bird's Hill, MA, 127D
Birdsall, NY, 114, NW
Bishops, CT, 146, SW
Black Creek (Erie), NY, 113, NE
Black Creek (PRR), NY, 113, NE
Black River, NY, 44, SW
Black Rock, NY, 93D
Black Rock Yard (Erie), NY, 93D
Black Siding, NJ, 158, SW
Black Walnut, PA, 137, SE
Blackinton, MA, 104, SE
Blackmount, NH, 51, SE
Blackstone, MA, 126, SE
Blackstone, ME, 5, NE
Blackwater, ME, 7, NW
Blackwell, PA, 135, SW
Blair, ME, 10, SE
Blair, NH, 68, NE
Blakeley, NY, 93, SE
Blake's, NH, 68, SE
Blakeslee, NY, 80, SW
Blanchard, ME, 19, NE
Blasdell, NY, 93B

Blauvelts, NY, 152, SW
Bleachery, MA, 108, SW
Bliss, NY, 94, SW
Blissville, PA, 132, NW
Blodgett, NH, 89, SW
Blodgett Mills, NY, 98, SE
Blood, NH, 107, NE
Bloomfield, CT, 144, NE
Bloomfield (DL&W; Erie), NJ, 158, NE
Bloomfield, NJ, 158, NE
Blooming Grove, NY, 151, NE
Bloomingdale, NJ, 151, SW
Bloomingdale, NY, 47, NE
Bloomville, NY, 120, NW
Blossburg, PA, 135, SE
Blossvale, NY, 80, NE
Blue Bell, NB, 6, NE
Blue Hill Ave., MA, 127E
Blue Point, NY, 160, SE
Bluestone, NY, 94, SE
Bluff Point, NY, 30, SW
Boardville, NJ, 151, SW
Bodfish, ME, 20, NW
Bogota (NYC; NYS&W), NJ, 158, NE
Bolivar, NY, 113, SE
Bolton, CT, 145, NW
Bolton, VT, 50, NW
Boltonville, VT, 51, SE
Bombay (BMRR; CN), NY, 27, NE
Bondsville, MA, 125, SW
Boonton, NJ, 158, NW
Boonville, NY, 81, NW
Boscawen, NH, 88, NE
Boston, North Station (B&M), MA, 127A
Boston, NY, 93, SE
Boston, South Station (BTCo.), MA, 127A
Boston Corners, NY, 122, SE
Boston Ranch, ME, 10, SW
Botanical Garden, NY, 159, NW
Botsford, CT, 153, NW
Bouckville, NY, 99, NE
Boundary, ME, 9, SE
Bourne, MA, 148, SE
Bow Jct., NH, 88, SE
Bowden, ME, 12, SE
Bowdoinham, ME, 56, SW
Bowman, NH, 53, NW
Bowmansville, NY, 93, NE
Boxford, MA, 109, SW

Boyce, NH, 88, NE
Boyd Lake, ME, 21, SW
Boyer, PA, 132, NE
Boylston, MA, 126, NE
Boylston St., MA, 127E
Brackett Jct., ME, 36, NW
Bradbury, ME, 70, SE
Bradbury's, ME, 1, SE
Bradford (B&O; Erie), PA, 132, NE
Bradford, MA, 108, NE
Bradford, ME, 21, SW
Bradford, NH, 88, NW
Bradford, RI, 156, NE
Bradford, VT, 67, NE
Bradford Shops, PA, 132, NE
Bradley Field, CT, 144, NE
Bragg, CT, 146, SW
Braggville, MA, 127, SW
Brainard, NY, 122, NE
Braintree, VT, 66, NE
Braintree Highlands, MA, 127, SE
Braleys, MA, 148, SW
Branch Jct., NJ, 158, SW
Branch Switch, ME, 55, NW
Branchville, CT, 153, NW
Branchville, NJ, 150, SE
Branchville Jct., NJ, 150, SE
Brandon, VT, 65, NE
Brandreth, NY, 62, NW
Brandt, PA, 138, NE
Brandy Brook, NY, 29, NW
Branford, CT, 154, NW
Brannen, ME, 5, SE
Brassua, ME, 11, SW
Brassua Station, ME, 11, SW
Brattle, MA, 127B
Brattleboro (B&M; CV), VT, 105, NE
Brattleboro (WEST), VT, 105, NE
Brayton, MA, 147, SE
Breakfast Hill, NH, 90, SW
Breesport, NY, 116, SE
Brentwood, NY, 160, NE
Bretton Woods (B&M), NH, 53, NW
Brewer Jct., ME, 39, NW
Brewerton, NY, 79, SE
Brewster, MA, 149, NE
Brewster (NH; NYC), NY, 152, NE
Briarcliff Farms, NY, 142, NE
Briarcliff Manor, NY, 152, SW

Brice, NY, 75, NE
Bridge St., CT, 145, SE
Bridge St., PA, 134, SE
Bridge St. (Port Washington), NY, 159, NW
Bridgehampton, NY, 162, NW
Bridgeport, CT, 153, SE, 153A
Bridgeport, Lower Yard, CT, 153A
Bridgeton, RI, 146, NE
Bridgewater, MA, 148, NW
Bridgewater, ME, 8, NW
Bridgewater, NH, 68, SE
Bridgewater, NY, 100, NE
Bridgton, ME, 54, SE
Bridgton Jct., ME, 70, NW
Brier Hill, NY, 25, SE
Briggs, NY, 142, SW
Briggs, NY, 45, SE
Brighton, MA, 127D
Brighton, NY, 76, SE
Brighton Ave., NJ, 158, NE
Brighton Ave., NY, 79, SE
Brightwood, MA, 124, SE
Brinckerhoff, NY, 142, SW
Brisben, NY, 118, NE
Bristol, CT, 144, SW
Bristol, NB, 8, NE
Bristol, NH, 68, SE
Bristol, RI, 147, SW
Bristol, VT, 49, SE
Brixment, NY, 75, SW
Broad Brook, CT, 144, NE
Broad Channel, NY, 159, SW
Broadalbin, NY, 83, SE
Broadalbin Jct., NY, 83, SW
Broadway (Fairlawn), NJ, 158, NE
Broadway, MA, 127C
Broadway, NY, 159, NW
Brockport, NY, 76, SW
Brockton, MA, 127, SE
Brockton (NKP), NY, 111, NW
Brockton (NYC), NY, 111, NW
Brockton (PRR), NY, 111, NW
Brockville (CN; CPR), ON, 25, SE
Bronx Park 180th St., NY, 159, NW
Bronxville, NY, 159, NW
Brookdale, NY, 76, SE
Brookfield, CT, 153, NW
Brookfield, MA, 125, SE
Brookfield, NH, 69, SE
Brookfield Jct., CT, 153, NW
Brookhaven, NY, 161, NW

Brookhaven Labs, NY, 161, NW
Brookhurst, NH, 69, SW
Brookland, PA, 134, NW
Brookline, MA, 127E
Brookline, NH, 107, SE
Brookline Hills, MA, 127D
Brooklyn Manor, NY, 159, SW
Brooks, MA, 126, NW
Brooks, ME, 38, SE
Brooks Ave., NY, 76A
Brookside, MA, 108, SE
Brookston, PA, 132, SW
Brookton, NY, 117, NW
Brookview, NY, 103, SE
Brown, MA, 108, SE
Brownfield, ME, 70, NW
Brownlee, PA, 135, SW
Browns, MA, 128, SW
Brownville, ME, 20, NE
Brownville, NY, 44, SW
Brownville Jct. (BAR; CPR), ME, 20, NE
Brundage, NY, 115, NW
Brunswick, ME, 72, NW
Brushton, NY, 27, NE
Bryant's Pond, ME, 54, NE
Bryn Mawr Park, NY, 159, NW
Buck, NY, 94, SW
Buckfield, ME, 55, NW
Buckland, CT, 144, NE
Buckland, MA, 105, SW
Bucksport, ME, 39, SW
Buffalo (DL&W), NY, 93A
Buffalo (Erie), NY, 93A
Buffalo (LV), NY, 93A
Buffalo, Central Station (NYC), NY, 93C
Buffalo, Coal Dumpers (DL&W), NY, 93A
Buffalo, Exchange St. (NYC), NY, 93A
Buffalo, Louisiana St. (PRR), NY, 93A
Buffalo, ME, 5, NW
Buffalo, Sycamore St. Jct., NY, 93C
Buffalo, Terrace (NYC), NY, 93A
Bugbee, ME, 5, NE
Bullville, NY, 141, SW
Burdett, NY, 116, NW
Burdicks, NY, 65, NW
Burgoyne, NY, 84, SE
Burke, NY, 28, NE
Burleigh, ME, 4, NE

Burleyville, NH, 69, SE
Burlington (CV) (R), VT, 49, NE
Burnham Jct., ME, 38, SW
Burns, NY, 114, NE
Burnside, CT, 144, NE, 144A
Burnside, NY, 151, NE
Burnwood, PA, 138, NE
Burpee, NB, 15, SW
Burr, ME, 2, SE
Burrage, MA, 128, SW
Burrows, PA, 134, SE
Burrows Lot (PRR), NY, 93A
Burrville, CT, 143, NE
Burt, NY, 74, NE
Burton's, VT, 85, NE
Burtt, MA, 108, SE
Burtville, PA, 133, NE
Bury, QC, 16, NW
Bushwick, NY, 159, SW
Bushwick Jct., NY, 159, SW
Buskirk, NY, 104, NW
Butler, CT, 146, NW
Butler, NJ, 151, SW
Buttonwoods, RI, 147, SW
Buxton, ME, 70, SE
Buzzards Bay, MA, 148, SE
Byams, MA, 108, SW
Byfield, MA, 109, NW
Byromtown, PA, 131, SE
Byron, ME, 35, SE
Byron, NY, 75, SE

C&C Jct., NY, 60, NE
C.J.I. Camps, NY, 150, NE
Cabin Run, PA, 136, SE
Cadiz, NY, 113, NW
Cadosia, NY, 139, NW
Cadyville, NY, 29, SE
Cairds, NB, 15, NW
Cairo, NY, 121, NE
Cairo Jct., NY, 122, NW
Calais (MEC), ME, 24, SW
Calais Jct., ME, 39, NW
Calciana, NY, 77, SE
Calcium, NY, 44, SW
Caldwell, NJ, 158, NW
Caledonia, NY, 95, NW
California Road, ME, 6, NW
Callicoon, NY, 139, NE
Calverton, NY, 161, NE
Cambria, NY, 74, SW
Cambridge, NY, 85, SW
Cambridge, VT, 31, SW

Cambridge Jct., VT, 31, SW
Camden (LV; NYC), NY, 80, NE
Camelot, NY, 142, SW
Cameron, NY, 115, SW
Cameron Mills, NY, 115, SW
Camillus, NY, 79, SW
Camp, MA, 149, SW
Camp 12, ME, 20, NW
Camp Benson, ME, 38, NW
Camp Edwards, MA, 148, SE
Camp Ellis, ME, 91, NW
Camp Ground, ME, 71, SW
Camp Jct., MA, 147, NE
Campbell (DL&W; Erie), NY, 115, SE
Campbell, ME, 6, NW
Campbell Hall (NYO&W), NY, 151, NW
Campbell Hall Jct. (Erie), NY, 151, NW
Campbells, ME, 24, SW
Campello, MA, 127, SE
Campgaw, NJ, 151, SE
Camps Mills, NY, 59, NE
Campton, NH, 68, NE
Campville, CT, 143, SE
Campville, NY, 117, SE
Canaan, CT, 123, SW
Canaan, NH, 67, SE
Canaan, NY, 123, NW
Canaan Road Crossover, ME, 37, SE
Canada Jct., NY, 29, NE
Canada Mills, MA, 126, NW
Canadian Jct., ME, 3, SW
Canadian Pacific Jct. (CN), NB, 2, NW
Canadohta, PA, 130, NW
Canajoharie, NY, 101, NE
Canal, NY, 81, SW, 81A
Canal Branch, NY, 81, SW, 81A
Canal Jct., MA, 125, SW
Canandaigua, NY, 96, NW
Canandea (PRR), NY, 113, NE
Canandea (W&B), NY, 113, NE
Canaseraga (Erie; PS&N), NY, 114, NW
Canastota (LV; NYC), NY, 80, SE
Candia, NH, 89, SW
Candor, NY, 117, SW
Canfield, NY, 116, NW
Canisteo (Erie; NY&P), NY, 114, NE

Cannondale, CT, 153, SW
Canobie Lake, NH, 108, NE
Canochet, RI, 156, NE
Canoe Camp, PA, 135, NE
Canterbury, NB, 15, NW
Canterbury, NH, 88, NE
Canton, CT, 144, NW
Canton, ME, 55, NW
Canton, NY, 26, SE
Canton, PA, 136, SW
Canton Jct., MA, 127, SE
Cape Elizabeth (South Portland), ME, 71A
Cape Jct., ME, 58, NW
Cape Jellison, ME, 58, NW
Cape Vincent, NY, 43, SW
Carbondale (DCH; NYO&W), PA, 138, SE
Carbondale (Erie), PA, 139, SW
Cardigan, NH, 68, SW
Cardinal, ON, 26, NW
Carey's Crossing, MA, 107, SW
Caribou (BAR; CPR), ME, 5, NE
Caribou, ME, 6, NW
Carle Place, NY, 159, SE
Carlson, PA, 132, SW
Carlstadt, NJ, 158, NE
Carlton, NY, 75, NE
Carlton Hill, NJ, 158, NE
Carltonville, MA, 109, SW
Carman, NY, 103, NW
Carmel, ME, 38, NE
Carmel, NY, 152, NE
Caroline, NY, 117, NW
Caron Brook, NB, 1, NE
Carrabassett, ME, 18, SE
Carrigain, NH, 53, SW
Carroll, NH, 52, NE
Carroll, NY, 113, SW
Carrollton, NY, 112, SE
Carson, ME, 5, NE
Carter, NY, 62, NW
Carteret, NJ, 158, SE
Carters, MA, 126, NE
Carthage, NY, 60, NE
Cartwrights, PA, 132, SE
Caryl, NY, 159, NW
Cary's Mills, ME, 8, SW
Caryville, MA, 127, SW
Casadaga, NY, 111, NW
Cascade, NH, 53, NE
Cashman's Crossing, MA, 107, SW

Castile, NY, 94, SE
Castle, NY, 80, SE
Castle Hill, MA, 109, SW
Castleton, VT, 65, SE
Castorland, NY, 60, NE
Catatonk, NY, 117, SW
Cathance, ME, 72, NW
Cato, NY, 78, SE
Catskill, NY, 122, SW
Catskill Landing, NY, 122, SW
Catskill Village, NY, 122, SW
Cattaraugus, NY, 112, NW
Catumet, MA, 148, SE
Caughdenoy, NY, 79, NE
Cavender, NH, 107, NW
Cavendish, VT, 86, NE
Cayuga, NY, 97, NE
Cayuga Jct., NY, 97, NE
Cayuta, NY, 116, NE
Caywood, NY, 97, SW
Cazenovia, NY, 99, NW
CD Crossover, PA, 130, NE
Cedar, MA, 127, SW
Cedar Ave., NY, 158, SE
Cedar Bluffs, NY, 84, SE
Cedar Grove, MA, 127E
Cedar Grove, ME, 39, SW
Cedar Grove, NJ, 158, NE
Cedar Hill Yard, CT, 154, NW, 154A
Cedar Manor, NY, 159D
Cedar Run, PA, 135, SW
Cedarcliff, NY, 142, SW
Cedarhurst, NY, 159, SE
Cedarville, NY, 100, NE
Cemetery, NY, 103, SE, 103A
Centennial Grove, MA, 109, SW
Center, ME, 39, NW
Center, NJ, 158, SW
Center, NY, 114, NW
Center Ave., NY, 159, SE
Center Barnstead, NH, 89, NW
Center Falls, NY, 85, SW
Center Moriches, NY, 161, NW
Center Rutland, VT, 65, SE
Center Village, MA, 128, SE
Center Village, NY, 118, SE
Centervale, VT, 51, NE
Centerville, NY, 113, NE
Centerville, NY, 59, SE
Centerville, PA, 130, SW
Central Ave., MA, 127E
Central Ave., NJ, 158, NE

Central Bridge, NY, 102, SW
Central Islip State Hospital, NY, 160, NE
Central Square (NYC; NYO&W), NY, 79, NE
Central Valley, NY, 151, NE
Central Village, CT, 146, SW
Centre, VT, 32, NW
Centredale, RI, 147, NW
Centreville, NB, 8, NE
Ceres, NY, 113, SW
Chadwick, NY, 81, SW
Chaffee, NY, 94, SW
Chaffee's, PA, 132, SW
Chaffin, MA, 126, NW
Chamberlains, ME, 22, NW
Chamberlains, NY, 59, NE
Chambers, NY, 116, NW
Chamcook, NB, 24, SE
Chamcook Lake, NB, 24, SE
Champlain, NY, 30, NW
Chandler, NH, 87, NW
Chapin, NY, 96, NE
Chapin Switch, NH, 87, SW
Chapman, ME, 5, SE
Chappaqua, NY, 152, SW
Charlemont, MA, 105, SW
Charlemont Crossovers, MA, 105, SW
Charles River, MA, 127, NW
Charles St. Engine Terminal, RI, 147A
Charlestown, NH, 87, SW
Charlotte (B&O; NYC), NY, 76, NE
Charlotte, ME, 24, SE
Charlotte, VT, 49, NW
Charlotteburg, NJ, 151, SW
Charlton, MA, 126, SW
Chase Mills, NY, 26, NE
Chasm Falls, NY, 28, NE
Chateauguay, NY, 28, NE
Chatham, MA, 149, SE
Chatham, NJ, 158, SW
Chatham, NY, 122, NE
Chatham Center, NY, 122, NE
Chaumont, NY, 43, SE
Chauncey, NY, 159, NW
Chautauqua, NY, 111, SW
Chazy, NY, 30, NW
Chazy Lake, NY, 29, SW
Cheapside, MA, 105, SE, 105A
Chelmsford, MA, 108, SW
Chelsea, MA, 127C

Chelsea, ME, 56, NE
Chelsea, NY, 142, SW
Chemung, NY, 116, SE
Chemung Sta., PA, 136, NE
Chenango Bridge, NY, 118, SW
Chenango Forks, NY, 118, SW
Chenango Jct., NY, 79, SE
Cherokee, ME, 14, SW
Cherry, CT, 144, NW
Cherry Brook, MA, 127, NW
Cherry Creek, NY, 111, SE
Cherry Plain, NY, 104, SW
Cherry Run, PA, 132, SW
Cherry Springs, PA, 134, SW
Cherry Valley, NY, 101, NW
Cherry Valley Jct., NY, 101, SE
Cherryfield, ME, 41, SW
Chesham, NH, 106, NE
Cheshire, CT, 144, SW
Cheshire, MA, 104, SE
Cheshire Harbor, MA, 104, SE
Chester, CT, 155, NW
Chester, MA, 124, NW
Chester, ME, 22, NW
Chester, NY, 151, NW
Chester, VT, 86, NE
Chester Heights, NY, 159, NW
Chestnut Hill, CT, 145, NW
Chestnut Hill, MA, 127D
Chichester, NH, 89, NW
Chichester, NY, 121, SW
Chicks, ME, 90, NE
Chicopee, MA, 124, SE
Chicopee Falls, MA, 124, SE
Chicopee Jct., MA, 124, SE
Childs, PA, 130, NE
Childwold, NY, 46, SE
Chili, NY, 76, SW
Chili Center, NY, 76, SW
China, ME, 57, NW
China Lake, ME, 56, NE
Chipmunk, NY, 112, SE
Chippawa, ON, 73, SE
Chittenango, NY, 80, SW
Chrome, NJ, 158, SE
Church's, ME, 8, NW
Churchville Jct., NY, 76, SW
Churchville, NY, 76, SW
Churubusco, NY, 29, NW
Cincinnatus, NY, 99, SW
City Mills, MA, 127, SW
City Point, ME, 57, NE
Clairs, NB, 1, NE

Clara, PA, 133, NE
Claremont, NH, 87, NW
Claremont Jct., NH, 87, NW
Claremont Park, NY, 159, NW
Clarence, NY, 93, NE
Clarence Center, NY, 74, SE
Clarendon, PA, 131, NE
Clarendon, VT, 66, SW
Clarendon Hills, MA, 127E
Clarendon Springs, VT, 65, SE
Clark, NJ, 158, SW
Clark Mills, NY, 81, SW
Clarks, ME, 5, SW
Clarks, ME, 56, NE
Clarks, NY, 46, NW
Claverack, NY, 122, SE
Clay, NY, 79, SE
Clayton, NY, 43, SE
Clayville, NY, 100, NW
Clear River, MA, 126, SW
Clematis Brook, MA, 127B
Clement, MA, 149, SW
Clemo, PA, 139, SW
Clemons, NY, 65, SW
Clermont (PS&N; PRR), PA, 133, SW
Clermont Jct., PA, 133, NW
Cleveland, MA, 126, SW
Cleveland, ME, 2, NW
Cleveland, NY, 80, SW
Clicquot, MA, 127, SW
Clifton, MA, 128, NW
Clifton, NJ, 158, NE
Clifton, NY, 158, SE
Clifton Mines, NY, 45, NE
Clifton Springs, NY, 96, NE
Cliftondale, MA, 127C
Clinton, CT, 154, NE
Clinton, MA, 126, NE
Clinton, ME, 37, SE
Clinton, NY, 81, SW
Clinton Ave., NJ, 158, SW
Clinton Corners, NY, 142, NW
Clinton Crossing, NY, 101, SW
Clinton Engine House, MA, 126, NE
Clinton Jct., MA, 126, NE
Clinton Mills, NY, 29, NW
Clinton Road, NY, 159, SE
Clinton St. (DL&W; LV), NY, 116, NE
Clintondale, NY, 141, SE
Clockville, NY, 80, SW

Closter, NJ, 159, NW
Clove Branch Jct., NY, 142, SW
Clove Valley, NY, 142, SE
Cloverdale, VT, 31, SW
Clyde, NY, 78, SW
Clyde, RI, 146, SW
Clymer, NY, 110, SE
CM Jct., MA, 126, NE
CM Jct., PA, 130, NE
Coal Yard, CT, 145, SE
Cobalt, CT, 144, SE
Cobb Bridge, VT, 66, NE
Cobham, PA, 131, SW
Cobleskill, NY, 102, SW
Cocheco, NH, 90, SW
Cochecton, NY, 139, SE
Cochessett, MA, 127, SE
Cochituate, MA, 127, NW
Coffins Mills, NY, 45, SE
Cohasset, MA, 128, SW
Cohocton, NY, 96, SW
Cohoes, NY, 103, NE
Cokertown, NY, 122, SW
Colburns, NY, 111, SW
Colby, ME, 5, NE
Colby, NH, 88, SW
Colchester, CT, 145, SW
Colchester, NY, 119, SE
Colchester, VT, 30, SE
Colchester, VT, 30, SW
Cold River, NH, 87, SW
Cold Spring, NY, 112, SW
Cold Spring, NY, 152, NW
Cold Spring, PA, 136, SE
Cold Spring Harbor, NY, 160, NW
Cold Water, NY, 76, SE
Coldbrook (B&A), MA, 125, NE
Coldbrook (B&M), MA, 125, NE
Colden, NY, 93, SE
Colebrook, CT, 143, NE
Colebrook, NH, 33, NE
Colegrove, PA, 133, SW
Colemans, NY, 142, NE
Colesburg, PA, 134, NW
Collamer, NY, 76, NW
College Point, NY, 159, NW
Colliers, NY, 120, NW
Collins, MA, 125, SW
Collins, ME, 2, SE
Collins, NY, 111, NW
Collins, NY, 45, SE
Collins St., MA, 109, SW
Collinsville, CT, 144, NW

Colonia, NJ, 158, SW
Colonie, NY, 103, SE, 103A
Colonie Shops (D&H), NY, 103, SE, 103A
Coltsville, MA, 123, NE
Columbia, CT, 145, SE
Columbia Bridge, NH, 33, NE
Columbia Falls, ME, 41, SE
Columbia X Roads, PA, 136, NW
Columbus, PA, 130, NE
Columbus Ave. (NH; NYW&B), NY, 159, NW
Columbus Jct., PA, 130, NE
Colza, PA, 130, NE
Communipaw (CNJ; LV), NJ, 158A
Comstock, NY, 85, NW
Concord, MA, 127, NW
Concord, NH, 88, SE
Concord, VT, 52, NW
Cones, NH, 33, NE
Conesus, NY, 95, SE
Conesus Lake Jct., NY, 95, NE
Coneville, PA, 133, NE
Conewango, NY, 111, SE
Congamond, MA, 124, SW
Congers, NY, 152, SW
Conicut, VT, 51, SE
Conifer, NY, 46, SE
Conklin, NY, 118, SW
Connors, NB, 1, SW
Connors Spur, NB, 2, NW
Conomo, MA, 109, SW
Constable, NY, 28, NW
Constantia, NY, 79, NE
Contoocook, NH, 88, NE
Convent, NJ, 158, NW
Converse, NH, 88, NW
Conway, MA, 105, SE
Conway, NH, 69, NE
Conway Centre, NH, 69, NE
Cook St., MA, 127D
Cooks, NY, 122, SE
Cook's Falls, NY, 140, NW
Coons, NY, 103, NE
Coopers (DL&W; Erie), NY, 115, SE
Coopers, NY, 84, NE
Coopers Mills, ME, 56, NE
Cooperstown, NY, 101, SW
Cooperstown Jct., NY, 120, NW
Coos Jct., NH, 33, SE
Copake, NY, 122, SE
Copake Falls, NY, 122, SE

Copenhagen, NY, 60, NE
Copenhagen Jct., NY, 60, NE
Copiague, NY, 160, SW
Copperville, NH, 34, SE
Corbett, NY, 119, SE
Corbin, NY, 103, SW
Cordage, MA, 148, NE
Cordaville, MA, 126, NE
Corfu, NY, 94, NW
Corinna, ME, 38, NW
Corinth, NY, 84, SW
Corning (DL&W; Erie; NYC), NY, 115, SE
Corning Yard (NYC), NY, 115, SE
Cornish, ME, 70, NW
Cornwall, NY, 151, NE
Cornwall Bridge, CT, 143, NW
Corona, NY, 159, SW
Corry (Erie; PRR), PA, 130, NE
Cortland, NY, 98, SE
Cortland Jct., NY, 117, NW
Corydon, PA, 132, NW
Coryville (PS&N; PRR), PA, 133, NW
Cos Cob, CT, 152, SE
Costello, PA, 133, SE
Costigan, ME, 21, SE
Cottage, ME, 1, SW
Cottage Farm, MA, 127A, 127E
Cottage Grove, CT, 144, NE
Cottekill, NY, 141, NE
Cotton Valley, NH, 69, SE
Cottrell, NB, 15, SW
Coudersport, PA, 133, NE
Country Club, NY, 122, SE
Country Life Press, NY, 159, SE
County Farm, NY, 101, SW
County Home, NY, 115, NW
County Home, NY, 82, SW
County House, NY, 114, NW
Courcelles, QC, 9, NW
Courchesne, NB, 1, NW
Coventry, NY, 118, NE
Coventry, RI, 146, SE
Coventry, VT, 32, NE
Covert, NY, 97, SE
Covington, PA, 135, SE
Cowanesque, PA, 134, NE
Cowley, PA, 136, SW
Coxsackie, NY, 122, NW
Crafts, NY, 152, NE
Craigs, NY, 95, NW
Cramers Bridge, NY, 84, SE

Cranberry Creek, NY, 83, SE
Cranberry Lake, NY, 46, SW
Cranford (CNJ; LV), NJ, 158, SW
Cranford, NJ, 158, SW
Craryville, NY, 122, SE
Crawford, NY, 113, NE
Crawford Jct., NY, 151, NW
Crawford Jct., PA, 132, NE
Crawford North, NH, 53, SW
Creamery Jct., MA, 125, NE
Creosoting Plant, NY, 80, SE
Crescent (NYC), NY, 103, NE
Crescent Beach, MA, 128, NW
Crescent Park, RI, 147, NW
Crescent St., MA, 127E
Cresskill, NJ, 159, NW
Crestwood, NY, 159, NW
Crittenden, NY, 94, NW
Crocketts, NY, 78, NE
Croghan, NY, 61, NW
Cromwell, CT, 144, SE
Crosby, PA, 133, SW
Cross Fork Jct., PA, 134, SW
Cross Roads, NY, 60, NE
Cross Roads, NY, 97, NE
Cross St., MA, 127B
Croton Falls, NY, 152, NE
Croton Heights, NY, 152, NW
Croton Lake, NY, 152, SW
Croton-on-Hudson, NY, 152, SW
Crouseville, ME, 5, NE
Crowleys, ME, 55, SE
Crown Point, NY, 65, NW
Croxton Yard, NJ, 158A
Crugers, NY, 152, SW
Crystal, ME, 13, NW
Crystal, NH, 34, SW
Crystal Lake, NJ, 151, SE
Crystal Lake, NY, 113, NW
Crystal Run, NY, 151, NW
Cuba (Erie), NY, 113, SW
Cuba (PRR), NY, 113, SW
Cuddleback, NY, 97, NW
Cumberland, ME, 71, NE
Cumberland Center, ME, 71, NW
Cumberland Mills (MEC; PTM), ME, 71, SW, 71A
Cumberland Mills, RI, 147, NW, 147B
Cummings, ME, 90, SW
Cummings, NY, 65, NW
Cumminsville, NY, 95, SE
Cupsuptic, ME, 17, SW

Currie, NB, 6, NE
Curriers, NY, 94, SW
Curtis, NY, 115, SE
Curtis Corner, ME, 55, NE
Curtis Crossing, MA, 128, SW
Cushings, ME, 1, SE
Cushman, MA, 124, NE
Custer City (B&O; Erie), PA, 132, NE
Cutchogue, NY, 155, SW
Cutler Summit, PA, 134, NW
Cuttingsville, VT, 86, NW
Cuyler, NY, 99, SW
Cuylerville, NY, 95, NW
CV, NY, 30, NW
CV Jct., PA, 135, NE
CV/B&M Jct., VT, 87, NW
Cyr Jct. (CN; CPR), NB, 3, SW

Daggett, ME, 6, SW
Daggetts, RI, 147, NW
Dahoga, PA, 132, SE
Dahoga Jct., PA, 132, SW
Dainsville, ON, 92, NW
Dairyman's League Creamery, NY, 120, NE
Dale, NY, 94, NE
Dallas, ME, 35, NE
Dalton, MA, 123, NE
Dalton, NH, 52, NE
Dalton, NY, 95, SW
Dalton, PA, 138, SE
Damariscotta Mills, ME, 56, SE
Damon Crossing, VT, 33, SW
Danbury, CT, 153, NW
Danbury, NH, 68, SW
Danby, VT, 86, NW
Danforth, ME, 14, SW
Danielson, CT, 146, NW
Dannemora, NY, 29, SE
Dansville, NY, 95, SE
Danvers, MA, 109, SW
Danversport, MA, 109, SW
Danville, VT, 51, NE
Danville Jct., ME, 55, SW
Darby, MA, 148, NE
Darien, CT, 153, SW
Darien Center, NY, 94, NW
Darlington, RI, 147B
Davenport Center (D&H), NY, 120, NW
Davenport Center (NYC), NY, 120, NW

Davidson, ME, 13, NW
Davis, NB, 3, SE
Davis, NB, 6, NE
Davis, PA, 134, NE
Dawson Run, PA, 131, SW
Day Jct., PA, 132, NW
Days Siding, ME, 90, NE
Daysville, NY, 59, SE
Dayton, NY, 112, NW
Dayville, CT, 146, NW
Dead River, ME, 17, SE
Deadwater, ME, 19, SW
Dean St., MA, 147, NE, 147C
Deans, ME, 21, NW
Deansboro, NY, 100, NW
Debec Jct., NB, 8, SE
Dedham, MA, 127, SE, 127F
Deep Creek, NB, 8, NE
Deep Cut, NY, 95, SW
Deep River, CT, 155, NW
Deer Lake, NB, 15, NW
Deer Park, NY, 160, NW
Deer River, NY, 60, NE
Deerfield, MA, 105, SE
Deerfield Jct., MA, 105, SE, 105A
Deering, ME, 71, SW, 71A
Deering Jct., ME, 71, SW, 71A
DeGolia, PA, 132, NE
DeKalb Jct., NY, 26, SW
DeKays, NJ, 151, SW
Delancey, NY, 120, SW
Delanson, NY, 102, NE
Delawanna, NJ, 158, NE
Delaware Packing Co., NY, 120, NE
Delaware Valley Creamery, NY, 120, NE
Delevan, NY, 113, NW
Delevan, NY, 81, NE
Delhi, NY, 120, NW
Dellwood, NY, 93, NE
Delmar, NY, 103, SW
Demarest, NJ, 159, NW
Denley, NY, 61, SW
Dennysville, ME, 42, NW
Dent, PA, 132, NE
Depew, NY, 93E
Deposit, NY, 119, SW
Derby, CT, 153, NE
Derby, ME, 21, SW
Derby, NY, 93, SW
Derrick, NY, 47, NW

Derrick City (WNY&P), PA, 132, NE
Derry, NH, 108, NW
DeRuyter, NY, 99, NW
Detroit, ME, 38, NW
Devereux, MA, 128, NW
Devon, CT, 153, SE
Deweys Mills, VT, 67, SW
DeWitt Yard, NY, 79, SE
Dexter, ME, 20, SW
Dexter, NY, 43, SE
Dexter Jct., NY, 43, SE
Diamond Crossing, NH, 52, NE
Diamond Crossing, VT, 87, SW
Diamond Hill, RI, 147, NW
Diana, NY, 45, SW
Dibblee, NB, 8, SE
Dickenson, NY, 27, SE
Dickenson Center, NY, 27, SE
Didell, NY, 142, SW
Dighton, MA, 147, NE
Dike St., RI, 147A
Dimmick, ME, 19, SW
Dimock, PA, 138, SW
Dimond, NH, 88, SW
Dineharts, NY, 115, NW
Ditchfield, QC, 9, SW
Dix, NY, 81, SW
Dixfield, ME, 36, SW
Dixon, NY, 96, NE
DL&W Jct., NY, 94, NE
Dobbs Ferry, NY, 152, SW
Dock, CT, 153, SW
Dock, NY, 152, SW
Dock Jct., NY, 76, SE
Dolby, ME, 12, SE
Dole Jct., NH, 106, NW
Dolgeville, NY, 82, SW
Donaldson, PA, 131, SE
Dongan Hills, NY, 158, SE
Dorchester, MA, 127E
Dorman, ME, 41, SW
Dorset Quarry, VT, 85, SE
Dotyville, PA, 130, SE
Douglas, NY, 49, NW
Douglas Jct., MA, 126, SW
Douglaston, NY, 159, NW
Dover, MA, 127, SW
Dover, NH, 90, SW
Dover Furnace, NY, 142, SE
Dover Jct., NH, 90, SW
Dover Plains, NY, 142, SE
Dover Point, NH, 90, SW

Dover-Foxcroft, ME, 20, SE
Downsville, NY, 120, SW
Dows, ME, 6, NW
Dresden, ME, 56, SW
Dresden, NY, 65, SW
Dresden, NY, 97, SW
Drew, ME, 13, SE
Drummond, NB, 3, SE
Drury, NH, 107, NW
Dryden, NY, 117, NW
Dublin St., CT, 143, SE
Dudley, ME, 7, SE
Duff (CN), ON, 93, NW
Duhring, PA, 132, SW
Dumbarton, NB, 24, NE
Dumfries, NB, 15, NE
Dummerston, VT, 105, NE
Dumont, NJ, 159, NW
Dundee, NY, 97, SW
Dunellen, NJ, 158, SW
Dunkirk (Erie), NY, 111, NW
Dunkirk (NKP), NY, 111, NW
Dunkirk (NYC), NY, 111, NW
Dunkirk (PRR), NY, 111, NW
Dunkirk Dock (Erie), NY, 111, NW
Dunns, ME, 71, NE
Dunraven, NY, 120, SE
Dunstable, MA, 108, SW
Dunwoodie, NY, 159, NW
Durant, NY, 152, SW
Durant City Jct., PA, 132, SW
Durham, NH, 90, SW
Durhamville, NY, 80, SE
Dushore, PA, 137, SW
Dutchess Jct., NY, 152, NW
Dwight, MA, 125, NW
Dye Plant, NH, 90, SW
Dyer Brook, ME, 7, SE
Dyes, NY, 118, SE
Dykemans, NY, 152, NE
Dyre Ave., NY, 159, NW

E. 3rd St., NY, 159, NW
E. 6th St., NY, 159, NW
Eagle, NY, 94, SW
Eagle Bay, NY, 62, NW
Eagle Bridge, NY, 104, NW
Eagle Harbor, NY, 75, SW
Eagle Lake, ME, 1, SE
Eagleville, CT, 145, NW
Earls, NY, 94, NW
Earlville, NY, 99, SE

East Alburgh, VT, 30, NE
East Alexander, NY, 94, NE
East Aurora, NY, 93, NE
East Barre, VT, 51, SW
East Bergen, NY, 76, SW
East Berkshire, VT, 31, NE
East Berlin, CT, 144, SE
East Bethany, NY, 94, NE
East Billerica, MA, 108, SE
East Blackstone, MA, 126, SE
East Boston (B&A; B&M), MA, 127E
East Boston, MA, 127E
East Bradford (B&O; Erie), PA, 132, NE
East Braintree, MA, 128, SW
East Branch, ME, 12, SE
East Branch, NY, 139, NE
East Bridgeport Yard, CT, 153, SE
East Bridgewater, MA, 128, SW
East Brighton, VT, 33, NW
East Brookfield, MA, 125, SE
East Buffalo, NY, 93C
East Buskirk, NY, 104, NW
East Cambridge, MA, 127A
East Canaan, CT, 123, SW
East Candia, NH, 89, SW
East Chatham, NY, 122, NE
East Chester, NY, 151, NW
East Clarence, NY, 74, SE
East Clarendon, VT, 66, SW
East Concord, NY, 93, SE
East Creek, NY, 82, SE
East Dedham, MA, 127, SE, 127F
East Deerfield East, MA, 105, SE, 105A
East Deerfield Yard, MA, 105, SE, 105A
East Dorset, VT, 85, SE
East Douglas, MA, 126, SE
East Dover, ME, 20, SE
East Everett, MA, 127A
East Fairfield, VT, 31, NW
East Farms, CT, 144, SW
East Foxboro, MA, 127, SE
East Freetown, MA, 148, NW
East Freetown, NY, 99, SW
East Gardner, MA, 107, SW
East Granby, CT, 144, NE
East Granville, VT, 50, SW
East Greenbush, NY, 103, SE
East Greenwich, NY, 85, SW
East Greenwich, RI, 147, SW

East Groton, NH, 107, SE
East Haddam, CT, 155, NW
East Hampton, NY, 162, NE
East Hanover, CT, 144, SW
East Hardwick, VT, 32, SW
East Hartford, CT, 144, NE, 144A
East Haven, CT, 154, NW
East Hebron, ME, 55, SW
East Hereford, QC, 16, SW
East Hickory, PA, 131, SW
East Highgate, VT, 31, NW
East Holliston, MA, 127, SW
East Homer, NY, 98, SE
East Honesdale, PA, 139, SE
East Hounsfield, NY, 60, NW
East Ithaca, NY, 117, NW
East Jaffrey, NH, 106, NE
East Jct., MA, 147, NW
East Kingston, NH, 108, NE
East Lexington, MA, 127B
East Lincoln Ave. (Mt. Vernon), NY, 159, NW
East Litchfield, CT, 143, NE
East Littleton, MA, 108, SW
East Longmeadow, MA, 124, SE
East Lyme, CT, 155, NE
East Lyndon, ME, 6, NW
East Machias, ME, 42, SW
East Manchester, NH, 108, NW
East Meredith, NY, 120, NW
East Middleboro, MA, 148, NW
East Milford, NH, 107, NE
East Millinocket, ME, 12, SE
East Milton, MA, 127E
East Moriches, NY, 161, NW
East New London (CV), CT, 155, NE, 155A
East New York, NY, 159, SW
East New York Mills, NY, 81, SW, 81A
East Newport, ME, 38, NE
East Northfield, MA, 106, SW
East Norwalk, CT, 153, SW
East Orange (DL&W; Erie), NJ, 158, NE
East Palmyra, NY, 77, SE
East Patterson, NJ, 158, NE
East Pembroke, NY, 94, NW
East Providence, RI, 147A
East Putney, VT, 106, NW
East River, CT, 154, NE
East Road, ME, 6, NW
East Rochester, NY, 77, SW

East Rockaway, NY, 159, SE
East Ryegate, VT, 51, SE
East Salamanca, NY, 112, SE
East Sandwich, MA, 149, SW
East Saratoga Jct., NY, 103, NE
East Saugus, MA, 128, NW
East Schaghticoke, NY, 103, NE
East Somerville, MA, 127A
East St. Johnsbury-Griswold, VT, 52, NW
East Steuben, NY, 81, NE
East Sudbury, MA, 127, NW
East Summit, PA, 134, NE
East Sumner, ME, 55, NW
East Swanton (CV; SJL), VT, 30, NE
East Switch, MA, 126, NE
East Syracuse, NY, 79, SE
East Taunton, MA, 147, NE
East Thompson, CT, 126, SW
East Vassalboro, ME, 56, NE
East View, NY, 152, SW
East Walden, NY, 141, SE
East Wallingford, CT, 154, NW
East Warren, RI, 147, SW
East Watertown, MA, 127D
East Waverly, NY, 116, SE
East Weare, NH, 88, SE
East Webster, MA, 126, SW
East Weymouth, MA, 128, SW
East Williamson, NY, 77, SE
East Williston, NY, 159, NE
East Wilton, ME, 36, SE
East Windsor, NY, 118, SE
East Windsor Hill, CT, 144, NE
East Worcester, NY, 101, SE
East X, NY, 97, NW
Eastham, MA, 149, NE
Easthampton, MA, 124, NE
Easton, MA, 127, SE
Easton, ME, 6, SW
Easton, NY, 85, SW
Eastondale, MA, 127, SE
Eastport, ME, 42, NE
Eastport, NY, 161, NE
Eastside, NH, 88, SE
Eastview, NH, 106, NE
Eastwood, ME, 90, NW
Eaton, ME, 14, SW
Eaton, NY, 99, NE
Eben, NY, 26, SE
Eden, NY, 93, SW
Eden Valley, NY, 93, SW

Edgar, NJ, 158, SW
Edgartown, MA, 157A
Edgemere, NY, 159, SW
Edgemont, NH, 87, NE
Edgewood, NY, 121, SE
Edgeworth, MA, 127C
Edison, NJ, 150, SE
Edmeston, NY, 100, SE
Edmundston (CN; CPR; TMC), NB, 2, NW
Edwards, NY, 114, SE
Edwards, NY, 45, NW
Edwards Park, NY, 123, NW
Egypt, MA, 128, SW
Elba, NY, 75, SE
Elberta, NY, 74, NW
Eldred, PA, 133, NW
Elgin, PA, 130, NE
Eliot, MA, 127D
Eliot, ME, 90, SW
Elizabeth (CNJ; PRR), NJ, 158, SE
Elizabeth Ave. (CNJ), NJ, 158, SE
Elizaville, NY, 122, SW
Elkhurst, VT, 32, NW
Elkland (B&O; NYC), PA, 135, NW
Ellenburgh, NY, 29, NW
Ellenton, PA, 136, SW
Ellenville, NY, 141, SW
Ellicottville, NY, 112, NE
Ellington, CT, 145, NW
Elliott, CT, 145, NE
Ellis, NY, 97, SW
Ellis Quarry, VT, 66, NE
Ellisburg, PA, 134, NW
Ellsworth, ME, 40, SE
Elm Grove, NY, 76, SE
Elm Park, NY, 158, SE
Elma, NY, 93, NE
Elmer, ME, 10, SW
Elmhurst, NY, 159, SW
Elmira (DL&W; Erie; LV; PRR), NY, 116, SW
Elmira Heights (DL&W; LV), NY, 116, SW
Elmora Ave., NJ, 158, SE
Elmsford, NY, 152, SW
Elmsville, NB, 24, NE
Elmwood, CT, 144, SE
Elmwood, ME, 55, SW
Elmwood, NB, 8, SE
Elmwood, NH, 107, NW
Elmwood, RI, 147A

Elnora, NY, 103, NW
Eltingville, NY, 158, SE
Elton, NY, 113, NW
Ely, VT, 67, NE
Embden, ME, 37, NW
Emerson, NH, 88, SW
Emerson, NJ, 158, NE
Emersons, VT, 66, NW
Emery, NH, 90, SW
Emeryville, NY, 45, NW
Emmons, NY, 79, SE
Empire Quarries, NY, 151, NW
Empire Road, ME, 55, SW
Emporium, PA, 133, SE
End of Branch, NY, 82, SW
Endeavor, PA, 131, SW
Endicott, MA, 127, SE, 127F
Endicott, NY, 117, SE
Enfield, MA, 125, NW
Enfield, ME, 21, NE
Enfield, NH, 67, SE
Englewood, NJ, 159, NW
Enner, NY, 96, NE
Ennishore, NB, 3, SE
Eno, NY, 142, NE
Enosburg Falls, VT, 31, NW
Ensenore, NY, 98, NW
Epping, NH, 89, SE
Epsom, NH, 89, SW
Erie Jct., NY, 76A
Erin, NY, 116, SE
Erskine, NJ, 151, SW
Erving, MA, 106, SW
Erwins, NY, 115, SE
Erwins Crossing, NY, 115, SE
Eschota, NY, 73, SE
Esmond, RI, 146, NE
Esopus, NY, 142, NW
Esperance, NY, 102, SW
Essex, CT, 155, NW
Essex, MA, 109, SW
Essex, NY, 49, NW
Essex Center, VT, 30, SE
Essex Falls, MA, 109, SW
Essex Fells (Erie; MT&E), NJ, 158, NW
Essex Jct., VT, 49, NE
Essex Siding, NY, 49, NW
Essex St., NJ, 158, NE
Essex-North Concord, VT, 52, NW
Etna, ME, 38, NE
Etna, NY, 117, NW
Eureka Springs, NY, 84, SW

Eustis Jct., ME, 17, SE
Evans Mills, NY, 44, SW
Evarts, VT, 67, SW
Everett, MA, 127A
Everett, NH, 88, SE
Everett, West St., MA, 127C
Exeter, NH, 109, NW
Ezarys, ME, 56, SE

Fabyan (B&M), NH, 53, NW
Fabyan-Bretton Woods (MEC), NH, 53, NW
Factoryville, PA, 138, SW
Fair Ground, NY, 79, SE
Fair Grounds, CT, 153, NW
Fair Grounds, ME, 55, SE
Fair Haven, NY, 78, NE
Fair Haven, VT, 65, SW
Fair Oaks, NY, 141, SW
Fair View, NY, 62, NW
Fairbanks, ME, 36, SE
Fairfield, CT, 153, SW
Fairfield, ME, 37, SE
Fairfield, VT, 31, NW
Fairgrounds, MA, 104, SE
Fairhaven, MA, 148, SW
Fairlee, VT, 67, NE
Fairmont, MA, 127E
Fairmont, NY, 79, SE
Fairmont, VT, 51, NW
Fairmont Ave., NJ, 158, NE
Fairmount, ME, 6, SW
Fairport, NY, 77, SW
Fairview, NJ, 158, NE
Fairview, NY, 94, SE
Fairville, NY, 77, SE
Falconer, NY, 111, SE
Fall River, MA, 147, SE
Fall River Wharf, MA, 147, SE
Falls Jct., NY, 73, SE, 73A
Falls View, ON, 73, SE
Falls Village, CT, 143, NW
Falls Village, MA, 147, NW
Fallsburgh, NY, 140, SE
Fallsburgh Tunnel, NY, 140, SE
Falmouth, MA, 148, SE
Falmouth, ME, 71, SE, 71A
Fancher, NY, 75, SE
Faneuil, MA, 127D
Fanwood, NJ, 158, SW
Far Rockaway, NY, 159, SE
Fargo, NY, 94, NW
Farley, MA, 106, SW
Farleys, CT, 144, NW
Farleys, NY, 97, NE
Farm Hill, MA, 127C
Farm St., MA, 127, SW
Farmers Valley (PS&N; PRR), PA, 133, NW
Farmersville, NY, 113, NW
Farmingdale, NY, 160, SW
Farmington, CT, 144, SW
Farmington, ME, 36, SE
Farmington, NH, 89, NE
Farmington, NY, 96, NW
Farnham, NY, 92, SE
Farnhams, MA, 104, SE
Farnumsville, MA, 126, SE
Farran's Point, ON, 26, NE
Farview, PA, 139, SW
Fassett, PA, 136, NW
Faulkner, MA, 127C
Fayetteville, NY, 79, SE
Federation, ME, 5, NE
Feeder Siding, ON, 92, NW
Feeding Hills, MA, 124, SE
Felts Mills, NY, 44, SW
Fenwick, CT, 155, NW
Fenwick, ON, 73, SW
Ferenbaugh, NY, 115, SE
Fernald, NH, 69, SE
Ferncroft, MA, 109, SW
Ferndale, NY, 140, NE
Ferndale, PA, 131, SE
Fernwood, NY, 79, NE
Ferrisburg, VT, 49, SW
Ferry St., MA, 147, SE
Fiberloid, MA, 124, SE
Fieldmore Springs, PA, 130, SE
Fields Corner, MA, 127E
Fields Point, RI, 147A
Fillmore (Erie), NY, 113, NE
Fillmore (PRR), NY, 113, NE
Findley, ON, 43, NW
Firthcliffe, NY, 151, NE
Fish Creek, NY, 80, SE
Fisher, ME, 6, SW
Fisher Hill, NY, 48, SE
Fishers, ME, 39, NE
Fishers, NY, 77, SW
Fishers, NY, 95, SW
Fishkill, NY, 142, SW
Fishkill Plain, NY, 142, SW
Fish's Eddy, NY, 139, NE
Fitchburg, MA, 107, SW
Fitches, VT, 33, SE
Fitchville, CT, 145, SE
Fitzgerald, NY, 45, SW
Fitzwilliam, NH, 106, NE
Flatbush Ave., NY, 159, SW
Fleetwood, NY, 159, NW
Fleischmann's, NY, 120, SE
Flemington, NB, 8, SE
Flemingville, NY, 117, SE
Flemming, NB, 3, NW
Fletcher, VT, 31, SW
Flint, NY, 96, NE
Floodwood, NY, 47, NW
Floral Park, NY, 159, SE
Florence, MA, 124, NE
Florence, VT, 65, SE
Florence Jct., VT, 65, SE
Florenceville, NB, 8, NE
Florentine Quarry, VT, 65, SE
Florida, NY, 151, NW
Flowerfield, NY, 160, NE
Flushing Main St., NY, 159, NW
Fly Summit, NY, 85, SW
Fogg Siding, NH, 34, SE
Folsom, VT, 33, SW
Folsom St., NH, 90, SW
Fonda, NY, 102, NW
Fonda, VT, 30, NE
Fonthill, ON, 73, SW
Forbes, MA, 127C
Forbes Ave., NY, 103, SE, 103A
Ford, NJ, 150, SE
Fordham, NY, 159, NW
Fore River, MA, 128, SW
Forest, ME, 14, SE
Forest, NY, 29, NW
Forest City (Erie; NYO&W), PA, 139, SW
Forest Hill, NJ, 158, NE
Forest Hills, MA, 127E
Forest Hills, NY, 159, SW
Forest Lake, MA, 125, SW
Forest Lake, NH, 106, NW
Forestdale, RI, 146, NE
Forestport, NY, 81, NE
Forestville, CT, 144, SW
Forestville, NY, 111, NE
Forge Village, MA, 108, SE
Forks Creek, ON, 92, NW
Forsyth, NY, 110, NE
Forsythe, ME, 19, NW
Fort Ann, NY, 85, NW
Fort Covington, NY, 28, NW
Fort Edward, NY, 84, NE
Fort Erie, ON, 93, NW
Fort Ethan Allen, VT, 30, SE
Fort Fairfield (BAR; CPR), ME, 6, NW
Fort Hunter, NY, 102, NW
Fort Johnson, NY, 102, NE
Fort Kent, ME, 1, NE
Fort Kent Pit, ME, 1, SE
Fort Kent Village, ME, 1, SE
Fort Montgomery, NY, 152, NW
Fort Plain, NY, 101, NE
Fort Richmond, NY, 158, SE
Fort Ticonderoga (D&H; R), NY, 65, NW
Fort Wadsworth, NY, 158, SE
Fort Washington, NY, 159, NW
Forthton, ON, 25, SW
Foster, ME, 8, NW
Foster, PA, 138, SW
Foundry, NH, 90, NW
Fournier, ME, 2, NW
Fowlerville, NY, 95, NW
Fox, ME, 5, SE
Fox Point, RI, 147A
Fox Ridge, NY, 78, SE
Foxboro, MA, 127, SE
Foxcroft, ME, 20, SE
Fraleighs, NY, 142, NW
Framingham, MA, 127, NW, 127G
Framingham Centre, MA, 127, NW
Frank Ave., NY, 159, SW
Frankfort, ME, 39, SW
Frankfort, NY, 81, SE
Franklin, CT, 145, SE
Franklin, MA, 127, SW
Franklin, ME, 40, SE
Franklin, NH, 88, NE
Franklin, NY, 119, NE
Franklin (DL&W; NYS&W), NJ, 150, SE
Franklin Ave. (Nutley), NJ, 158, NE
Franklin Falls, NH, 88, NE
Franklin Jct., MA, 127, SW
Franklin Jct., NH, 88, NE
Franklin Park, MA, 127C
Franklin Road, ME, 40, SW
Franklin Springs, NY, 81, SW
Franklinville, NY, 113, NW
Fraser Jct. (CN), NB, 2, NW
Frasers, NY, 120, SW
Free Bridge, NH, 88, SE

Freedom, NY, 113, NW
Freedonia, NY, 111, NW
Freeman, NY, 115, SW
Freeman, PA, 132, SE
Freemans, CT, 124, SE
Freeman's Point, NH, 90, SW
Freeport, ME, 71, NE
Freeport, NY, 159, SE
Freeville, NY, 98, SW
French Mountain, NY, 84, NE
Frenchville, ME, 2, NW
Fresh Pond, MA, 127B
Fresh Pond, NY, 159C
Fresh Pond Jct. (LI; NYCN), NY, 159C
Frewsburg, NY, 111, SE
Friendship (Erie), NY, 113, SE
Friendship (PS&N), NY, 113, SE
Fruitland, NY, 77, SW
Frye, ME, 35, SE
Fryeburg, ME, 54, SW
Fuera Bush, NY, 103, SW
Fuller, ME, 56, NW
Fullers, NY, 103, SW
Fulton, NY, 79, NW
Fultonville, NY, 102, NW
Furniss, NY, 78, NE

G&O Jct., NY, 45, NW
G&W Jct., NY, 95, NW
Gabriels, NY, 47, NE
Gaines, PA, 134, NE
Gaines Jct., PA, 134, SE
Gainesville, NY, 94, SE
Gale, NH, 88, NW
Galena, NY, 99, SE
Galeton, PA, 134, SE
Gallatin, NY, 122, SE
Gananoque, ON, 43, NE
Gananoque Jct., ON, 43, NE
Gang Mills, NY, 115, SE
Gansevoort, NY, 84, SE
Garbutt, NY, 76, SW
Garden City, NY, 159, SE
Gardenville Yard (NYC), 93, NW, 93E
Gardiner, ME, 56, SW
Gardiner, NY, 141, SE
Gardner, MA, 107, SW
Gardner East, MA, 107, SW
Gardners, ME, 42, NW
Garfield, NJ, 158, NE

Garland (NYC; PRR), PA, 131, NW
Garrison, NH, 88, SE
Garrison, NY, 152, NW
Garwood, NJ, 158, SW
Garwoods (Erie; PS&N), NY, 114, NW
Gasport, NY, 74, SE
Gassetts, VT, 86, NE
Gates, NY, 84, SE
Gates, PA, 132, NW
Gawango, PA, 132, NW
Gaylordsville, CT, 143, SW
Gaysville, VT, 66, NE
GD, NY, 73, SE
Gedney Way, NY, 152, SW
Genasco, NJ, 158, SW
Genesee (B&O; NY&P), PA, 134, NW
Genesee Dock, NY, 76, SE
Genesee River, NY, 76, NE
Genesee Springs, PA, 134, NW
Geneseo, NY, 95, NW
Geneva, NY, 97, NW
Georges, NH, 33, NE
Georges River, ME, 57, SE
Georgetown, CT, 153, NW
Georgetown, MA, 109, SW
Georgia, VT, 30, SE
Georgiaville, RI, 146, NE
Germania, PA, 134, SE
Germantown, NY, 122, SW
Germonds, NY, 152, SW
Gerrish, NH, 88, NE
Gerry, NY, 111, SE
Getty Square (Yonkers), NY, 159, NW
Getzville, NY, 74, SW
Ghent, NY, 122, NE
Gibbs, CT, 145, SE
Gibson, NY, 159, SE
Giffords, NY, 98, NW
Gilbert, ME, 12, SE
Gilbert, NY, 97, SW
Gilbertville (B&A), MA, 125, NE
Gilbertville (B&M), MA, 125, NE
Gilbertville, ME, 55, NW
Gilboa, NH, 106, NW
Gilead, ME, 54, NW
Gilford, ME, 21, NE
Gilford, NH, 69, SW
Gill, MA, 106, SW
Gillett, PA, 136, NW

Gillette, NJ, 158, SW
Gilman, ME, 2, NW
Gilman, VT, 52, NE
Gilson, PA, 131, SE
Glade, PA, 131, NE
Glastenbury, VT, 104, NE
Gleason Jct., MA, 126, NE
Gleasondale, MA, 126, NE
Glen, NH, 53, SE
Glen Cove, NY, 159, NE
Glen Hazel, PA, 132, SE
Glen Head, NY, 159, NE
Glen Ridge (DL&W; Erie), NJ, 158, NE
Glen Rock, NJ, 158, NE
Glen St. (Glen Cove), NY, 159, NE
Glenbrook, CT, 152, SE
Glenburn, ME, 39, NW
Glenburn, PA, 138, SE
Glencliff, NH, 68, NW
Glendale, MA, 123, NW
Glendale, NH, 69, SW
Glendale, NY, 159, SW
Glendale, RI, 146, NE
Glendyne, NB, 1, NW
Glenfield (G&W; NYC), NY, 61, SW
Glenford, NY, 121, SE
Glenham, NY, 142, SW
Glenmount, NY, 103, SW
Glens, NJ, 151, SW
Glens Falls, NY, 84, NE
Glenwood, MA, 126, SW
Glenwood, MA, 127C
Glenwood, NJ, 151, NW
Glenwood, NY, 159, NW
Glenwood, NY, 93, SE
Glenwood Jct., NY, 151, NW
Gloucester, MA, 109, SE
Gloversville, NY, 83, SW
Glynden, PA, 130, NE
Godefroy's, NY, 150, NE
Goffs Falls, NH, 108, NW
Goffstown, NH, 88, SE
Golah, NY, 95, NE
Gold, PA, 134, NW
Golden's Bridge, NY, 152, NE
Golders, ME, 55, SE
Golf Grounds, NY, 162, NW
Gonic, NH, 90, NW
Good Will Farm, ME, 37, SE
Goodmans, NJ, 158, SW
Goodrich, ME, 6, NW

Goodwin, ME, 6, NW
Goodwins, ME, 40, SE
Gordon, ME, 13, SW
Gore, ME, 11, SW
Gorham, ME, 71, SW
Gorham, NH, 53, NE
Gorham, NY, 96, NE
Goshen, NY, 151, NW
Goulds, RI, 156, NE
Gouverneur, NY, 45, NW
Gowanda, NY, 111, NW
Grafton, MA, 126, SE
Grafton, NH, 68, SW
Graham, NY, 150, NE
Graham, NY, 152, SW
Granby, CT, 144, NW
Grand Beach, ME, 71, SW
Grand Central Terminal (New York) (NH; NYC), NY, 159, NW
Grand Falls (CN; CPR), NB, 3, SE
Grand Gorge, NY, 120, NE
Grand Hotel Station, NY, 121, SW
Grand Isle, ME, 2, NE
Grand Isle, VT, 30, SW
Grand River, NB, 3, NW
Grand Valley, PA, 130, SE
Grand View, NY, 152, SW
Granger, NY, 96, NE
Granite Springs, NY, 152, NW
Graniteville, MA, 108, SE
Graniteville, VT, 51, SW
Grant Ave., NJ, 158, SW
Grant City, NY, 158, SE
Grants, CT, 143, NE
Granville, NY, 85, NW
Grasmere, NH, 88, SE
Grasmere, NY, 158, SE
Grasmere Jct., NH, 88, SE
Grasse River Club, NY, 46, SE
Grasselli, NJ, 158, SE
Gratwick, NY, 74, SW
Gravel Pit, VT, 51, NW
Gravesville, NY, 81, NE
Gray, ME, 71, NW
Great Barrington, MA, 123, SW
Great Bend, NY, 44, SE
Great Bend, PA, 138, NE
Great Kills, NY, 158, SE
Great Neck, NY, 159, NE
Great Notch, NJ, 158, NE
Great River, NY, 160, SE
Great Valley, NY, 112, SE
Great Works, ME, 39, NE

Great Works, ME, 90, SW
Greece, NY, 76, NE
Green Harbor, MA, 128, SE
Green Haven, NY, 142, SE
Green Island, NY, 103, SE, 103A
Green Lake, ME, 39, SE
Green Pond Jct., NJ, 151, SW
Green River, NB, 2, NE
Green Road, NB, 8, SW
Greenbush, MA, 128, SE
Greendale, NY, 122, SW
Greendale, PA, 132, SW
Greene, ME, 55, SE
Greene, NY, 118, NW
Greene, RI, 146, SE
Greenfield, MA, 105, SE, 105A
Greenfield, NH, 107, NW
Greenfield East, MA, 105, SE, 105A
Greenfield Siding, NY, 84, SW
Greenfield West, MA, 105, SE
Greenhurst, NY, 111, SW
Greenland, NH, 90, SW
Greenlaw, ME, 4, NE
Greenlawn, NY, 160, NW
Greenport, NY, 155, SW
Greens Corners, NY, 59, NE
Greens Corners, VT, 30, NE
Greens Farms, CT, 153, SW
Green's Farm, ME, 17, SE
Greensboro, VT, 32, SW
Greenvale, NY, 159, NE
Greenville (BAR; CPR), ME, 19, NE
Greenville (CNJ; PRR), NJ, 158A
Greenville, MA, 107, NW
Greenway, NY, 80, SE
Greenwich, CT, 152, SE
Greenwich, MA, 125, NW
Greenwich Jct., NY, 85, SW
Greenwich Lake, MA, 125, NW
Greenwich Valley, MA, 125, NW
Greenwood, MA, 127C
Greenwood, NY, 114, SE
Gregorytown, NY, 119, SE
Grey Oaks, NY, 159, NW
Greycourt, NY, 151, NW
Greyhound, MA, 147, NE
Greylock, MA, 104, SE
Greystone, NH, 69, SW
Greystone, NY, 159, NW
Griffins, CT, 144, NE
Griffiths, PA, 132, SE

Grimes Mill, ME, 6, NW
Grindstone, ME, 12, SE
Griswold, ME, 7, NW
Griswolds, NY, 94, NW
Groton, MA, 107, SE
Groton, NY, 98, SW
Groton, VT, 51, SE
Groton Ferry Wharf, CT, 155, NE
Grout, VT, 87, SW
Grove St., NJ, 158, NE
Grove Station, ME, 91, NW
Groveland, MA, 108, NE
Groveland, NY, 95, SW
Grover, PA, 136, SW
Groveton (B&M; CN), NH, 33, SE
Guffey, PA, 132, NE
Guilderland Center, NY, 103, SW
Guildhall, VT, 33, SE
Guilford, CT, 154, NE
Guilford, ME, 20, SW
Guilford, NY, 119, NW
Guion, ME, 6, SW
Gulf, RI, 147A
Gulf Head, NY, 61, SW
Gulf Summit, NY, 118, SE
Gulfport, NY, 158, SE
Gun Hill Road, NY, 159, NW
Guthrie, NB, 6, NW

Haberman, NY, 159, SW
Hackensack, NJ, 158, NE
Hacketts Switch, ME, 55, SW
Haddam, CT, 154, NE
Hadley, MA, 124, NE
Hadley, NH, 107, NW
Hadley, NY, 84, NW
Haggett, MA, 108, SE
Hailesboro, NY, 45, NW
Haines Falls, NY, 121, SE
Halcottville, NY, 120, SE
Halcyon, NH, 88, NW
Hale, ME, 35, SE
Hale, NB, 8, NE
Hales Eddy, NY, 119, SW
Half Acre, NY, 97, NE
Half Way, ME, 38, SW
Halfway, ME, 41, SW
Halfway, NY, 79, SW
Halifax, MA, 128, SW
Hall, NY, 96, NE
Hallon, NY, 115, SE
Hallowell, ME, 56, NW
Hallstead, PA, 138, NE

Halsey, NJ, 150, SW
Halstead, NY, 122, SE
Hamburg, NJ, 150, SE
Hamburg, NY, 93, SW
Hamden, NY, 120, SW
Hamilton, NY, 99, NE
Hamilton & Wenham, MA, 109, SW
Hamilton Beach, NY, 159, SW
Hamlet, RI, 126, SE
Hamlin, NY, 76, NW
Hamlin, PA, 133, SW
Hammel, NY, 159, SW
Hammond, NY, 44, NE
Hammond, PA, 135, NE
Hammondsport, NY, 115, NE
Hammondville, NY, 64, NE
Hampden, ME, 39, SW
Hampden Jct., MA, 125, SW
Hampshire Road, NH, 108, SE
Hampstead, NH, 108, NE
Hampton, CT, 145, NE
Hampton, NH, 109, NW
Hampton Bays, NY, 161, NE
Hampton Falls, NH, 109, NW
Hampton Mills, MA, 124, NE
Hamptonburgh, NY, 151, NE
Hancock (Erie; NYO&W), NY, 139, NW
Hancock, NH, 107, NW
Hanford, ME, 5, NW
Hanford, NJ, 150, NE
Hankins, NY, 139, NE
Hannawa Falls, NY, 27, SW
Hannibal, NY, 78, NE
Hanover, MA, 128, SW
Hanover, NJ, 158, NW
Hanover Springs, CT, 153, NW
Harbor Creek (NKP), PA, 110, SW
Harbor Creek (NYC), PA, 110, SW
Harbor View, MA, 127C
Hardings, ME, 72, NW
Hardwick, VT, 32, SW
Hardy Pond, ME, 21, NW
Hardys, NY, 94, SE
Harkness, NY, 29, SE
Harlem River, NY, 159A
Harmon, NY, 152, SW
Harmony, ME, 37, NE
Harpursville, NY, 118, SE
Harriman, NY, 151, NE
Harrington, ME, 41, SW

Harrington Park, NJ, 159, NW
Harris, RI, 146, SW
Harrison, ME, 54, SE
Harrison (NH; NYW&B), NY, 159, NE
Harrison Square, MA, 127E
Harrison St., NJ, 158, NE
Harrison Valley, PA, 134, NE
Harrisville, NH, 106, NE
Harrisville, NY, 45, SW
Harrisville, RI, 146, NE
Harters, NY, 60, NE
Hartford, CNE Sta., CT, 144, NE, 144A
Hartford, CT, 144, NE, 144A
Hartford, ME, 55, NW
Hartford, VT, 67, SW
Hartford Yard, CT, 144A
Hartland, ME, 38, NW
Hartland, NB, 8, NE
Hartland, VT, 67, SW
Hartsdale, NY, 152, SW
Hartwick Seminary, NY, 101, SW
Hartwood, NY, 140, SE
Harvard, MA, 107, SE
Harvard, NY, 119, SE
Harvard St., MA, 127E
Harvey, ME, 8, NW
Harvey, NB, 15, SE
Harwards, ME, 56, SW
Harwich, MA, 149, SE
Harwich Centre, MA, 149, SE
Hasbrouck Heights, NJ, 158, NE
Haskell, NJ, 151, SW
Hastings, MA, 127, NW
Hastings, MA, 149, NE
Hastings, NY, 159, NW
Hatfield, MA, 124, NE
Hathorne, MA, 109, SW
Haven, NY, 140, SE
Haverhill, MA, 108, NE
Haverhill, NH, 51, SE
Haverhill Engine House, MA, 108, NE
Haverstraw (Erie; NYC), NY, 152, SW
Hawkins, ME, 7, NW
Hawk's Mountain, NY, 139, NE
Hawkshaw, NB, 15, NE
Hawleyville, CT, 153, NW
Haworth, NJ, 159, NW
Hawthorne, NJ, 158, NE
Hawthorne, NY, 152, SW

Hayden Row, MA, 126, SE
Haydens, CT, 144, NE
Haydenville, MA, 124, NE
Hayes, NH, 90, NW
Haynes, NY, 118, NE
Haystack, CT, 123, SE
Hayts Corners, NY, 97, SW
Hazardville, CT, 144, NE
Hazelhurst, PA, 132, SE
Hazelton Mills, PA, 132, NE
Hazelwood, MA, 127E
Hazen, NH, 52, NE
Head Tide, ME, 56, SE
Healdville, VT, 86, NW
Heart Lake, PA, 138, NW
Heath St., MA, 127E
Heathcote, NY, 159, NW
Hebronville, MA, 147, NW
Hector, NY, 97, SW
Hedding, NH, 89, SE
Helena (CN; NYC), NY, 27, NE
Hemlock, NY, 95, NE
Hemlock, PA, 131, NE
Hemlock Lake, NY, 95, NE
Hempstead, NY, 159, SE
Hempstead Gardens, NY, 159, SE
Henniker, NH, 88, SW
Henniker Jct., NH, 88, SW
Henrietta, NY, 76, SE
Henry's Mill, PA, 131, SE
Herkimer, NY, 82, SW
Hermitage, NY, 115, NW
Hermon, NY, 45, NE
Hermon Pond, ME, 39, NW
Herrick Center, PA, 139, SW
Herrings, NY, 44, SE
Heuvelton, NY, 26, SW
Hewitt, NB, 24, NE
Hewitt, NJ, 151, SW
Hewlett, NY, 159, SE
Heywood, MA, 107, SW
Hibernia, NJ, 158, NW
Hibernia, NY, 142, NW
Hickory Grove, PA, 138, NE
Hickox, PA, 134, NW
Hicksville, NY, 159, NE
Higbie Ave., NY, 159, SW
Higganum, CT, 144, SE
High Bridge, NY, 159, NW
High Cut, ME, 19, NE
High Falls, NY, 141, NE
High St. Jct., CT, 144, NW
High View, NY, 141, SW

Highgate, VT, 30, NE
Highgate Spring, VT, 30, NE
Highland, CT, 144, SE
Highland, MA, 127D
Highland, NY, 142, SW
Highland Ave., NJ, 158, NE
Highland Falls, NY, 152, NW
Highland Jct., CT, 143, SE
Highland Lake, MA, 127, SW
Highland Mills, NY, 151, NE
Highlands, MA, 128, NW
Highlands, NH, 53, NW
Highpine, ME, 90, NE
Highwood, CT, 154, NW
Hill, NH, 68, SE
Hill Crossing, MA, 127B
Hill Top, MA, 126, SW
Hill Yard, MA, 107, SE
Hillburn, NY, 151, SE
Hillman, ME, 7, SE
Hillman, NB, 8, SE
Hills, ME, 8, NW
Hills Grove, RI, 147, SW
Hillsboro, NH, 88, SW
Hillsdale, NJ, 151, SE
Hillsdale, NY, 122, SE
Hillside, ME, 71, NE
Hillside, NB, 6, NE
Hillside, NJ, 158, SE
Hillside, NY, 159D
Hilton, NY, 76, NW
Himbaugh, PA, 130, NW
Himrod, NY, 97, SW
Himrods Jct., NY, 97, SW
Hinckley, ME, 37, SE
Hinckley, NY, 81, NE
Hingham, MA, 128, SW
Hinman, NY, 116, NE
Hinsdale (Erie), NY, 113, SW
Hinsdale (PRR), NY, 113, SW
Hinsdale, MA, 123, NE
Hinsdale, NH, 106, NW
Hiram, ME, 70, NW
Hoadley's, PA, 139, SW
Hobart, NY, 120, NE
Hoboken, NJ, 158A
Hockenhall, ME, 6, NW
Hoffmans, NY, 102, NE
Hogtown, ME, 39, NW
Ho-Ho-Kus, NJ, 158, NE
Holban Yard, NY, 159D
Holbrook, MA, 127, SE
Holbrook, NY, 160, NE

Holcomb, NY, 96, NW
Holden, MA, 126, NW
Holden, ME, 39, SE
Holeb, ME, 10, SW
Holiday, PA, 135, NE
Holland, NY, 159, SW
Holland, NY, 93, SE
Holland Patent, NY, 81, SW
Holley, NY, 75, SE
Hollis, MA, 107, SE
Hollis, NY, 159D
Hollister, VT, 65, SE
Holliston, MA, 127, SW
Holmes, NY, 142, SE
Holmesville, NY, 100, SW
Holton, NH, 88, SW
Holts Hill, CT, 144, SW
Holtsville, NY, 160, NE
Holwood, NH, 107, NE
Holyoke, MA, 124, SE
Homer, NY, 98, SE
Honeoye Falls, NY, 95, NE
Honeoye Jct., NY, 95, NE
Honesdale (D&H; Erie), 139, SW
Honeydale, NB, 24, NE
Hooksett, NH, 89, SW
Hoosac Tunnel, MA, 105, SW
Hoosick, NY, 104, NW
Hoosick Falls, NY, 104, NW
Hoosick Jct., NY, 104, NW
Hop City, ME, 35, SE
Hope, RI, 146, SW
Hope Ave., MA, 126, SW
Hope Valley, RI, 146, SE
Hopedale, MA, 126, SE
Hopewell Jct., NY, 142, SW
Hopkins, CT, 153, SW
Hopkins, ME, 6, NW
Hopkins, VT, 85, NE
Hopkinton, MA, 126, SE
Hornell (Erie; PS&N), NY, 114, NE
Horse Meadow, NH, 51, SE
Horse Shoe, NY, 46, SE
Horseback, ME, 8, SW
Horseheads (DL&W; Erie; LV; PRR), NY, 116, SW
Horton's, NY, 139, NE
Hoskins, CT, 144, NW
Hotham, ME, 5, SW
Houghton, ME, 35, NE
Houghton, NY, 113, NE
Houghton Farm, NY, 151, NE

Houghtonville, ME, 6, NW
Houlton, ME, 8, SW
Housatonic, MA, 123, NW
Houseville, NY, 61, SW
Houstons, NY, 95, NW
Houstons-North Lincoln, ME, 22, NW
Hovers, PA, 131, SE
Howard (B&O; Erie), PA, 132, NE
Howard, RI, 147, SW
Howard Beach, NY, 159, SW
Howards, NY, 49, SW
Howarths, MA, 126, SW
Howe, MA, 109, SW
Howe Brook, ME, 7, NE
Howells, NY, 151, NW
Howe's Cave, NY, 102, SW
Howland, ME, 21, SE
Hoxies, ME, 37, SE
Hoyts, ME, 56, NW
Hoytville, PA, 135, SW
Hubbard, NH, 108, NE
Hubbard, NY, 93, NE
Hubbardston, MA, 126, NW
Hubbardsville, NY, 100, NE
Hudson, MA, 126, NE
Hudson, ME, 21, SW
Hudson, NH, 108, SW
Hudson, NY, 122, NW
Hudson Ave., NJ, 159, NW
Hudson Falls, NY, 84, NE
Hudson Jct., NY, 151, NW
Hudson Upper, NY, 122, SW
Hugenot Park, NY, 158, SE
Huguenot, NY, 150, NE
Humaston, NY, 80, NE
Humberstone, ON, 92, NW
Hume, NY, 113, NE
Hunter, NY, 121, SE
Hunters Point Ave., NY, 159B
Huntington, MA, 124, SW
Huntington, NY, 160, NW
Huntington Ave. (B&A), MA, 127A
Hurlbut, MA, 123, NW
Hurley, NY, 141, NE
Hutchins, PA, 132, SE
Hyannis, MA, 149, SW
Hyannis Wharf, MA, 149, SW
Hyatt, NY, 45, NW
Hyde Park, MA, 127D
Hyde Park, NY, 142, NW
Hyde Park, VT, 31, SE

Hydetown, PA, 130, SE
Hydeville, VT, 65, SE
Hydro, NB, 3, SE
Hyndesville, NY, 101, SE
Hyper-Humus, NJ, 150, SE

I.N.R. Jct. (CN; VB), NB, 3, SW
Ice House, NH, 107, SE
Ice House Jct., NY, 77, SW
Ice Pond, NY, 142, NE
Ilion, NY, 81, SE
Indian Castle, NY, 82, SW
Indian Crossing, PA, 133, NW
Indian Orchard, MA, 124, SE
Indian Point, ME, 24, SW
Indian Pond, ME, 19, NW
Industry, NY, 76, SE
Ingalls, MA, 128, NW
Ingalls, ME, 12, SW
Ingall's Crossing, MA, 108, SE
Ingall's Road, ME, 70, NE
Ingells, NY, 79, NW
Inghams, NY, 82, SW
Inman, NB, 6, SE
Instanter, PA, 132, SE
Interlaken, NY, 97, SE
Intervale, NH, 53, SE
Inverness, NY, 95, NW
Inwood, NY, 159, NW
Inwood, NY, 159, SE
Inwood, VT, 51, NE
Iona Island, NY, 152, NW
Ionia, NY, 96, NW
Ipswich, MA, 109, SW
Ira, NY, 78, SE
Iron Works, NY, 103, SE, 103A
Irona, NY, 29, NE
Irondale, NY, 82, SW
Ironton, NY, 27, NE
Ironville, NY, 64, NE
Iroquois, ON, 26, NW
Irvineton (NYC; PRR), PA, 131, NW
Irving, NY, 92, SE
Irvington, NJ, 158, SE
Irvington, NY, 152, SW
Iselin, NJ, 158, SW
Ishua, NY, 113, SW
Island Creek, MA, 128, SE
Island Falls, ME, 7, SW
Island Park, MA, 108, NE
Island Park, NY, 159, SE
Island Pond, VT, 33, NW

Isle La Motte, VT, 30, NW
Islington, MA, 127, SE
Islip, NY, 160, SE
Italy Yard, VT, 30, NE
Itaska, NY, 118, NW
Ithaca (DL&W; LV), NY, 116, NE

J&B Jct., PA, 132, SE
Jackman, ME, 10, SW
Jackson, NY, 142, SW
Jackson Ave., NJ, 158A
Jackson Corners, NY, 122, SE
Jackson Summit, PA, 135, NE
Jacksonville, ME, 42, NW
Jacobs, ME, 5, NE
Jamaica, NY, 159D
Jamaica, VT, 86, SW
Jamaica Plain, MA, 127E
James City, PA, 132, SW
Jamesport, NY, 161, NE
Jamestown (Erie), NY, 111, SE
Jamestown (JW&NW), NY, 111, SE
Jamesville, MA, 126, SW
Jamesville, NY, 98, NE
Jamison, PA, 131, SW
Jamison, PA, 134, SW
Jamison Road, NY, 93, NE
Jardine Brook, NB, 3, NE
Jasper, ON, 25, NW
Java Center, NY, 94, SW
Jay, ME, 36, SE
Jayville, NY, 45, SE
Jefferson, MA, 126, NW
Jefferson, NH, 53, NW
Jefferson Ave., NY, 158, SE
Jeffersonville, VT, 31, SW
Jemtland, ME, 2, SE
Jericho, VT, 31, SW
Jermyn (D&H; NYO&W), PA, 138, SE
Jersey City (CNJ; Erie; LV; PRR), NJ, 158A
Jewett, ME, 90, SW
Jewett City, CT, 146, SW
Jewettville, NY, 93, SE
JN, NY, 79, NW
JO Sidings, PA, 130, NW
Johnson, NY, 150, NE
Johnson, VT, 31, SE
Johnson City (DL&W; Erie), NY, 118, SW
Johnsonburg, NY, 94, SW

Johnsonville, NY, 103, NE
Johnstown, NY, 102, NW
Johnstown, NY, 83, SW
JoJo Jct., PA, 132, SW
Jones Point, NY, 152, NW
Jonesboro, ME, 41, SE
Jonesville, VT, 50, NW
Jordan, NY, 79, SW
Jordan, ON, 73, SW
Joslin, NH, 106, NW
Journal Sq. (H&M), NJ, 158A
Judd's Bridge, CT, 143, SW
Junius, NY, 97, NW

Kaaterskill, NY, 121, SE
Kaaterskill Jct., NY, 121, SE
Kalurah, NY, 45, SE
Kamankeag, ME, 35, NW
Kane, PA, 132, SW
Kane Jct., PA, 132, SW
Kanesholm, PA, 132, SE
Kanona (DL&W; Erie; Prat), NY, 115, NW
Karter, NY, 44, SE
Kasoag, NY, 80, NW
Kasson, PA, 132, SE
Kast Bridge, NY, 82, SW
Katahdin Iron Works, ME, 20, NE
Katama, MA, 157A
Katonah, NY, 152, NE
Kaufman's, NY, 120, SW
Kayuta, NY, 81, NE
Keag (MEC), ME, 13, SW
Keane, PA, 139, SW
Keating Summit (B&O; PRR), PA, 133, SE
Keegan, ME, 3, SW
Keene, NH, 106, NW
Keenes, NY, 44, NE
Keepawa, NY, 62, NW
Keeseville, NY, 30, SW
Kellettville, PA, 131, SW
Kellogg, PA, 137, SW
Kelly's Corners, NY, 120, SE
Kellyville, NH, 87, NE
Kenberma, MA, 128, NW
Kendaia, NY, 97, SW
Kendall, NY, 75, NE
Kendall, VT, 67, NE
Kendall Green, MA, 127, NW
Kenilworth, NJ, 158, SW
Kennebago, ME, 17, SW
Kennebec, ME, 56, NE

Kennebunk, ME, 90, NE
Kennebunk Beach, ME, 90, NE
Kennebunkport, ME, 91, NW
Kennedy, NY, 111, SE
Kensico Cemetery, NY, 152, SW
Kent, CT, 143, SW
Kent Jct., ME, 1, NE
Kenwood, NY, 80, SE
Kenyons, RI, 156, NE
Keogh, ME, 9, SE
Kerhonkson, NY, 141, NW
Kerry's, NY, 119, SW
Ketner, PA, 132, SE
Kettle Point, RI, 147A
Keuka Mills, NY, 96, SE
Kew Gardens, NY, 159, SW
Kidder's, ME, 58, NW
Kilbourne, PA, 134, NE
Kilburn, NB, 6, SE
Kildare, NY, 46, NE
Killawog, NY, 117, NE
Kimball, VT, 32, SE
Kineo Station, ME, 11, SE
Kingfield, ME, 36, NE
Kingman, ME, 13, SE
Kings Bridge, NY, 159, NW
Kings Ferry, NY, 97, SE
Kings Park, NY, 160, NW
Kings Park State Hosp., NY, 160, NW
Kings Siding, NY, 84, SW
Kingsbridge Road, NY, 159, NW
Kingsland, NJ, 158, NE
Kingsley, PA, 138, NW
Kingston, MA, 148, NE
Kingston, NY, 141, NE
Kingston, ON, 43, SW
Kingston (NAP), RI, 156, NE
Kingston (NH), RI, 156, NE
Kingston Jct., ON, 43, NW
Kingston Point, NY, 142, NW
Kinzua, PA, 132, NW
Kinzua Viaduct, PA, 132, NE
Kipps, NY, 151, NW
Kirby, ME, 13, SW
Kirkland, NY, 81, SW
Kirkville, NY, 80, SW
Kirkville Jct., NY, 80, SW
Kirkwood, NY, 118, SW
Kitchawan, NY, 152, SW
Kittery Jct., ME, 90, SW
Kittery Navy Yard, ME, 90, SE
Kittery Point, ME, 90, SE

Klondike, CT, 146, NW
Knapp, NY, 103, SW
Knapps, NY, 27, NW
Knights, ME, 21, NW
Knowlesville, NY, 75, SW
Knox, ME, 38, SE
Knox, NY, 102, SE
Knoxville (B&O; NYC), PA, 135, NW
Kodak Switch, NY, 76, SE
Kortright Station, NY, 120, NW
Kowenhoven Road, NY, 159, SW
Kushequa, PA, 132, NE
Kyleton, ME, 19, NE
Kyserike, NY, 141, NE

L&O Jct., NY, 74, SE
La Fargeville, NY, 44, SW
La Grange, ME, 21, SW
La Mont, PA, 132, SW
La Plume, PA, 138, SE
Labby, ME, 1, SE
Lac Baker, NB, 1, NE
Laceyville, PA, 137, SE
Lackawanna, NY, 93B
Lackawaxen, PA, 150, NW
Lacona, NY, 59, SE
Laconia, NH, 69, SW
Laddsburg, PA, 137, SW
Lafayette, NJ, 150, SE
Lafayette, PA, 132, SE
LaGrange, NY, 142, SW
Lake, NY, 151, NW
Lake, VT, 33, NW
Lake Austin, ME, 19, SW
Lake Bonaparte, NY, 45, SW
Lake Carey, PA, 138, SW
Lake Clear Jct., NY, 47, NE
Lake George, NY, 84, NE
Lake Grinnell, NJ, 150, SE
Lake Katrine, NY, 142, NW
Lake Kushaqua (D&H; NYC), NY, 28, SE
Lake Mahopac, NY, 152, NE
Lake Moxie, ME, 19, NW
Lake Placid, NY, 48, NW
Lake Pleasant, MA, 105, SE
Lake Ridge, NY, 97, SE
Lake St., MA, 127B
Lake View, ME, 21, NW
Lake View, NJ, 158, NE
Lake View, NY, 159, SE
Lake View, NY, 93, SW

Lakeport, NH, 69, SW
Lakeside, VT, 51, NW
Lakeville, CT, 143, NW
Lakeville, MA, 148, NW
Lakeville, NB, 8, NE
Lakeville, NY, 151, SW
Lakeville, NY, 95, NE
Lakewood, ME, 7, NE
Lakewood, NH, 69, NE
Lakewood, NY, 111, SW
Lakewood, PA, 139, NW
Lakewood, RI, 147, NW
Lambert Lake, ME, 14, SE
Lamson, NY, 79, SW
Lancaster, MA, 126, NE
Lancaster, NY, 93, NE
Lancaster (B&M), NH, 52, NE
Landers, ME, 19, NW
Landrus, PA, 135, SE
Landsdowne, ON, 43, NE
Lanesboro, PA, 138, NE
Lanesboro, VT, 51, NW
Lanesville, NY, 121, SW
Lang, NH, 88, SE
Langdon, NY, 118, SW
Langdons, VT, 65, SE
Lansing, NY, 97, SE
Lansingburg, NY, 103, NE
Laona, NY, 111, NW
Lapham's Mills, NY, 29, SE
Laquin, PA, 136, SE
Larabee, PA, 133, NW
Larabees Point, VT, 65, NW
Larchmont (NH; NYW&B), NY, 159, NW
Larchmont Gardens, NY, 159, NW
LaSalle (Erie; NYC), NY, 74, SW
Laurel, CT, 144, SE
Laurel, NY, 161, NE
Laurel House Station, NY, 121, SE
Laurelton, NY, 159, SW
Lawrence, MA, 108, SE
Lawrence, NB, 24, NE
Lawrence, NY, 159, SE
Lawrence Jct., MA, 108, SE
Lawrence Mills-South Gardner, ME, 56, SW
Lawrenceville (Erie; NYC), PA, 135, NE
Lawrenceville, NY, 122, SW
Lawton, CT, 146, SW
Lawtons, NY, 93, SW

Lebanon, CT, 145, SE
Lebanon, NH, 67, SE
Ledges, ME, 1, SE
Ledges, NB, 1, SE
Lee, MA, 123, NW
Lee, NH, 89, SE
Leeds, MA, 124, NE
Leeds, NY, 122, NW
Leeds Jct., ME, 55, SE
Leetes Island, CT, 154, NE
Lehigh Valley Jct., NY, 97, NW
Leicester, NY, 95, NW
Leicester Jct., VT, 65, NE
Lenox, MA, 123, NE
Leolyn, PA, 136, SW
Leominster, MA, 107, SW
Leonardsville, NY, 100, NW
Leonia, NJ, 159, NW
LeRoy (B&O; Erie; NYC), NY, 95, NW
Les Etroits, NB, 1, NW
Leverett, MA, 124, NE
Lewis Center, ME, 55, NE
Lewis Run (B&O; Erie), PA, 132, NE
Lewisburg, NJ, 150, SE
Lewiston, ME, 55, SE
Lewiston, ME, 55, SW
Lewiston Lower, ME, 55, SE
Lewiston-Water St., NY, 73, SE
Lexington, MA, 127B
Libby's Pit, ME, 55, NE
Liberty, NY, 140, NW
Liberty, PA, 133, SE
Liberty Corners, NY, 150, NE
Liebig's Lane, NJ, 158, SE
Lille, ME, 2, NE
Lily Pond, ME, 20, SW
Lima, NY, 95, NE
Lime, NY, 95, NW
Lime Rock, CT, 143, NW
Lime Rock, NY, 95, NW
Lime Rock Crossing, ME, 57, SE
Limerick, NY, 43, SE
Limestone, ME, 6, NW
Limestone, NB, 6, NE
Limestone, NY, 112, SE
Lincoln, MA, 127, NW
Lincoln, ME, 21, NE
Lincoln, NH, 52, SE
Lincoln, NY, 159, NW
Lincoln Mills, ME, 38, NW

Lincoln Park, NJ, 158, NW
Lincoln Park, NY, 76A
Lincoln Square, MA, 126, NW
Lincolnville, PA, 130, NW
Linden, MA, 127C
Linden, NJ, 158, SW
Linden, NY, 94, NE
Linden Jct., NJ, 158, SE
Lindenhurst, NY, 160, SW
Lindenwood, MA, 127C
Lindley, NY, 115, SE
Lindsay, NB, 8, SE
Linlithgo, NY, 122, SW
Linore, PA, 133, SE
Linwood, NY, 95, NW
Lisbon, ME, 55, SE
Lisbon, NH, 52, SW
Lisbon, NY, 26, SW
Lisbon Falls, ME, 71, NE
Lisle, NY, 117, NE
Litchfield, CT, 143, SE
Litchfield, NY, 117, SW
Little Falls (DL&W; Erie), NJ, 158, NE
Little Falls, NY, 82, SW
Little Ferry (NYC; NYS&W), NJ, 158, NE
Little Genesee, NY, 113, SE
Little Neck, NY, 159, NE
Little Rapids, NY, 62, NW
Little River Mills, NB, 1, SW
Little Valley, NY, 112, NW
Little Yard, NY, 98, SE
Littlefield, ME, 55, SW
Littlefield, NH, 90, SW
Littleton, MA, 107, SE
Littleton, ME, 8, SW
Littleton & Bethlehem, NH, 52, NW
Livermore Falls, ME, 55, NE
Livermore Falls, NH, 68, NE
Liverpool, NY, 79, SE
Livingston, NY, 158, SE
Livingston Manor, NY, 140, NW
Livonia, NY, 95, NE
Llewellyn, NJ, 158, NE
Lochmere, NH, 88, NE
Lock Berlin, NY, 78, SW
Locke, NY, 98, SW
Locke's Mills, ME, 54, NE
Lockport, NY, 74, SE
Lockport Jct., NY, 74, SW
Lockwood, NY, 116, SE

Locust Manor, NY, 159, SW
Locust Valley, NY, 159, NE
Lodi, NY, 97, SW
Logue, PA, 134, SW
Londonderry, NH, 108, NW
Long A, ME, 12, SW
Long Beach, ME, 90, SE
Long Beach, NY, 159, SE
Long Eddy, NY, 139, NE
Long Flat, NY, 119, SE
Long Hill, CT, 153, NE
Long Pond, ME, 10, SE
Long Swamp, QC, 16, NW
Long Valley, PA, 136, SE
Long Valley Station, PA, 136, SE
Longview, MA, 148, SE
Longwood, MA, 127A, 127E
Longwood, NY, 94, NW
Lonsdale, RI, 147, NW, 147B
Lookout Jct., PA, 138, SE
Loon Cove, NH, 89, NW
Loon Lake (D&H; NYC), NY, 28, SE
Lordville, NY, 139, NE
Loring AFB, ME, 6, NW
Lorraine, NJ, 158, SE
Lottsville, PA, 131, NW
Louden Hill, PA, 138, NW
Lounsberry, NY, 117, SW
Loveland, NY, 93, SE
Lovell, PA, 130, NE
Lowbanks, ON, 92, NW
Lowell, MA, 108, SW
Lowell Jct., MA, 108, SE
Lowell St., MA, 108, SE
Lowelltown, ME, 9, SE
Lower Switch, NY, 117, NW
Lower Warner, NH, 88, NW
Lower Yard, ME, 35, SE
Lowerre, NY, 159, NW
Lowerton (Erie), NY, 74, SE
Lowman, NY, 116, SE
Lowrey, PA, 136, SW
Low's Bridge, ME, 20, SW
Lowville (LBR; NYC), NY, 61, NW
Lucerne-in-Maine, ME, 39, SE
Ludlow, MA, 125, SW
Ludlow, ME, 7, SE
Ludlow, NY, 159, NW
Ludlow, PA, 132, SW
Ludlow, VT, 86, NW
Ludlowville, NY, 97, SE
Lunenberg, VT, 52, NE

Luzon, NY, 140, SE
Lycoming, NY, 79, NW
Lyford's Siding, NH, 68, NE
Lyme, CT, 155, NW
Lyn Jct., ON, 25, SW
Lynbrook, NY, 159, SE
Lynch, PA, 131, SE
Lyndhurst, NJ, 158, NE
Lyndon, VT, 32, SE
Lyndonville, NY, 75, NW
Lyndonville, VT, 33, SW
Lynn, MA, 128, NW
Lynn, ON, 25, SW
Lynn Common, MA, 128, NW
Lynnfield, MA, 108, SE
Lynnfield Center, MA, 108, SE
Lyon Mountain, NY, 29, SW
Lyons, CT, 153, SE
Lyons, NY, 48, NW
Lyons, NY, 78, SW
Lyons Falls, NY, 61, SW

M&U Jct., NJ, 150, NE
Macedon, NY, 77, SW
Machias (B&O), NY, 113, NW
Machias (PRR), NY, 113, NW
Machias, ME, 4, SE
Machias, ME, 42, SW
Machiasport, ME, 42, SW
Machine Shop, MA, 108, SE
Mack Point, ME, 58, NW
Mackamp, ME, 10, SE
MacKenzie, NB, 1, NE
Macy Jct., ME, 35, NW
Madawaska, ME, 2, NW
Madawaska, NY, 28, SW
Madbury, NH, 90, SW
Madge, PA, 132, SW
Madison, CT, 154, NE
Madison, ME, 37, NW
Madison, NH, 69, NE
Madison, NJ, 158, NW
Madison St., NY, 103, SE, 103A
Madrid, ME, 36, NW
Madrid, NY, 26, SE
Madrid Jct., ME, 36, NW
Madrid Village, ME, 36, NW
Magaguadavic, NB, 15, SE
Magee, PA, 131, SW
Magnolia, MA, 109, SE
Magowan, VT, 32, NW
Mahopac, NY, 152, NE
Mahwah, NY, 151, SE

Maines, ME, 6, NW
Mainstream, ME, 37, NE
Maitland, ON, 25, SE
Malba, NY, 159, NW
Malden (E), MA, 127C
Malden, MA, 127C
Malden-on-Hudson, NY, 122, SW
Mallory, NY, 79, NE
Mallorytown, ON, 44, NW
Malone (NYC; R), NY, 28, NW
Malone Jct., NY, 28, NW
Maltbys, CT, 143, NW
Malvern, NY, 159, SE
Malvina, QC, 16, SW
Mamakating, NY, 141, SW
Mamaroneck (NH; NYW&B), NY, 159, NE
Mamaroneck Ave., NY, 152, SW
Manchester, CT, 144, NE
Manchester, MA, 109, SW
Manchester, NH, 108, NW, 108A
Manchester, NY, 96, NE
Manchester, VT, 85, SE
Manchester Bridge, NY, 142, SW
Manhassett, NY, 159, NE
Manhattan Beach, NY, 159, SW
Manitou, NY, 152, NW
Manlius, NY, 80, SW
Mann's Crossing, MA, 107, SW
Mannsville, NY, 59, SE
Manorville, NY, 161, NW
Mansfield, CT, 145, NW
Mansfield, MA, 127, SE
Mansfield, PA, 135, NE
Manton, RI, 147, NW
Manville, RI, 147, NW
Mapes, NY, 114, SW
Maple Grove, MA, 104, SE
Maple Grove, ME, 6, SW
Maple Shade Farm, NY, 84, SE
Maple Springs, NY, 111, SW
Maple View, NY, 79, NE
Maples, CT, 153, NE
Mapleton, ME, 5, SE
Mapleton, NH, 33, SE
Mapleton, NY, 74, SW
Maplewood (DL&W; RV), NJ, 158, SW
Maplewood, MA, 127C
Maplewood, ME, 36, SE
Maplewood, NY, 76A
Maplewood Yard (NYC), NY, 76A
Maquam, VT, 30, NE

Maranacook, ME, 56, NW
Marathon, NY, 117, NE
Marble Hill, NY, 159, NW
Marble Ridge, MA, 108, SE
Marbles, ME, 35, NE
Marcellus, NY, 98, NW
Marcellus Falls, NY, 79, SW
Marcy, NY, 81, SW
Margaretville, NY, 120, SE
Margison, ME, 5, NE
Marietta, NY, 98, NW
Marilla, NY, 93, NE
Mariner's Harbor, NY, 158, SE
Marion, MA, 148, SW
Marion, ME, 42, NW
Marion, NY, 77, SE
Marion Jct., NJ, 158A
Markhams, NY, 111, NW
Marlboro (B&M), MA, 126, NE
Marlboro (NH), MA, 126, NE
Marlboro, NH, 106, NE
Marlborough, NY, 142, SW
Maromas, CT, 144, SE
Marrs, ME, 11, SW
Mars Hill, ME, 6, SW
Marsh, ME, 56, SE
Marsh Creek, PA, 135, NW
Marshalls, NY, 115, NW
Marshfield, MA, 128, SE
Marshfield, NH, 53, NW
Marshfield, VT, 51, NW
Marshville, ON, 92, NW
Martin (CN; CPR), NB, 3, SW
Martin, ME, 2, SE
Martindale, NY, 122, SE
Martins, NJ, 150, SE
Martinsville, NY, 74, SW
Martisco (MOL; NYC), NY, 79, SW
Martville, NY, 78, NE
Marvindale, PA, 132, SE
Maryland, NY, 101, SW
Marysville, NY, 98, NW
Masaris, ME, 5, SW
Mascoma, NH, 67, SE
Mason, MA, 107, SW
Masons, NH, 33, SE
Maspeth, NY, 159, SW
Massabesic, NH, 108, NW
Massachusetts Ave., MA, 127A
Massapequa, NY, 160, SW
Massapequa Park, NY, 160, SW
Massena, NY, 27, NW
Massey, NY, 60, NW

Masspeag, CT, 155, NE
Mast Yard, NH, 88, NE
Masten, PA, 136, SW
Masthope, PA, 139, SE
Mastic, NY, 161, NW
Matfield, MA, 128, SW
Matoon, NY, 44, NE
Mattapan, MA, 127E
Mattapoisett, MA, 148, SW
Mattawamkeag (CPR; MEC), ME, 13, SW
Mattewan, NY, 142, SW
Matthews, NH, 69, SE
Mattituck, NY, 161, NE
Mattocks, ME, 70, NE
Maxeys, ME, 56, NE
Maxwell, NB, 24, SW
Maybrook (NH), NY, 151, NE
Maybrook Jct. (L&HR), NY, 151, NE
Maybrook Yard, NY, 141, SE
Mayburg, PA, 131, SE
Maybury's, NY, 98, SE
Mayfield, NY, 83, SW
Mayfield, PA, 138, SE
Mayfield Yard, PA, 138, SE
Maynard, MA, 127, NW
Maynard, ME, 6, NW
Mayo-East Concord, VT, 52, NW
Maysville, ME, 6, NW
Mayville (JW&NW), NY, 111, NW
Mayville (PRR), NY, 111, NW
Mayville Jct., NY, 111, NW
Maywood, NJ, 158, NE
Maywood, NY, 119, NW
McAdam, NB, 15, SW
McAfee, NJ, 150, SE
McConnellsville (LV; NYC), NY, 80, NE
McDonald, NY, 47, NW
McDonald's "Y", PA, 131, SW
McDougall, NY, 97, NW
McGeorges, ME, 42, NW
McGraw, ME, 6, NW
McGraw, NY, 98, SE
McIndoes, VT, 51, NE
McIntyre, NY, 142, NE
McKeagan, NB, 8, NE
McKeever, NY, 61, SE
McKenna, NB, 8, SE
McKinley, PA, 132, SW
McKinneys, NY, 116, NE
McLean, NY, 98, SW

McNally, ME, 4, NE
McShea, ME, 6, NW
MD Siding, NY, 65, NW
Mead, NY, 98, SW
Meadow Brook, ME, 13, SE
Meadow Brook, NY, 151, NE
Meadowbrook, NY, 159, SE
Meadows Yard, NJ, 158A
Meadowview, MA, 108, SE
Mechanic Falls, ME, 55, SW
Mechanicville, NY, 103, NE
Mechanicville Yard, NY, 103, NE
Medfield, MA, 127, SW
Medfield Jct., MA, 127, SW
Medford, ME, 21, SW
Medford, NY, 160, NE
Medford, Park St., MA, 127C
Medford Jct., MA, 127C
Medina, NY, 75, SW
Meductic, NB, 15, NW
Medway, MA, 127, SW
Megantic, QC, 9, SW
Mehoopany, PA, 137, SE
Mellenville, NY, 122, NE
Mellville, MA, 147, SW
Melrose, CT, 144, NE
Melrose, MA, 127C
Melrose, NY, 103, NE
Melrose, NY, 159A
Melrose, PA, 138, NE
Melrose Highlands, MA, 127C
Meltcalf, NY, 78, NE
Melvin, NH, 88, NW
Memphis, NY, 79, SW
Menands, NY, 103, SE, 103A
Mendon, NY, 95, NE
Menlo Park, NJ, 158, SW
Meno, NY, 28, SW
Meredith, NH, 68, SE
Meriden, CT, 144, SW
Meriden, West Main St., CT, 144, SW
Meriden Jct., CT, 143, SE
Merillon Ave., NY, 159, SE
Merriam, NY, 49, NW
Merrick, NY, 159, SE
Merrickville, NY, 119, NE
Merrickville, ON, 25, NW
Merrimac, MA, 108, NE
Merrimack, NH, 108, NW
Merritton, ON, 73, SE
Mertensia, NY, 96, NW
Meshoppen, PA, 137, SE

Messalonskee-North Belgrade, ME, 37, SW
Messengerville, NY, 117, NE
Metcalfs, MA, 127, SW
Methuen, MA, 108, SE
Metuchen (LV; PRR; RDG), NJ, 158, SW
Mexico, NY, 79, NE
Michaud, ME, 1, SE
Michigan Ave. (BCK), NY, 93A
Middle Falls, NY, 84, SE
Middle Granville, NY, 85, NW
Middle Kiln, NY, 28, SE
Middleboro, MA, 148, NW
Middlebury, PA, 135, NW
Middlebury, VT, 49, SE
Middlefield, MA, 123, NE
Middleport, NY, 75, SW
Middlesburgh, NY, 102, SW
Middlesex, MA, 108, SW
Middlesex, NJ, 158, SW
Middlesex, NY, 96, SW
Middlesex, VT, 50, NE
Middlesex Jct., MA, 127, NW
Middleton, MA, 108, SE
Middletown (Erie; M&U; NYO&W), NY, 151, NW
Middletown, CT, 144, SE
Middletown, MA, 147, SW
Middletown-Lincoln Center, ME, 22, NW
Middleville, NY, 82, SW
Midland, MA, 127, SW
Midland Park, NJ, 158, NE
Midmont, PA, 132, SE
Midway, CT, 155, NE
Midway, ME, 5, SW
Midway, VT, 87, SW
Milan, PA, 136, NE
Miles Pond, VT, 52, NW
Miles River, MA, 109, SW
Milford, CT, 153, SE
Milford, MA, 126, SE
Milford, ME, 39, NE
Milford, NH, 107, NE
Milford, NY, 101, SW
Military Road (NYC), NY, 93D
Milk House, CT, 145, SE
Mill Brook, ME, 20, NE
Mill Creek-South Orrington, ME, 39, SW
Mill Neck, NY, 159, NE
Mill Plain, CT, 152, NE

Mill Rift, PA, 150, NE
Mill Village, PA, 130, NW
Mill Yard, ME, 20, SW
Millbrook, NY, 142, NE
Millburn, NJ, 158, SW
Millbury, MA, 126, SW
Millbury Jct., MA, 126, SE
Milldale, CT, 144, SW
Miller Farm, PA, 130, SE
Miller Place, NY, 161, NW
Millers, CT, 123, SW
Millers, NY, 75, NW
Millers, PA, 130, NW
Millers, PA, 131, SE
Miller's Falls, MA, 106, SW
Millerton, NY, 142, NE
Millerton, PA, 136, NW
Millicete, NB, 6, NE
Millinocket, ME, 12, SE
Millis, MA, 127, SW
Millport, NY, 116, NW
Millport, PA, 133, NE
Mills, NY, 117, NE
Mills, PA, 134, NE
Milltown, ME, 24, SW
Milltown Jct. (CPR; MEC), NB, 24, SW
Millville, MA, 126, SE
Millwood, NY, 152, SW
Milo, ME, 21, NW
Milo Mills, NY, 96, SE
Milton, MA, 127E
Milton, NH, 90, NW
Milton, NY, 142, SW
Milton, VT, 30, SE
Mina, PA, 133, SE
Mineola, NY, 159, SE
Minetto, NY, 79, NW
Mineville, NY, 48, SE
Minister, PA, 131, SE
Minnehaha, NY, 61, SE
Minnowbrook, NY, 62, NW
Minoa, NY, 79, SE
Misery, ME, 11, SW
Mitchell Field, NY, 159, SE
Model City, NY, 74, SW
Modena, NY, 141, SE
Modoc Ledge, CT, 146, NW
Mohawk, NY, 81, SE
Mohola, NJ, 150, SE
Moira (BMRR; R), NY, 27, NE
Moira Yard, NY, 27, NE
Monmouth, ME, 55, SE

Monponsett, MA, 128, SW
Monroe, NJ, 150, SE
Monroe, NY, 151, NE
Monroe Bridge, MA, 105, SW
Monroe St., NY, 97, NE
Monroe-Cedar Knoll, NJ, 158, NW
Monroeton (LV; S&NY), PA, 137, SW
Monsey, NY, 151, SE
Monsey Heights, NY, 151, SE
Monson, MA, 125, SW
Monson, ME, 20, NW
Monson Jct. (BAR; Monson), ME, 20, SW
Montague, MA, 105, SE
Montague, ON, 92, NW
Montague City, MA, 105, SE
Montauk, NY, 156, SW
Montclair (DL&W; Erie), NJ, 158, NE
Montclair, MA, 127E
Montclair Heights, NJ, 158, NE
Montello, MA, 127, SE
Monteola, NY, 60, SE
Montezuma, NY, 78, SE
Montgomery, NY, 141, SE
Monticello, ME, 8, NW
Monticello, NY, 140, SE
Montour Falls, NY, 116, NW
Montowese, CT, 154, NW
Montpelier, VT, 50, NE
Montpelier Jct., VT, 50, NE
Montrose (DL&W; LV), PA, 138, NW
Montrose, MA, 108, SE
Montrose, NY, 152, NW
Montrose, ON, 73, SE
Montrose Jct., ON, 73, SE
Montserrat, MA, 109, SW
Montsweag, ME, 72, NE
Montvale, MA, 127B
Montvale, NJ, 151, SE
Montville, CT, 155, NE
Monument Beach, MA, 148, SE
Moodys, ME, 38, NW
Mooers (D&H; R), NY, 29, NE
Mooers Forks, NY, 29, NE
Mooers Jct., NY, 29, NE
Moonachie, NJ, 158, NE
Moons, NY, 111, NW
Moore's Camp, ME, 20, NW
Moore's Mills, NB, 24, NW
Moores Mills, NY, 142, SW

Moorhead (NYC), PA, 110, SW
Moorheads (NKP), PA, 110, SW
Moorland, MA, 129, SE
Moosehead, ME, 11, SE
Moosup, CT, 146, SW
Moraine, NY, 114, NE
Moravia, NY, 98, SW
Morgan Crossing, MA, 125, NW
Moriah, NY, 49, SW
Moritz, NY, 112, SE
Morkill, ME, 20, NW
Morrill, NB, 6, NE
Morrills, MA, 127, SE
Morris, PA, 135, SW
Morris Heights, NY, 159, NW
Morris Park (NH; NYW&B), NY, 159, NW
Morris Park, NY, 159D
Morris Park Shops, NY, 159D
Morris Plains, NJ, 158, NW
Morris Run, PA, 135, SE
Morrisburg, ON, 26, NE
Morrison, PA, 132, NW
Morrisonville, NY, 29, SE
Morrissania, NY, 159, NW
Morristown (DL&W; MT&E), NJ, 158, NW
Morristown, NY, 25, SE
Morrisville, NY, 99, NE
Morrisville, VT, 31, SE
Morrow Road, ME, 6, NW
Morse, NY, 79, NE
Morsemere, NJ, 158, NE
Mortimer (Erie), NY, 76A
Mortimer (LV), NY, 76A
Mortimer (NYC), NY, 76A
Morton, NY, 76, NW
Morton St., MA, 127E
Mottville, NY, 98, NW
Moulin, NY, 62, SW
Mount Arab, NY, 46, SE
Mount Auburn, MA, 127B
Mount Bowdoin, MA, 127E
Mount Holly, VT, 86, NW
Mount Hope, MA, 127E
Mount Hope, NY, 159, NW
Mount Jewett (B&O; Erie), PA, 132, SE
Mount Kisco, NY, 152, SE
Mount Morris, NY, 95, SW
Mount Pleasant, MA, 148, SW
Mount Pleasant, NY, 152, SW
Mount Riga, NY, 142, NE

Mount St. Vincent, NY, 159, NW
Mount Tabor, NJ, 158, NW
Mount Upton, NY, 119, NW
Mount Vernon (NH; NYC), NY, 159, NW
Mount Washington, NH, 53, NW
Mount Whittier, NH, 69, NE
Mountain, NJ, 158, NW
Mountain Ave., NJ, 158, NE
Mountain Dale, NY, 140, SE
Mountain Lakes, NJ, 158, NW
Mountain Spring, NY, 150, NE
Mountain View (DL&W; Erie), NJ, 158, NW
Mountain View, NY, 28, SE
Mountainview, NH, 69, SE
Mountainville, NY, 151, NE
Mountorne, NH, 52, NE
Mountville, NJ, 158, NW
Mt. Abram, ME, 36, NW
Mt. Abram Jct., ME, 36, NE
Mt. Alton, PA, 132, NE
Mt. Carmel, CT, 154, NW
Mt. Desert Ferry, ME, 40, SE
Mt. Hermon, MA, 106, SW
Mt. Ivy, NY, 151, SE
Mt. Marion, NY, 122, SW
Mt. Pleasant, NY, 121, SW
Mt. Riga, NY, 122, SE
Mt. Ross, NY, 122, SE
Mt. Tom, MA, 124, NE
Mulford, NJ, 150, SE
Mumford, NY, 95, NW
Munns, NY, 99, NE
Munroe, MA, 127B
Murphy Road, ME, 6, NW
Murray Hill, NJ, 158, SW
Murray Hill, NY, 159, NW
Muschopauge, MA, 126, NW
Muscongus Bay, ME, 57, SW
Myobeach, PA, 137, SE
Mystic, CT, 156, NW
Mystic Jct., MA, 127A
Mystic, PA, 133, NE

NJ & NY Jct., NJ, 158, NE
N.Y. State Fish Hatchery, NY, 115, NW
Nagog (North Acton), MA, 108, SW
Nahor, NH, 107, NW
Nansen, PA, 132, SW
Nantasket, MA, 128, NW

Nantucket, MA, 157B
Nanuet, NY, 151, SE
Napanoch, NY, 141, SW
Napiers, NY, 113, NW
Naples, NY, 96, SW
Narragansett Pier, RI, 157, NW
Narrows, MA, 148, SE
Narrowsburg, NY, 139, SE
Nash Road, MA, 148, SW
Nashoba, MA, 108, SW
Nashua, NH, 108, NW
Nashua Union Station, NH, 108, NW
Nasonville, RI, 146, NE
Nassau, NY, 158, SE
Nassau Boulevard, NY, 159, SE
Nastasket Jct., MA, 128, SW
Natick, MA, 127, NW
Natick, RI, 147, SW
Natural Bridge, NY, 45, SW
Naugatuck, CT, 153, NE
NE Junction, NY, 111, SW
Needham, MA, 127D
Needham Jct., MA, 127D
Neelytown, NY, 151, NE
Nehasane, NY, 46, SW
Nelson, NY, 61, SE
Nelson, PA, 135, NE
Nemasket, MA, 148, NW
Nepera Park, NY, 159, NW
Neponset, MA, 127E
Nepperhan, NY, 159, NW
Netherwood, NJ, 158, SW
New Albany, PA, 137, SW
New Baltimore, NY, 122, NW
New Bedford, MA, 148, SW
New Bedford Wharf, MA, 148, SW
New Berlin, NY, 100, SW
New Berlin Jct. (NYO&W; UV), NY, 119, NW
New Boston, NH, 107, NE
New Braintree, MA, 125, NE
New Bremen, NY, 61, NW
New Brighton, NY, 158, SE
New Britain, CT, 144, SW
New Canaan, CT, 153, SW
New Castle, ME, 56, SE
New City, NY, 152, SW
New Denmark, NB, 6, NE
New Dorp, NY, 158, SE
New Durham, NH, 89, NE
New Gloucester, ME, 71, NW

New Hamburg, NY, 142, SW
New Hampton, NY, 151, NW
New Hartford, CT, 144, NW
New Hartford, NY, 81, SW, 81A
New Haven, Belle Dock, CT, 154, NW, 154A
New Haven, CT, 154, NW, 154A
New Haven, NY, 79, NW
New Haven, Silver St., CT, 154, NW, 154A
New Haven, VT, 49, SE
New Haven Jct., VT, 49, SE
New Hempstead, NY, 151, SE
New Hyde Park, NY, 159, SE
New Lebanon, NY, 123, NW
New Lenox, MA, 123, NE
New Limerick, ME, 8, SW
New London, CT, 155, NE, 155A
New Market, NJ, 158, SW
New Milford, CT, 143, SW
New Milford, NJ, 158, NE
New Milford, NY, 151, SW
New Milford, PA, 138, NE
New Paltz, NY, 141, SE
New Providence, NJ, 158, SW
New Rochelle, NY, 159, NW
New Salem, MA, 125, NW
New Scotland, NY, 103, SW
New Sweden, ME, 5, NE
New Woodstock, NY, 99, NW
Newark, NY, 77, SE
Newark Valley, NY, 117, SE
Newburg, NB, 8, SE
Newburgh (Erie), NY, 151, NE
Newbury, NH, 87, NE
Newbury, VT, 51, SE
Newburyport, MA, 109, NW
Newcomb, NB, 15, SW
Newells, ME, 56, NE
Newfane, VT, 105, NE
Newfield, NY, 116, NE
Newfield, PA, 134, NW
Newfield Jct. (B&O; CPA), PA, 134, NW
Newfields, NH, 90, SW
Newfoundland, NJ, 151, SW
Newhall, ME, 71, SW
Newington, CT, 144, SE
Newington, NH, 90, SW
Newmarket, NH, 90, SW
Newport, NH, 87, NE
Newport, NY, 81, SE
Newport, RI, 157, NW

Newport, VT, 32, NE
Newport Jct., ME, 38, NW
Newton, MA, 127D
Newton, NH, 108, NE
Newton, NJ, 150, SE
Newton, PA, 131, NW
Newton Centre, MA, 127D
Newton Falls, NY, 46, SW
Newton Highlands, MA, 127D
Newton Hook, NY, 122, NW
Newton Jct., NH, 108, NE
Newton Upper Falls, MA, 127D
Newtonville, MA, 127D
Newtown, CT, 153, NW
Niagara, MA, 123, NE
Niagara Falls, JH Freight Office, NY, 73, SW, 73A
Niagara Falls, NY, 73, SE, 73A
Niagara Falls, ON, 73, SE, 73A
Niagara Falls, ON, 73, SW, 73A
Niagara Jct., NY, 73, SE
Niagara-on-the-Lake, ON, 73, NE
Nichols, NY, 115, SW
Nichols, NY, 117, SW
Nicholson, PA, 138, SW
Nicolin, ME, 39, SE
Nightingale, NY, 98, NW
Nile, NY, 113, SE
Niles, NY, 119, NW
Niles Valley, PA, 135, NW
Nineveh, NY, 118, SE
Niobe, NY, 111, SW
Niskayuna, NY, 103, NW
Niverville, NY, 122, NE
Nixon, ME, 4, NE
Nixon, NB, 8, NE
No. 6, ME, 36, NW
Noank, CT, 156, NW
Nobadeer, MA, 157B
Nobles, PA, 130, SW
Noblesboro, ME, 57, SW
Noone, NH, 107, NW
Norcross, ME, 12, SW
Norfolk, CT, 143, NE
Norfolk, MA, 127, SW
Norfolk, NY, 27, NW
Norfolk Downs, MA, 127E
Norlands, ME, 55, NE
Norman, MA, 149, SE
Noroton Heights, CT, 153, SW
Norowottuck, MA, 125, NW
Norridgewock, ME, 37, SW
North Abington, MA, 128, SW

North Acton, MA, 108, SW
North Adams, MA, 104, SE
North Adams Jct., MA, 123, NE
North Alexander, NY, 94, NE
North Andover, MA, 108, SE
North Anson, ME, 37, NW
North Attleboro, MA, 147, NW
North Bangor, ME, 39, NW
North Bay, NY, 80, SE
North Bennington, VT, 104, NE
North Bergen (NNJ; NYS&W), NJ, 158, NE
North Bergen (NYC; NYS&W), NJ, 158A
North Berwick (E), ME, 90, NE
North Berwick (W), ME, 90, NE
North Beverly, MA, 109, SW
North Billerica, MA, 108, SW
North Boston, NY, 93, SW
North Branford Quarry, CT, 154, NW
North Bridgton, ME, 54, SE
North Brookfield, MA, 125, NE
North Brookfield, NY, 100, NE
North Brookline, NH, 107, NE
North Carver, MA, 148, NW
North Charlestown, NH, 87, NW
North Chelmsford, MA, 108, SW
North Chittenango, NY, 80, SW
North Clarendon, VT, 66, SW
North Collins, NY, 93, SW
North Conway, NH, 53, SE
North Creek (D&H; NLI), NY, 64, SW
North Croghan, NY, 44, SE
North Cromwell, CT, 144, SE
North Dana, MA, 125, NW
North Dartmouth, MA, 147, SE
North Dighton, MA, 147, NE
North Dorset, VT, 85, NE
North East (NKP), PA, 110, SW
North East (NYC), PA, 110, SW
North Easton, MA, 127, SE
North Elizabeth, NJ, 158, SE
North Enosburg, VT, 31, NE
North Fair Haven, NY, 78, NE
North Ferrisburg, VT, 49, NW
North Frankfort, NY, 81, SE
North Germantown, NY, 122, SW
North Grafton, MA, 126, SE
North Grosvenordale, CT, 146, NW
North Hackensack, NJ, 158, NE

North Hampton, NH, 109, NW
North Hanson, MA, 128, SW
North Harford, NY, 117, NE
North Hatfield, MA, 124, NE
North Haven, CT, 154, NW
North Haven Cabin, CT, 154, NW
North Hawthorne, NJ, 158, NE
North Hero, VT, 30, NW
North Hoosick, NY, 104, NW
North Ilion, NY, 81, SE
North Java, NY, 94, SW
North Jay, ME, 36, SE
North Jct., VT, 30, NE
North Lawrence, MA, 108, SE
North Lawrence, NY, 27, NE
North Leeds, ME, 55, NE
North Leominster, MA, 107, SE
North Leroy, NY, 76, SW
North Lexington, MA, 127B
North Macedon, NY, 77, SW
North Memphis, NY, 79, SW
North Oxford Mills, MA, 126, SW
North Pelham, NY, 159, NW
North Pepperell, MA, 107, SE
North Petersburg, NY, 104, NW
North Port Byron, NY, 78, SE
North Pownal, VT, 104, NW
North Rahway, NJ, 158, SW
North Reading, MA, 108, SE
North River, MA, 128, SE
North River, NY, 63, SE
North Rose, NY, 78, SW
North Scituate, MA, 128, SW
North Sheldon, VT, 31, NW
North Somerville, MA, 127A
North Spencer, NY, 116, NE
North St., MA, 109, SW
North Stephentown, NY, 104, SW
North Stoughton, MA, 127, SE
North Stratford, NH, 33, NE
North Tonawanda, NY, 74, SW, 74A
North Troy, VT, 32, NW
North Twin, ME, 12, SW
North Underhill, VT, 31, SW
North Vassalboro, ME, 56, NE
North Walpole Yard, NH, 87, SW
North Warren, PA, 131, NE
North Weare, NH, 88, SE
North Weedsport, NY, 78, SE
North Whitefield, ME, 56, SE
North Wilbraham, MA, 125, SW
North Wilmington, MA, 108, SE

North Windham, CT, 145, NE
North Woburn, MA, 108, SE
North Woodstock, NH, 52, SE
North Worcester, MA, 126, NW
Northampton, MA, 124, NE
Northboro, MA, 126, NE
Northboro, VT, 67, NE
Northbridge, MA, 126, SE
Northern Maine Jct. (BAR; MEC), ME, 39, NW
Northfield, MA, 106, SW
Northfield, NH, 88, NE
Northfield, VT, 50, SE
Northfield Farms, MA, 106, SW
Northford, CT, 154, NW
Northport, NY, 160, NW
Northport Jct., NY, 160, NW
Northrups, NJ, 150, SE
Northumberland, NH, 33, SE
Northvale, NJ, 152, SW
Northvale, NJ, 159, NW
Northville, NH, 87, NE
Northville, NY, 83, SE
Norton, MA, 147, NE
Norton Hollow, NY, 114, SE
Nortons, NY, 111, NW
Norton's Siding, NY, 85, NW
Norwalk Mills, CT, 153, SW
Norway, ME, 54, SE
Norwich, CT, 145, SE
Norwich, NY, 99, SE
Norwich, VT, 67, SW
Norwichtown, CT, 145, SE
Norwood (N&STL; R), NY, 27, NW
Norwood Central, MA, 127, SE
Norwood, MA, 127, SE
Norwood, NJ, 159, NW
Nostrand Ave., NY, 159, SW
Notchland, NH, 53, SW
Notre Dame, ME, 2, NE
Nowland, ME, 5, SW
Noyack Road, NY, 162, NW
Nunda, NY, 95, SW
Nunda Jct., NY, 95, SW
Nutley, NJ, 158, NE
Nutt's Pond, NH, 108, NW
Nyack, NY, 152, SW
NYC Jct., NY, 79, NW
Nypen, NY, 116, SW

Oak, NB, 8, SE
Oak Bay, NB, 24, SE
Oak Bluffs, MA, 157A
Oak Grove, MA, 127C
Oak Hill, ME, 71, SW
Oak Island, MA, 128, NW
Oak Lawn, RI, 147, NW
Oak Point Yard, NY, 159A
Oak St., MA, 124, SE
Oak Summit, NY, 142, NE
Oak Tree, NJ, 158, SW
Oakdale, MA, 126, NW
Oakdale, NY, 160, SE
Oakdale Park, NH, 88, NE
Oakfield, ME, 7, SE
Oakfield, NY, 75, SW
Oakland, MA, 127C
Oakland, ME, 37, SE
Oakland, NJ, 151, SE
Oakland, RI, 146, NE
Oakland, VT, 30, SE
Oakland Beach, RI, 147, SW
Oakland Centre, RI, 146, NE
Oakland Farm, ME, 90, SE
Oakley, NY, 114, NE
Oakridge, NJ, 151, SW
Oaks Corners, NY, 96, NE
Oakville, NY, 44, NE
Oakwood, NY, 97, NE
Oakwood Heights, NY, 158, SE
Ocean Ave., CT, 155, NE
Ocean Park, ME, 91, NW
Ocean Side, ME, 90, SE
Ocean Spray, MA, 128, NW
Oceanside, NY, 159, SE
Odessa, NY, 116, NW
Ogdensburg (NYC; R), NY, 26, SW
Ogdensburg, NY, 25, SE
Ogren Road, ME, 5, NE
Olamon, ME, 21, SE
Old Chatham, NY, 122, NE
Old Clarendon, PA, 131, NE
Old Forge, NY, 62, SW
Old Furnace, MA, 125, NE
Old Greenwich, CT, 152, SE
Old Line Jct., NY, 29, SW
Old Northport, NY, 160, NW
Old Orchard Beach, ME, 71, SW
Old Saybrook, CT, 155, NW
Old Town (BO&M), ME, 39, NE
Old Town, NY, 158, SE
Old Wye, NY, 78, NE
Oldtown (BAR), ME, 39, NE
Olean (Erie), NY, 113, SW
Olean (PRR), NY, 113, SW

Olean (PS&N), NY, 113, SW
Oliver, NY, 61, SW
Oliverian, NH, 52, SW
Olmstead, PA, 133, NE
Olneyville, RI, 147, NW, 147A
Onatavia, NY, 98, NE
Onawa, ME, 20, NW
Onchiota (D&H), NY, 47, NE
Onchiota (NYC), NY, 47, NE
Ondawa, NY, 84, SE
Oneco, CT, 146, SW
Oneida (NYC; NYC), NY, 80, SE
Oneida Castle, NY, 80, SE
Onekio, NY, 61, SE
Oneonta, NY, 119, NE
Onoville, NY, 112, SW
Onset, MA, 148, NE
Ontario, NY, 77, SW
Onway Lake, NH, 89, SE
Oquaga, NY, 119, SW
Oquossoc, ME, 35, NW
Oradell, NJ, 158, NE
Oramel, NY, 113, NE
Orange (DL&W; Erie), NJ, 158, NE
Orange, CT, 153, NE
Orange, MA, 106, SW
Orangeburg (Erie; NYC), NY, 152, SW
Orangeburg, NJ, 150, SE
Orchard Park, NY, 93, NE
Orchard St., NJ, 158, NE
Orcutts, CT, 145, NW
Ordway, MA, 127, NW
Orient Heights, MA, 127C
Oriskany, NY, 81, SW
Oriskany Falls, NY, 100, NW
Orleans, MA, 149, NE
Orleans, NY, 96, NE
Orleans, VT, 32, NE
Orleans Corners, NY, 44, SW
Ormsbee Road, PA, 130, NE
Ormsby, PA, 132, NE
Orono, ME, 39, NE
Orrington, ME, 39, SW
Orson, PA, 139, NW
Ortonville, NB, 6, NE
Orwell, VT, 65, NW
Osborntown, CT, 143, SE
Oscawanna, NY, 152, SW
Osceola (B&O; NYC), PA, 135, NW
Ossining, NY, 152, SW

Ossipee, NH, 69, SE
Oswayo, PA, 133, NE
Oswegatchie, NY, 45, SE
Oswego (DL&W), NY, 78A
Oswego, NY, 79, NW
Oswego (NYC), NY, 78A
Oswego (NYO&W), NY, 78A
Oswego Frt. Ho. (NYC), NY, 78A
Otsego, NY, 119, NE
Otis, NB, 15, NE
Otis, NY, 76A
Otis Hill, ME, 37, SE
Otis Jct., NY, 121, NE
Otisco Lake, NY, 98, NW
Otisville, NY, 150, NE
Otter Lake, NY, 61, SE
Otter River, MA, 106, SE
Otterson St., NH, 108, NW
Outlet, ME, 11, SE
Outlet, NY, 96, NE
Ovid, NY, 97, SW
Owasko Lake, NY, 97, NE
Owego (DL&W; Erie; LV), NY, 117, SW
Owens, NJ, 150, NE
Owls Head, NY, 28, SE
Oxford (DL&W; NYO&W), NY, 118, NE
Oxford, CT, 153, NE
Oxford, MA, 126, SW
Oxford, ME, 55, SW
Oxford, NY, 151, NE
Oxford, ON, 25, NE
Oyster Bay, NY, 159, NE
Ozone Park, NY, 159, SW

Pacific Ave., NJ, 158A
Packard, ME, 21, NW
Paddlefords, NY, 96, NW
Page, NY, 60, SE
Painted Post (DL&W; Erie), NY, 115, SE
Palatine Bridge, NY, 101, NE
Palenville, NY, 121, SE
Palermo, ME, 57, NW
Palisades Park, NJ, 158, NE
Palmer, MA, 125, SW
Palmer Falls, NY, 84, SW
Palmers Cove, CT, 155, NE
Palmertown, CT, 155, NE
Palmyra, NY, 77, SE
Pamelia, NY, 60, NW
Panama, NY, 110, SE

Paper Mill, MA, 108, NE
Parent, ME, 2, SE
Paris, NY, 100, NE
Parish, NY, 79, NE
Park, NY, 116, SE
Park Hill, NY, 159, NW
Park Ridge, NJ, 151, SE
Park Siding, ME, 5, SE
Park Square, MA, 127A
Parker, MA, 106, SE
Parker, NH, 88, SE
Parker, NY, 119, NW
Parkers, ME, 21, SW
Parkers, NY, 85, SW
Parker's Crossover, ME, 37, SE
Parker's Glen, PA, 150, NW
Parkhurst, ME, 6, SW
Parks, CT, 146, NW
Parkside, NY, 159, SW
Parksville, NY, 140, NW
Parkville, CT, 144, SE, 144A
Parrot, ME, 19, NE
Parsons, ME, 90, NE
Pascoag, RI, 146, NE
Passadumkeag, ME, 21, SE
Passaic (DL&W; Erie), NJ, 158, NE
Passaic Jct., NJ, 158, NE
Passaic Park, NJ, 158, NE
Passumpsic, VT, 51, NE
Patchin, NY, 93, SE
Patchogue, NY, 160, SE
Pattee, NH, 67, SE
Patten, ME, 13, NW
Patten Jct., ME, 13, NW
Patterson (DL&W; Erie), NJ, 158, NE
Patterson, NY, 142, SE
Patterson Ave., NJ, 158A
Patterson Broadway, NJ, 158, NE
Patterson City, NJ, 158, NE
Pattersonville, NY, 102, NE
Paul, NY, 79, NW, 78A
Pauls, ME, 5, NE
Pauls, NY, 97, NE
Pavillion, NY, 94, NE
Pawling, NY, 142, SE
Pawtucket, RI, 147B
Payns, NY, 122, NE
Pea Cove, ME, 39, NE
Peabody, MA, 109, SW
Peace Dale, RI, 157, NW
Pearl Creek, NY, 94, NE

Pearl River, NY, 151, SE
Pebble Dell, PA, 131, SE
Pecks Bridge, MA, 123, NE
Pecksport, NY, 99, NE
Peconic, NY, 155, SW
Pecowsic, MA, 124, SE
Peekskill, NY, 152, NW
Peel, NB, 8, NE
Pejepscot Mills, ME, 71, NE
Pekin, NY, 74, SW
Pelham, NY, 159, NW
Pelham Manor, NY, 159, NW
Pelham Parkway, NY, 159, NW
Pelhamwood, NY, 159, NW
Pelletier, ME, 2, NW
Pellettown, NJ, 150, SE
Pemberton, MA, 128, NW
Pembroke, ME, 42, NE
Pembroke, NY, 75, SW
Penacook, NH, 88, NE
Pendleton, NY, 74, SW
Penn Yan, NY, 96, SE
Pennellville, NY, 79, NW
Pennsylvania Station (New York) (LI; LV; NH; PRR), NY, 159, SW
Penny Bridge, NY, 159, SW
Penobscot, ME, 22, NW
Pepacton, NY, 120, SW
Pepper, CT, 153, NE
Pepperell, MA, 107, SE
Pequannock, NJ, 158, NW
Percy, NH, 34, SW
Perham, ME, 5, NE
Perham Jct., ME, 36, NW
Perham Road (AV), ME, 5, NE
Perkins, ME, 12, SW
Perkinsville, NY, 95, SE
Perley's Mill, ME, 70, NE
Perry, MA, 128, SW
Perry, ME, 42, NE
Perry, NY, 93, SE
Perry, NY, 94, SE
Perry, ON, 92, NW
Perrys, NY, 94, SW
Perrysburg, NY, 112, NW
Perth Amboy (CNJ; LV), NJ, 158, SW
Perth Jct., NB, 6, SE
Peru, ME, 36, SW
Peru, NY, 29, SE
Peruton, NY, 98, SW
Peterboro, NH, 107, NW

Petersburg, NY, 104, NW
Petersburg Jct., NY, 104, NW
Petroleum, NJ, 158, SE
Petroleum Center, PA, 130, SE
Phair, ME, 6, SW
Phelps, NY, 96, NE
Phelps Jct. JC (PRR), NY, 96, NE
Philadelphia, NY, 44, SE
Phillips (B&O; NYC), PA, 135, NW
Phillips, ME, 36, NW
Phillips, NB, 8, SE
Phillips Beach, MA, 128, NW
Phillipsdale, RI, 147, NW
Phillipse Manor, NY, 152, SW
Phillipsport, NY, 141, SW
Philmont, NY, 122, NE
Phoenicia, NY, 121, SW
Phoenix, NY, 79, SW
Phoenix Mills, NY, 101, SW
Pickering, NH, 90, SW
Piercefield, NY, 46, SE
Pierces, ME, 39, SW
Pierces Bridge, MA, 127B
Piermont, NY, 152, SW
Piermont, VT, 67, NE
Pierre, ME, 1, SE
Pierrepont Manor, NY, 59, SE
Piffard, NY, 95, NW
Pike, NH, 51, SE
Pilgrim, MA, 148, NW
Pine, PA, 135, SW
Pine Aire, NY, 160, NW
Pine Brook, NY, 159, NW
Pine Bush, NY, 141, SW
Pine Camp, NY, 44, SW
Pine City, NY, 116, SW
Pine Hill, NY, 121, SW
Pine Island, NY, 151, NW
Pine Island Jct. (Erie; LNE), NY, 151, NW
Pine Meadow, CT, 144, NW
Pine Orchard, CT, 154, NW
Pine Plains, NY, 142, NE
Pine Plains Jct., NY, 142, NE
Pine Point, ME, 71, SW
Pine Ridge, MA, 108, SW
Pine Valley, NY, 116, SW
Pinelawn, NY, 160, SW
Piscataqua, NH, 90, SW
Pit Four, NY, 81, NE
Pittsfield (NYC; PRR), PA, 131, NW
Pittsfield, MA, 123, NW

Pittsfield, ME, 38, NW
Pittsfield, NH, 89, NW
Pittsford, NY, 76, SE
Pittsford, VT, 65, SE
Pixley, NY, 113, NW
Plainfield, CT, 146, SW
Plainfield, NJ, 158, SW
Plainfield, VT, 51, NW
Plains, NJ, 150, SE
Plainville, CT, 144, SW
Plainville, MA, 127, SW
Plaistow, NH, 108, NE
Plandome, NY, 159, NE
Plantsville, CT, 144, SW
Plattsburg, NY, 30, SW
Plauderville, NJ, 158, NE
Playstead, MA, 128, NW
Pleasant Hill, MA, 127C
Pleasant Mount, PA, 139, SW
Pleasant Plains, NY, 158, SE
Pleasant Point, ME, 42, NE
Pleasant Point, NY, 79, NW
Pleasant St., MA, 127C
Pleasant St., MA, 128, NW
Pleasant Valley, NY, 115, NW
Pleasant Valley, NY, 120, SW
Pleasant Valley, NY, 142, NW
Pleasantville, NY, 152, SW
Plimptonville, MA, 127, SE
Plum Brook, NY, 27, NW
Plumadore, NY, 28, SE
Plumadore Jct., NY, 28, SE
Plymouth, MA, 148, NE
Plymouth, NH, 68, NE
Pocantico Hills, NY, 152, SW
Pocasset, MA, 148, SE
Point of Pines, MA, 128, NW
Pokiok, NB, 15, NE
Poland, NY, 81, SE
Pollard Brook, ME, 21, NE
Pomfret, CT, 146, NW
Pomona, NY, 151, SE
Pompton Jct., NJ, 151, SW
Pompton Lakes, NJ, 151, SW
Pompton Plains, NJ, 158, NW
Pond Eddy, PA, 150, NW
Pondville, MA, 127, SW
Ponemah, NH, 107, NE
Pontiac, RI, 147, SW
Poolville, NY, 99, NE
Popes Hill, MA, 127E
Port Allegany (CPA; PRR), PA, 133, NW

Index of Passenger and Non-passenger Stations / 243

Port Byron, NY, 78, SE
Port Chautauqua, NY, 111, SW
Port Chester (NH; NYW&B), NY, 152, SE
Port Colborne, ON, 92, NW
Port Crane, NY, 118, SW
Port Ewen, NY, 142, NW
Port Gibson, NY, 77, SE
Port Henry, NY, 49, SW
Port Jefferson, NY, 160, NE
Port Jervis (Erie; NYO&W), NY, 150, NE
Port Kent (D&H; KACL), NY, 30, SW
Port Leyden, NY, 61, SW
Port Morris, NY, 159A
Port Orange, NY, 150, NE
Port Reading (CNJ; RDG), NJ, 158, SW
Port Robinson, ON, 73, SE
Port Washington, NY, 159, NE
Port Weller, ON, 73, SE
Portage, ME, 5, NW
Portage, NY, 94, SE
Portageville, NY, 94, SE
Portland (CN), ME, 71, SE, 71A
Portland, CT, 144, SE
Portland (NKP), NY, 111, NW
Portland (NYC), NY, 111, NW
Portland (old), 71, SW, 71A
Portland, Preble St., ME, 71A
Portland & Rochester Jct., ME, 71A
Portland Point, NY, 97, SE
Portland Union Station (MEC; PTM), ME, 71, SW, 71A
Portlandville, NY, 101, SW
Portsmouth, MA, 147, SW
Portsmouth, NH, 90, SW
Portville (PRR), NY, 113, SW
Portville (PS&N), NY, 113, SW
Post Creek, NY, 116, SW
Potsdam, NY, 27, SW
Potter, NJ, 158, SW
Potter Brook, PA, 134, NE
Potter Place, NH, 88, NW
Poughkeepsie, NY, 142, SW
Poughkeepsie Jct., NY, 142, SW
Poughquag, NY, 142, SE
Poultney, VT, 65, SE
Pownal, ME, 71, NE
Pownal, VT, 104, NE
Powwow River, NH, 108, NE

Poyntelle, PA, 139, NW
Pratt, MA, 107, NW
Pratts Jct., MA, 126, NE
Prattsburgh, NY, 96, SW
Preble, NY, 98, SE
Prebles, ME, 56, SE
Pre-Emption, NY, 96, NE
Prescott (CPR; CN), ON, 25, SE
Presho, NY, 115, SE
Presque Isle (BAR; CPR), ME, 5, SE
Presque Isle Jct., ME, 5, SE
Preston Park, PA, 139, NW
Prides, MA, 109, SW
Prides, ME, 7, NW
Primrose, RI, 146, NE
Primus, MA, 107, SE
Prince William, NB, 15, SE
Princes Bay, NY, 158, SE
Princeton, MA, 126, NW
Princeton, ME, 23, SE
Print Works, RI, 147, NW
Pritchard, PA, 135, NE
Proctor, MA, 109, SW
Proctor, VT, 65, SE
Proctorsville, VT, 86, NE
Produce Siding, NY, 96, NE
Profile House, NH, 52, SE
Prompton, PA, 139, SW
Prospect, CT, 144, SE
Prospect, ME, 39, SW
Prospect, NJ, 158, NE
Prospect, NY, 110, SE
Prospect, NY, 81, NE
Prospect Ave., NJ, 158, NE
Prospect Jct., NY, 81, NE
Prosser, NY, 113, SW
Protection, NY, 94, SW
Providence, RI, 147, NW, 147A
Provincetown, MA, 129, SE
PS&P Jct., ME, 71, SW
Pulaski, NY, 59, SE
Pulvers, NY, 122, NE
Purdy's, NY, 152, NE
Putnam, CT, 146, NW
Putnam, NY, 65, SW
Putnamville, MA, 109, SW
Putney, VT, 105, NE
Pyrites, NY, 26, SE

Quaker Bridge, NY, 112, SW
Quaker Ridge, NY, 159, NW
Quakish, ME, 12, SW

Quarry, ME, 19, NE
Quarry Jct., CT, 144, SW
Quarryville, NJ, 150, NE
Quebec Jct., NH, 52, NE
Quechee, VT, 67, SW
Queens Village, NY, 159, SE
Queenston, ON, 73, SE
Quidneck, RI, 146, SE
Quimby, ME, 2, SE
Quinapoxet, MA, 126, NW
Quincy, MA, 127E
Quincy, NH, 68, NW
Quinebaug, MA, 126, SW
Quinnwood, PA, 132, SE
Quisibus, NB, 2, NE
Quoddy, ME, 42, NE
Quogue, NY, 161, NE
Quonset Point Naval Air Sta., RI, 147, SW

Radburn (Fairlawn), NJ, 158, NE
Raddin, MA, 128, NW
Rahway, NJ, 158, SW
Rainbow Lake, NY, 47, NE
Ralston, PA, 136, SW
Ramapo, NY, 151, SE
Ramsey, NJ, 151, SE
Rand Cove, ME, 21, NW
Randall, NY, 102, NW
Randallsville, NY, 99, NE
Randolph, MA, 127, SE
Randolph, ME, 56, SW
Randolph, NH, 53, NW
Randolph, NY, 112, SW
Randolph, VT, 66, NE
Rands, ME, 5, SE
Rangeley, ME, 35, NE
Rankins Mills, ME, 70, NW
Ransomville, NY, 74, SW
Raquette Lake, NY, 62, NE
Rasselas, PA, 132, SE
Rathbone, NY, 115, SW
Ravena, NY, 122, NW
Raybrook, NY, 47, NE
Raymond, NH, 89, SE
Raymond, PA, 134, NW
Raymondville, NY, 27, NW
Raynham, MA, 147, NE
Reades, CT, 146, SW
Readfield, ME, 56, NW
Reading, MA, 108, SE
Reading Center, NY, 116, NW
Reading Highlands, MA, 108, SE

Readsboro, VT, 105, NW
Readville (Dedham Br.), MA, 127F
Readville (Main Line), MA, 127F
Readville (Midland), MA, 127F
Readville Shops, MA, 127F
Readville Transfer, MA, 127F
Realty, ME, 17, SW
Red Bridge, MA, 125, SW
Red Creek, NY, 78, NE
Red Hook, NY, 142, NW
Red House (Erie), NY, 112, SW
Red House (PRR), NY, 112, SW
Red Onion Switch, NY, 151, NW
Red Rapids, NB, 6, NE
Red Rock Siding, NY, 141, NE
Red School, MA, 106, SE
Redding, CT, 153, NW
Redington, ME, 36, NW
Redstone, NH, 53, SE
Redwood, NY, 44, NW
Reeder, NY, 97, NW
Reeds, NY, 96, NE
Reed's Ferry, NH, 108, NW
Reeds Gap, CT, 154, NE
Reformatory Station, MA, 127, NW
Rego Park, NY, 159, SW
Relius, NY, 97, NE
Relyas, NY, 141, SE
Remsen, NY, 81, NE
Renchans, NY, 115, NW
Renfrew, MA, 104, SE
Rensselaer, NY, 103, SE
Rensselaer Falls, NY, 26, SW
Republic, NY, 160, SW
Reservoir, MA, 127D
Reservoir Switch, MA, 126, NE
Retsof, NY, 95, NW
Retsof Jct., NY, 95, NW
Revere, MA, 127C
Revere St., MA, 128, NW
Rexville, NY, 114, SE
Reynolds, NY, 103, NE
Reynolds Bridge, CT, 143, SE
Reynoldsville, NY, 142, SE
Rheims, NY, 115, NW
Rhinebeck, NY, 142, NW
Rhinecliff, NY, 142, NW
Riccars, ME, 55, SW
Rices, NY, 60, NW
Riceville, PA, 130, NW
Richburg, NY, 113, SE

Richfield Jct., NY, 100, NE
Richfield Springs, NY, 101, NW
Richford (CPR; CV), VT, 31, NE
Richford, NY, 117, NE
Richland, NY, 59, SE
Richmond, MA, 123, NW
Richmond, ME, 56, SW
Richmond, VT, 50, NW
Richmond Furnace, MA, 123, NW
Richmond Hill, NY, 159D
Richmond Summit, MA, 123, NW
Richmond Valley, NY, 158, SE
Richmondville, NY, 101, SE
Richville, NY, 45, NW
Rickers, VT, 51, SE
Rideau, ON, 43, NW
Riders, NY, 122, NE
Riderville, PA, 132, NE
Ridge Road, NY, 93B
Ridgefield, CT, 153, NW
Ridgefield, NJ, 158, NE
Ridgefield Park (NYC; NYS&W), NJ, 158, NE
Ridgeland, NY, 76A
Ridgeway, NY, 152, SW
Ridgeway, ON, 92, NE
Rifle Range, NY, 76, SE
Rigby Yard, ME, 71, SW, 71A
Rileys, ME, 36, SE
Rindgemere, NH, 90, NW
Ringwood, NJ, 151, SW
Ringwood Jct., NJ, 151, SW
Ripley, NY, 110, NE
Rippleton, NY, 99, NW
Rising, MA, 123, SW
Rista, ME, 5, NE
Ritchie, NB, 15, NW
River de Chute, NB, 6, SE
River Edge, NJ, 158, NE
River Falls, CT, 145, SE
River Forks, NY, 100, NE
River Point, RI, 146, SE
River St., MA, 126, SE
River St., MA, 127E
River St., NJ, 158, NE
Riverdale, NH, 88, SE
Riverdale, NJ, 158, NW
Rivergate, NY, 44, SW
Riverhead, NY, 161, NE
Riverhill, NH, 88, SE
Riverside, CT, 152, SE
Riverside, MA, 124, SE
Riverside, ME, 56, NE

Riverside, NJ, 158, NE
Riverside, NY, 118, SW
Riverside, NY, 159, NW
Riverside, NY, 64, SW
Riverside, RI, 147, NW
Riverside, VT, 87, SW
Riverside Jct., NY, 112, SE
Riverton, NH, 52, NW
Riverton, VT, 50, SE
Riverview, MA, 127D
Riverview-Anson, ME, 37, NW
Roaring Branch, PA, 136, SW
Robbins, ME, 1, NE
Robbins, ON, 92, NE
Roberts, MA, 127, NW
Roberts, ME, 6, NW
Robinson, ME, 42, NW
Robinson, ME, 8, NW
Robinwood, NY, 46, SW
Rochdale, MA, 126, SW
Rochelle Park, NJ, 158, NE
Rochester (B&O), NY, 76, SE, 76A
Rochester (Erie; LV; PRR), NY, 76, SE, 76A
Rochester, NH, 90, NW
Rochester (NYC), NY, 76A
Rochester, VT, 66, NW
Rochester Belt Line Jct., NY, 76A
Rochester Jct., NY, 95, NE
Rochester-State St. (NYC), NY, 76A
Rock, MA, 148, NW
Rock, NY, 45, SW
Rock Glen, NY, 94, SE
Rock Rift, NY, 119, SE
Rock Tavern, NY, 151, NE
Rockaway Park, NY, 159, SW
Rockdale, NY, 119, NW
Rockfall, CT, 144, SE
Rockingham, NH, 90, SW
Rockingham Park, NH, 108, NE
Rockland, MA, 128, SW
Rockland, ME, 57, SE
Rockland, NY, 49, NW
Rockport, MA, 109, SE
Rockport, ME, 57, SE
Rockstream, NY, 116, NW
Rockville, CT, 145, NW
Rockville, NY, 113, NE
Rockville Center, NY, 159, SE
Rockwells Mills, NY, 119, NW
Rocky Hill, CT, 144, SE

Rocky Hill, MA, 127, SW
Rocky Point, NY, 161, NW
Rodbourne, NY, 116, SE
Rogers, NY, 44, SE
Rogers, NY, 48, NE
Rogersville, NY, 95, SE
Roix Road, NB, 24, NE
Rolling Dam, NB, 24, NE
Rollins Farm, NH, 90, SW
Rollinsford, NH, 90, SW
Rollstone St., MA, 107, SW
Rome, NY, 48, NE
Rome, NY, 81, SW
Rome Jct., NY, 81, SW
Romulus, NY, 97, NW
Rondaxe, NY, 62, NW
Ronkonkoma, NY, 160, NE
Rooseveltown, NY, 27, NE
Roots, NY, 44, SW
Rosborough, NB, 15, NE
Roscoe, NY, 140, NW
Rose Hill, NY, 98, NW
Rose Lake, PA, 134, NW
Rosebank, NY, 158, SE
Rosedale, NY, 159, SE
Roseland, NJ, 158, NW
Roselle, NJ, 158, SW
Roselle Park, NJ, 158, SW
Rosendale, NY, 141, NE
Roseton, NY, 142, SW
Roseville Ave., NJ, 158, NE
Rosiere, NY, 43, SE
Roslindale, MA, 127D
Roslyn, NY, 159, NE
Ross' Crossing, NY, 95, SW
Ross Run Jct., PA, 131, SW
Rossburg, NY, 113, NE
Rotterdam Jct., NY, 102, NE
Rotterdam Yard, NY, 102, NE
Roulette, PA, 133, NE
Round Lake, NY, 103, NW
Round Top, PA, 135, SW
Roundout, NY, 142, NW
Rouses Point, NY, 30, NW
Rowayton, CT, 153, SW
Rowena, NB, 6, NE
Rowes Corner, NH, 89, SW
Rowley, MA, 109, SW
Roxbury, CT, 143, SW
Roxbury, MA, 127E
Roxbury, ME, 35, SE
Roxbury, NY, 120, NE
Roxbury, VT, 50, SE

Royal, MA, 149, SW
Royalston, MA, 106, SE
Royalton, VT, 66, NE
Roys, NJ, 150, SE
Rumford, ME, 35, SE
Rumford, RI, 147, NW
Rumford Jct., ME, 55, SW
Rummerfield, PA, 137, SW
Rumney, NH, 68, NW
Rupert, VT, 85, NE
Rush, NY, 95, NE
Rushville, NY, 96, NE
Russell, MA, 124, SW
Russell, PA, 131, NE
Russell City, PA, 132, SW
Russells, NY, 98, NW
Russia, NY, 29, SW
Rutherford, NJ, 158, NE
Rutland (R), VT, 66, SW
Rutland, MA, 126, NW
Rutland-West St. (D&H), VT, 66, SW
Rye (NH; NYW&B), NY, 159, NE

Sabattus, ME, 55, SE
Sabinsville, PA, 134, NE
Sacandaga Park, NY, 83, SE
Sacandaga Siding, NY, 84, NW
Sackets Harbor, NY, 59, NE
Saco (E), ME, 71, SW
Saco, MA, 127D
Saco, ME, 91, NW
Sag Harbor, NY, 162, NW
Sagamore, MA, 148, NE
Sailors Snug Harbor, NY, 158, SE
Salamanca (B&O), NY, 112, SE
Salamanca (Erie), NY, 112, SE
Salamanca (PRR), NY, 112, SE
Salem, MA, 109, SW
Salem, ME, 36, NW
Salem, NH, 108, NE
Salem, NY, 85, SW
Salem Jct., MA, 108, SE
Salisbury, CT, 143, NW
Salisbury, MA, 109, NW
Salisbury, NY, 159, SE
Salisbury, VT, 65, NE
Salisbury Center, NY, 82, SW
Salisbury Mills, NY, 151, NE
Salmon Falls (E), ME, 90, SW
Salmon Falls (W), NH, 90, SW
Salmon Falls, ME, 24, SW
Salt Point, NY, 142, NW

Index of Passenger and Non-passenger Stations / 245

Sanborn, NY, 74, SW
Sanborns, ME, 72, NW
Sanbornville, NH, 69, SE
Sand Pit, ME, 12, SE
Sanders, ME, 36, NW
Sandersdale, MA, 125, SE
Sandusky, NY, 113, NW
Sandwich, MA, 149, NW
Sandy Creek, ME, 54, SE
Sandy Hook, CT, 153, NW
Sandy Point, ME, 39, SW
Sandy River, ME, 35, NE
Sanford & Springvale, ME, 90, NW
Sanford Lake Mines, NY, 47, SE
Sangerville, ME, 20, SW
Sanitaria Springs, NY, 118, SW
Santa Clara, NY, 28, SW
Saranac Inn, NY, 47, NW
Saranac Lake, NY, 47, NE
Saranac Lake Jct., NY, 47, NE
Saratoga Lake, NY, 84, SE
Saratoga Springs, NY, 84, SW
Sardinia, NY, 93, SE
Sargent, NH, 88, SE
Saugerties, NY, 122, SW
Saugus, MA, 127C
Saundersville, MA, 126, SE
Sauquoit, NY, 81, SW
Savannah, NY, 78, SW
Savona (DL&W; Erie), NY, 115, NE
Sawyers, NH, 90, SW
Sawyers River, NH, 53, SW
Saxonville, MA, 127, NW
Saylesville, RI, 147B
Sayre, PA, 136, NE
Sayville, NY, 160, SE
SC Jct., PA, 136, NE
Scantic, CT, 144, NE
Scarboro Beach, ME, 71, SW
Scarborough, NY, 152, SW
Scarsdale, NY, 152, SW
Scarsdale, NY, 159, NW
Schaghticoke, NY, 103, NE
Schenectady, NY, 103, NW, 103B
Schenevus, NY, 101, SW
Schodack Landing, NY, 122, NW
Schoharie, NY, 102, SW
Schoharie Jct., NY, 102, SW
Scholes, NY, 114, NW
Schoodic, ME, 21, NW
Schoodic, ME, 40, SE

Schoodic Stream Jct., ME, 12, SE
Schuyler Jct., NY, 84, SE
Schuylerville, NY, 84, SE
Scio (Erie; W&B), NY, 114, SW
Sciota, NY, 29, NE
Scituate, MA, 128, SE
Scotchman's Cut, NY, 142, SW
Scotia, NY, 103, NW
Scott, ME, 14, NE
Scott, NH, 52, NE
Scotts, ME, 6, SW
Scotts, NH, 52, NE
Scottsville, NY, 76, SW
Scottsville, NY, 95, NW
Sea Cliff, NY, 159, NE
Sea St., ME, 42, NE
Sea View, MA, 128, SE
Seabury, ME, 90, SE
Seaford, NY, 160, SW
Searsport, ME, 58, NW
Seaside, NY, 159, SW
Sebago Lake, ME, 70, NE
Sebattis, NY, 46, SE
Seboois, ME, 21, NE
Secaucus, NJ, 158A
Secaucus Yard, NJ, 158A
Sedgwick Ave., NY, 159, NW
Seeley Creek, NY, 116, SW
Seeleyville, PA, 139, SW
Selkirk, NY, 103, SW
Selkirk Yard, NY, 103, SW
Seneca Castle, NY, 96, NE
Seneca Falls, NY, 97, NW
Seneca Jct., NY, 97, NW
Seneca Mills, NY, 96, SE
Seneca River, NY, 78, SE
Sergeant, PA, 132, SW
Setauket, NY, 160, NE
Severance, NH, 89, SW
Seward, NY, 101, SE
Sewaren, NJ, 158, SW
Seymour, CT, 153, NE
Shady Hill, MA, 127, NW
Shaftsbury, VT, 85, SE
Shaker, CT, 124, SE
Shandaken, NY, 121, SW
Shannock, RI, 156, NE
Sharon, MA, 127, SE
Sharon, NY, 142, NE
Sharon, VT, 67, NW
Sharon Center, PA, 133, NE
Sharon Heights, MA, 127, SE
Sharon Springs, NY, 101, NE

Sharps, ME, 8, NW
Sharp's Mill, ME, 5, NW
Shattucks, ME, 56, SE
Shavertown, NY, 120, SW
Shaw, ME, 5, NE
Shaw, NY, 113, SW
Shawmut, MA, 127E
Shawmut, ME, 37, SE
Shawmut Jct., MA, 127E
Shawsheen, MA, 108, SE
Sheds Corners, NY, 99, NW
Sheepscot, ME, 56, SE
Sheffield, MA, 123, SW
Sheffield (S&T; PRR; TIV), PA, 131, SE
Sheffield Ave., NJ, 159, NW
Sheffield Farms Creamery, NY, 120, NW
Sheffield Jct., PA, 132, SW
Shekomeko, NY, 142, NE
Shelburne, NH, 53, NE
Shelburne, VT, 49, NE
Shelburne Falls, MA, 105, SE
Shelburne Falls East, MA, 105, SE
Shelburne Falls West, MA, 105, SE
Sheldon, VT, 31, NW
Sheldon Jct. (CV; SJL) VT, 31, NW
Sheldon Springs, VT, 31, NW
Sheldrake Springs, NY, 97, SW
Shelton, CT, 153, NE
Shepaug, CT, 153, NW
Sherborn, MA, 127, SW
Sherburne, NY, 99, SE
Sherburne Four Corners, NY, 99, SE
Sheridan, ME, 5, SW
Sheridan, NY, 111, NE
Sherks, ON, 92, NE
Sherman, ME, 13, NW
Sherman, NY, 110, SE
Sherman, NY, 49, SW
Sherrill, NY, 80, SE
Shetucket, CT, 145, SE
Shields, NY, 116, SE
Shinglehouse, PA, 133, NE
Shinhopple, NY, 119, SE
Shinnecock Hills, NY, 162, NW
Shirley, MA, 107, SE
Shirley, ME, 19, NE
Shogomoc, NB, 15, NW

Shohola, PA, 150, NW
Shongo, NY, 113, NE
Shongo, NY, 114, SW
Shore Line Crossing, NB, 24, NE
Shoreham, NY, 161, NW
Shoreham, VT, 65, NW
Shoreys, ME, 7, SE
Short Falls, NH, 89, SW
Short Hills, NJ, 158, SW
Shortsville, NY, 96, NE
Shunpike, NY, 142, NE
Shurtleffs, NY, 46, NE
Shushan, NY, 85, SW
Shuy, ME, 55, NE
Siasconset, MA, 157B
Siberia, ME, 13, NW
Siding 4, NY, 141, NE
Sidney (D&H; NYO&W), NY, 119, NW
Siegas (CN; CPR), NB, 3, SW
Sierks, NY, 94, NW
Silver Creek, NY, 92, SE
Silver Hill, MA, 127, NW
Silver Lake, MA, 108, SE
Silver Lake, MA, 128, SW
Silver Lake, NJ, 158, NE
Silver Lake, NY, 94, SE
Silver Lake Jct., NY, 94, SE
Silver Springs, NY, 94, SE
Silverdale, ON, 73, SW
Silvernails, NY, 122, SE
Silvers Mills, ME, 20, SE
Simonton Corner, ME, 57, SE
Simsbury, CT, 144, NW
Sinclairville, NY, 111, NW
Singac, NJ, 158, NE
Sink Hole, PA, 138, NE
Sisson, NY, 27, SW
Sizerville, PA, 133, SE
Skaneateles, NY, 98, NW
Skaneateles Falls, NY, 98, NW
Skaneateles Jct., NY, 79, SW
Skensowane, NY, 62, NW
Skerry Brook, ME, 5, SW
Skinner, ME, 9, SE
Skinners, NY, 139, SE
Skinners Eddy, PA, 137, SE
Skowhegan, ME, 37, NE
Slate Hill, NJ, 151, NW
Slater, RI, 147, NW
Slatersdale, RI, 146, NE
Sloatsburg, NY, 151, SE
Slocums, RI, 146, SE

Smethport (B&O; PRR; PS&N), PA, 133, NW
Smith, NY, 81, NW
Smithboro (Erie; LV), NY, 117, SW
Smithfield, RI, 146, NE
Smith's, MA, 125, NW
Smith's Basin, NY, 85, NW
Smith's Ferry, MA, 124, NE
Smith's Mills, NY, 111, NE
Smith's Point, NH, 69, SW
Smith's Run, PA, 132, SE
Smithtown, NY, 160, NE
Smyrna, NY, 99, SE
Smyrna Mills, ME, 7, SE
Snyder, NY, 117, SW
Soapstone, MA, 105, SW
Soapstone, MA, 125, NW
Sockanossett, RI, 147, NW
Sodus, NY, 77, SE
Sodus Center, NY, 77, SE
Sodus Point, NY, 78, NW
Soldier Pond, ME, 1, SE
Solon, ME, 37, NW
Solon, NY, 98, SE
Solsville, NY, 99, NE
Solvay, NY, 79, SE
Somerset, MA, 147, NE
Somerset, ME, 11, SW
Somerset Jct., MA, 147, NE
Somerset Jct., ME, 11, SW
Somersworth, NH, 90, NW
Somerville, MA, 127A
Somerville Highlands, MA, 127A
Somerville Jct., MA, 127A
Sonyea, NY, 95, SW
South Acton, MA, 127, NW
South Albion, ME, 38, SW
South Amsterdam, NY, 102, NE
South Ashburnham, MA, 107, SW
South Athol, MA, 106, SW
South Barre, MA, 125, NE
South Barre, VT, 50, SE
South Bay, NY, 65, SW
South Bay, NY, 80, SE
South Beach, MA, 157A
South Beach, NY, 158, SE
South Beach Jct., MA, 157A
South Bedford, NH, 107, NE
South Bennington, NH, 107, NW
South Berwick, ME, 90, SW
South Bethlehem, NY, 103, SW
South Billerica, MA, 108, SW

South Bolton, MA, 126, NE
South Bombay, NY, 27, NE
South Boston, MA, 127A
South Bound Brook, NJ, 158, SW
South Braintree, MA, 127, SE
South Branch, PA, 137, SW
South Brewer, ME, 39, NW
South Bridgewater, MA, 148, NW
South Britain, CT, 153, NE
South Brookline, NH, 107, SE
South Byron, NY, 75, SE
South Cairo, NY, 122, NW
South Cambridge, NY, 104, NW
South Charlestown, NH, 87, SW
South Chatham, MA, 149, SE
South China, ME, 56, NE
South Clinton, MA, 126, NE
South Clyde, NY, 78, SW
South Columbia, NY, 100, NE
South Corinth, NY, 84, SW
South Coventry, CT, 145, NW
South Dayton, NY, 111, NE
South Deerfield, MA, 124, NE
South Dennis, MA, 149, SE
South Easton, MA, 127, SE
South Edmeston, NY, 100, SW
South Elizabeth, NJ, 158, SE
South Farmingdale, NY, 160, SW
South Fort Plain, NY, 101, NE
South Franklin, VT, 31, NW
South Gilboa, NY, 120, NE
South Granby, NY, 79, NW
South Hanson, MA, 128, SW
South Harwich, MA, 149, SE
South Hero, VT, 30, SW
South Jct., NY, 30, SW
South Keene (Spur), NH, 106, NE
South Kortright, NY, 120, NE
South La Grange, ME, 21, SW
South Lawrence, MA, 108, SE
South Lima, NY, 95, NE
South Little Falls, NY, 82, SW
South Livonia, NY, 95, NE
South Londonderry, VT, 86, SW
South Lowell, MA, 108, SW
South Lyndeboro, NH, 107, NW
South Manchester, CT, 144, NE
South Manchester, NH, 108, NW, 108A
South Merrimack, NH, 107, NE
South Middleboro, MA, 148, NW
South Middleton, MA, 108, SE
South Milford, NH, 107, NE

South Monson, MA, 125, SW
South Montrose, PA, 138, NW
South Nashua, NH, 108, SW
South New Berlin, NY, 100, SW
South Norwalk, CT, 153, SW
South Nyack, NY, 152, SW
South Orange, NJ, 158, SW
South Orangeburg, NJ, 150, SE
South Palmyra, NY, 77, SE
South Paris, ME, 54, SE
South Patterson, NJ, 158, NE
South Peabody, MA, 109, SW
South Plainfield, NJ, 158, SW
South Portland, ME, 71, SE, 71A
South Presque Isle, ME, 6, SW
South Rangeley, ME, 35, NW
South River, MA, 105, SE
South Royalton, VT, 66, NE
South Ryegate, VT, 51, SE
South Schenectady (D&H), NY, 102, NE
South Schenectady (NYC), NY, 103, NW, 103B
South Sebec, ME, 20, SE
South Shaftsbury, VT, 104, NE
South Spencer, MA, 125, SE
South Stoughton, MA, 127, SE
South Strong, ME, 36, NE
South Sudbury, MA, 127, NW
South Swansea, MA, 147, SE
South Switch, ME, 39, NW
South Unadilla, NY, 119, NW
South Union, ME, 57, SW
South Utica, NY, 81, SW, 81A
South Uxbridge, MA, 126, SE
South Vandalia, NY, 112, SE
South Wales, NY, 93, SE
South Wallingford, VT, 86, NW
South Walpole, MA, 127, SW
South Wareham, MA, 148, NE
South Weymouth, MA, 128, SW
South Willington, CT, 145, NW
South Wilmington, MA, 108, SE
South Windham, CT, 145, SE
South Windham, ME, 71, SW
South Woodbury, VT, 51, NW
South Worcester, MA, 126, SW, 126A
Southampton, MA, 124, SE
Southampton, NY, 162, NW
Southboro, MA, 126, NE
Southbridge, MA, 125, SE
Southbury, CT, 153, NE

Southern Inlet, ME, 42, NW
Southfields, NY, 151, SE
Southington, CT, 144, SW
Southington Road, CT, 144, SW
Southold, NY, 155, SW
Southport, CT, 153, SW
Southport, NY, 116, SW
Southport Jct., NY, 116, SW
Southville, MA, 126, NE
Southwick, MA, 124, SW
Sparkill, NY, 152, SW
Sparrowbush, NY, 150, NE
Sparta, NJ, 150, SE
Sparta Jct. (L&HR), NJ, 150, SE
Spartansburg, PA, 130, NE
Spaulding, ME, 5, NE
Spears, ME, 57, SW
Spellmans, NY, 30, NW
Spencer, NY, 116, SE
Spencer, NY, 117, SW
Spencerport, NY, 76, SW
Spencerville, ON, 25, NE
Speonk, NY, 161, NE
Sprague's, MA, 104, SE
Sprakers, NY, 101, NE
Spring Brook, CT, 144, SE
Spring Brook, NY, 93, NE
Spring Cove, NY, 28, SW
Spring Creek, PA, 130, NE
Spring Farm, ME, 18, SE
Spring Glen, NY, 141, SW
Spring St., MA, 127D
Spring St., NJ, 158, SE
Spring Valley, NY, 151, SE
Springdale, CT, 152, SE
Springdale Cemetery, CT, 152, SE
Springfield, MA, 124, SE, 124A
Springfield, NJ, 158, SW
Springfield, VT, 87, NW
Springfield Gardens, NY, 159, SW
Springfield Station, NH, 87, NW
Springsville, NY, 93, SE
Springville, PA, 138, SW
Springville Jct., NY, 93, SE
Springwater, NY, 95, SE
Spuyten Duyvil, NY, 159, NW
Squa Pan, ME, 5, SW
Squanocook Jct., MA, 107, SE
Squaw Brook, ME, 11, SE
St. Albans, NY, 159D
St. Albans, VT, 30, NE
St. Andrews, NB, 24, SE
St. Andrews North, NB, 24, SE

St. Basil, NB, 2, NE
St. Catherines, ON, 73, SW
St. Cecile, QC, 9, SW
St. Croix, ME, 7, NE
St. Croix Jct., ME, 24, SW
St. David, ME, 2, NE
St. David, ON, 73, SE
St. Elmo, NY, 141, SE
St. Evariste, QC, 9, NW
St. Francis, ME, 1, SW
St. Froid, ME, 4, NE
St. George, NY, 158, SE
St. Isadore, QC, 16, SW
St. James, NY, 160, NE
St. Jaques, NB, 2, NW
St. John, ME, 1, SW
St. Johns Park, NY, 158, SE
St. Johnsbury, VT, 51, NE
St. Johnsville, NY, 101, NE
St. Joseph's, NY, 140, SE
St. Leonard (CN; CPR), NB, 3, SW
St. Luce, ME, 2, NW
St. Regis Falls, NY, 27, SE
St. Samuel, QC, 9, SW
St. Sebastien, QC, 9, NW
St. Stephen (CPR), NB, 24, SW
Staatsburg, NY, 142, NW
Stacyville, ME, 13, NW
Stafford, CT, 145, NW
Stafford, NY, 94, NE
Stafford, VT, 86, NW
Stafford Bridge, NY, 84, SE
Stamford, CT, 152, SE
Stamford, NY, 120, NE
Stamford, ON, 73, SE
Stanard, NY, 114, SW
Standing Stone, PA, 137, SW
Standish, MA, 128, SE
Standish, MA, 147, NE
Standish, NY, 29, SW
Stanfordville, NY, 142, NE
Stanhope, QC, 33, NW
Stanley, MA, 128, SW
Stanley, NY, 96, NE
Stapleton, NY, 158, SE
Starbirds, ME, 20, SE
Starbrick (NYC; PRR), PA, 131, NE
Stark, NH, 34, SW
Starkey, NY, 97, SW
Starlight, PA, 132, SE
Starlight, PA, 139, NW

Starrucca, PA, 139, NW
State Bridge, NY, 80, SE
State Hospital, NY, 142, SE
State Line, CT, 143, NW
State Line, MA, 123, NW
State Line, MA, 125, SW
State Line, NH, 106, SE
State Line, PA, 110, SW
State Line, VT, 65, SW
State Road, ME, 5, SE
State School, NY, 142, NE
State School, NY, 151, NW
State St., CT, 144A
Steamburg, NY, 112, SW
Stebbins, ME, 6, NW
Steeles, CT, 145, NW
Steep Falls, ME, 70, NE
Stellaville, NY, 45, NE
Stelton, NJ, 158, SW
Stephentown, NY, 104, SW
Stepney, CT, 153, NE
Sterling, CT, 146, SW
Sterling, MA, 126, NW
Sterling, NY, 78, NE
Sterling Forest, NJ, 151, SW
Sterling Jct., MA, 126, NW
Sterlington (Erie; SMR), NY, 151, SE
Stevens, MA, 108, SE
Stevens, VT, 33, SE
Stevens, VT, 33, SW
Stevens Point, PA, 138, NE
Stevenson, CT, 153, NE
Stevenson, PA, 137, SW
Stevensville, ON, 92, NE
Stewart Manor, NY, 159, SE
Stickney, NB, 8, NE
Stickneys, NY, 115, NW
Stiles, NY, 79, SW
Still River, MA, 126, NE
Stillwater (BO&M; MEC), ME, 39, NE
Stillwater, NJ, 150, SW
Stillwater, NY, 103, NE
Stillwater Centre, NY, 103, NE
Stimsons, ME, 5, SW
Stirling, NJ, 158, SW
Stissing, NY, 142, NE
Stissing Jct., NY, 142, NE
Stittville, NY, 81, SW
Stockbridge, MA, 123, NW
Stockbridge, VT, 66, NW
Stockholm, ME, 2, SE

Stockholm, NJ, 150, SE
Stockport, NY, 122, NW
Stockton, ME, 58, NW
Stokesdale, PA, 135, NW
Stone, PA, 135, SW
Stone Haven, MA, 127, SE, 127F
Stoneham (PRR; TIV), PA, 131, NE
Stoneham, MA, 127C
Stonington, CT, 156, NW
Stonington Wharf, CT, 156, NW
Stony Beach, MA, 128, NW
Stony Brook, MA, 127, NW
Stony Brook, NY, 160, NE
Stony Brook Glen, NY, 95, SE
Stony Creek, CT, 154, NE
Stony Creek, NY, 84, NW
Stony Ford, NY, 151, NW
Stony Hollow, NY, 141, NE
Stony Point, NY, 152, SW
Stormville, NY, 142, SE
Stoughton Jct., MA, 127, SE
Stowell, NH, 107, NE
Straiton Ave., NY, 159, SW
Stratford, CT, 153, SE
Stratham, NH, 90, SW
Strattons, CT, 144, NW
Stricklands, ME, 55, NE
Strong, ME, 36, NE
Stroughs, NY, 44, SW
Struthers, PA, 131, NE
Stuyvesant, NY, 122, NW
Submarine Base, CT, 155, NE
Sudbury, MA, 127, NW
Suffern, NY, 151, SE
Sugar Brook, NB, 15, SW
Sugar Hill, NH, 52, SW
Sugar Loaf, NY, 151, NW
Sugar Run, PA, 132, NW
Sullivan, NY, 80, SW
Summerdale, NY, 110, SE
Summit (DL&W; RV), NJ, 158, SW
Summit, CT, 143, NE
Summit, CT, 144, SW
Summit, MA, 126, NE
Summit, MA, 126, NW
Summit, ME, 12, NE
Summit, ME, 35, NE
Summit, ME, 36, NE
Summit, NH, 106, NW
Summit, NH, 68, SW
Summit, NY, 113, SE

Summit, NY, 114, SW
Summit, NY, 117, NW
Summit, NY, 118, NE
Summit, NY, 62, NW
Summit, NY, 94, NW
Summit, RI, 146, SE
Summit, VT, 32, SE
Summit, VT, 33, NW
Summit, VT, 86, NW
Summit Park, NY, 151, SE
Summit Siding, MA, 126, NW
Summit Siding, NH, 90, NW
Summitville, NY, 141, SW
Summit—Young's Gap, NY, 140, NE
Sunapee, NH, 87, NE
Suncook, NH, 89, SW
Sunderland, VT, 85, SE
Sundown, NH, 108, NE
Sunnyside Jct., NY, 159B
Surfside, MA, 128, NW
Surfside, MA, 157B
Suspension Bridge (Erie), NY, 73, SW, 73A
Suspension Bridge (NYC), NY, 73, SE, 73A
Suspension Bridge Yard (LV), NY, 73, SW, 73A
Suspension Bridge Yard (NYC), NY, 73, SW, 73A
Susquehanna, PA, 138, NE
Susquehanna Transfer (NNJ; NYS&W), NJ, 158
Sussex (LNE; NYS&W), NJ, 150, SE
Sussex Jct., NJ, 150, SE
Sutton, VT, 32, SE
Swainboro, NH, 68, NW
Swains (CNYW; Erie; PS&N), NY, 114, NW
Swampscott, MA, 128, NW
Swanton (CV; SJL), VT, 30, NE
Swanzey, NH, 106, NW
Swartswood, NJ, 150, SW
Swartswood, NY, 116, SE
Swartswood Jct., NJ, 150, SW
Sweden, ME, 5, NE
Sweets, NY, 100, SW
Sycamore St. Jct., NY, 93C
Sylvan Beach (LV; NYC), NY, 80, SE
Sylvan Jct., NY, 80, SE
Sylvan Lake, NY, 142, SE

Syosset, NY, 159, NE
Syracuse (DL&W), NY, 79, SE, 79A
Syracuse, Erie Boulevard (NYC), NY, 79, SE, 79A
Syracuse, Fayette St. (NYC), NY, 79A
Syracuse, WS (NYC), NY, 79A

Taconic Siding, NY, 104, SW
Taftville, VT, 67, SW
Tahawus, NY, 63, NE
Tainters, PA, 132, NE
Talbot, MA, 126, NE
Talcottville, CT, 145, NW
Talcville, NY, 45, NW
Talcville, VT, 66, NW
Tallmans, NY, 151, SE
Tally Ho, PA, 132, NE
Talmadge Hill, CT, 153, SW
Tanners, NY, 122, SE
Tannersville, NY, 121, SE
Tannery, ME, 7, SW
Tapleyville, MA, 109, SW
Tappan (NNJ; NYC), NY, 152, SW
Tarbell, NH, 107, NW
Tariffville, CT, 144, NW
Tarkiln, RI, 146, NE
Tarrantine, ME, 11, SW
Tarrytown, NY, 152, SW
Tarrytown Heights, NY, 152, SW
Taughannock Falls, NY, 97, SE
Taunton, MA, 147, NE, 147C
Taunton, ME, 11, SW
Taylor, NY, 95, NW
Teaneck, NJ, 158, NE
Teed's Mill, NB, 8, SE
Temple, NB, 15, NW
Templeton, MA, 106, SE
Tenafly, NJ, 159, NW
Terminus, NY, 159, SE
Terrace Hill, NH, 69, SW
Terryville, CT, 144, SW
Tewksbury Centre, MA, 108, SE
Tewksbury Jct., MA, 108, SE
Texas, MA, 126, SW
Thamesville, CT, 145, SE
Thayer, MA, 126, NE
Thayers Mills, NH, 53, NW
The Falls, ME, 40, SE
The Glen, NY, 64, SW
The Heights, ME, 19, SW
The Raunt, NY, 159, SW

The Wye, ME, 55, NW
Thendara, NY, 61, SE
Theresa, NY, 44, SW
Theriault, NB, 2, NE
Thetford, VT, 67, NE
Thiells, NY, 151, SE
Thomaston, CT, 143, SE
Thomaston, ME, 57, SE
Thompson, CT, 146, NW
Thompson, PA, 138, NE
Thompson Ridge, NY, 141, SW
Thompsons, ME, 38, NW
Thompsons, NY, 78, SW
Thompsonville, CT, 144, NE
Thomson, NY, 84, SE
Thorndike, MA, 125, SW
Thorndike, ME, 38, SW
Thornton, MA, 128, NW
Thornton, NH, 68, NE
Thornton's Ferry, NH, 108, NW
Thornwood, NY, 152, SW
Thorold, ON, 73, SE
Three Mile Bay, NY, 43, SE
Three River Point, NY, 79, SW
Three Rivers, MA, 125, SW
Throop, NY, 97, NE
Thurman, NY, 84, NW
Tiadaghton, PA, 135, SW
Ticonderoga, VT, 65, NW
Ticonderoga Jct., VT, 65, NW
Tidioute, PA, 131, SW
Tie Plant, NH, 108, NW
Tiffany, PA, 138, NW
Tifft Jct. (LV), NY, 93C
Tifft Terminal (LV), NY, 93A
Tifft Yard (NKP), NY, 93A
Tilly Foster, NY, 152, NE
Tilton, NH, 88, NE
Timoney, ME, 7, SE
Tin Bridge, PA, 136, NW
Tinker, NB, 6, NW
Tioga (Erie; NYC), PA, 135, NE
Tioga Center (Erie; LV), NY, 117, SW
Tioga Jct., PA, 135, NE
Tiona (PRR; TIV), PA, 131, SE
Tip Top, NY, 114, SW
Tirrell Hill, NH, 107, NE
Titusville (NYC; PRR), PA, 130, SE
Tiverton, RI, 147, SE
Tivoli, NY, 122, SW
Tobique Narrows, NB, 6, NE

Todds Farm, ME, 14, SE
Togus, ME, 56, NE
Tomah, ME, 14, SE
Tomkins, PA, 135, NE
Tomkins Cove, NY, 152, NW
Tomkinsville, NY, 158, SE
Tonawanda, NY, 74, SW, 74A
Tonawanda Jct., NY, 74, SW, 74A
Topsfield, MA, 109, SW
Topsham, ME, 72, NW
Torpedo, PA, 131, NW
Torrington, CT, 143, NE
Totmans, ME, 37, NW
Tottenville, NY, 158, SW
Touisset, MA, 147, SE
Towaco, NJ, 158, NW
Towanda, PA, 137, NW
Towanda-Washington St., PA, 137, NW
Towantic, CT, 153, NE
Tower Hill, MA, 127, NW
Tower Hill, NY, 152, SW
Tower Hill, NY, 158, SE
Town Line, NY, 93, NE
Towers (NH; NYC), NY, 152, NE
Townley, NJ, 158, SE
Townsend, MA, 107, SE
Townsend Harbor, MA, 107, SE
Townshend, VT, 86, SE
Trafton, ME, 5, SW
Transit, NY, 74, SE
Transit Bridge, NY, 113, NE
Trap Rock, MA, 124, SE
Trap Rock, ME, 5, SE
Tremley, NJ, 158, SE
Tremont, MA, 148, NW
Tremont, NY, 159, NW
Trenton Chasm, NY, 81, NE
Trenton Falls, NY, 81, NE
Tribes Hill, NY, 102, NW
Trinity Place (B&A), MA, 127A
Trionda, NY, 84, SE
Trout Brook, NY, 139, NE
Troutdale, ME, 19, NW
Trowbridge, PA, 136, NW
Troy, NH, 106, NE
Troy, PA, 136, NW
Troy Union Station, NY, 103, SE, 103A
Troy Yard (B&M), NY, 103A
Trudell, QC, 9, SW
Truemans, PA, 131, SE
Trumansburg, NY, 97, SE

Trumbull, CT, 153, SE
Trunkeyville, PA, 131, SW
Truro, MA, 149, NE
Truxton, NY, 98, SE
Tryonville, PA, 130, SW
Tuckahoe, NY, 159, NW
Tully, NY, 98, NE
Tunk Lake, ME, 40, SE
Tunkhannock (LV), PA, 138, SW
Tunnel, NY, 118, SE
Tunnel Switch, VT, 87, SW
Tupper, VT, 66, NW
Tupper Lake, NY, 47, SW
Turkey Brook, CT, 153, NE
Turners Falls Jct., MA, 105A
Turnpike, MA, 108, SW
Turtle Point, PA, 133, NW
Tuscarora, NY, 95, SW
Tusten, NY, 139, SE
Tuxedo, NY, 151, SE
Twin Lakes, CT, 123, SW
Twin Mountain, NH, 52, NE
Tyler, NH, 88, NE
Tyler City, CT, 154, NW
Tyngsboro, MA, 108, SW
Tyter, MA, 106, SW
Tyter West, MA, 106, SW

Ulster, PA, 136, NE
Ulster Park, NY, 142, NW
Ulysses (CPA; NYC), PA, 134, NW
Unadilla, NY, 119, NW
Uncas Road, NY, 62, NW
Underhill, VT, 31, SW
Union, ME, 57, SW
Union, NH, 89, NE
Union City (Erie; PRR), PA, 130, NW
Union Grove, NY, 120, SW
Union Hall St., NY, 159D
Union Hill, NY, 77, SW
Union Market, MA, 127D
Union Springs, NY, 97, NE
Union Square, MA, 127A
Union Village, RI, 146, NE
Uniondale, PA, 139, SW
Unionville, CT, 144, NW
Unionville, MA, 127, SW
Unionville, ME, 41, SW
Unionville, NY, 103, SW
Unionville, NY, 150, NE
Unionville, NY, 27, SW
Unity, ME, 38, SW

University Heights, NY, 159, NW
Uphams Corner, MA, 127E
Upper Kent, NB, 6, SE
Upper Montclair, NJ, 158, NE
Upper Switch, NY, 117, NW
Upper Woodstock, NB, 8, SE
Upton, MA, 126, SE
Upton, NY, 161, NW
Upton Lake, NY, 142, NW
Uptonville, NY, 76, SE
Ushers, NY, 103, NW
Utica (Genesee St.), NY, 81, SE, 81A
Utica, NY, 81, SE, 81A
Utica (Washington St.), NY, 81, SE, 81A
Uxbridge, MA, 126, SE

Vail Mills, NY, 83, SE
Vails Gate Jct., NY, 151, NE
Valcour, NY, 30, SW
Valhalla, NY, 152, SW
Valley, NB, 8, SE
Valley Cottage, NY, 152, SW
Valley Falls, NY, 103, NE
Valley Falls, RI, 147, NW, 147B
Valley Jct., NY, 150, NE
Valley Mills, NY, 80, SE
Valley Stream, NY, 159, SE
Valley View, NY, 96, SW
Valois, NY, 97, SW
Van Buren, ME, 3, SW
Van Buren (NYC), NY, 111, NW
Van Buren (PRR), NY, 111, NW
Van Buren Pit, ME, 3, SW
Van Cortlandt, NY, 159, NW
Van Deusenville, MA, 123, SW
Van Etten, NY, 116, SE
Van Etten Jct., NY, 116, SE
Van Fleet, NY, 115, SW
Van Hoesen, NY, 103, SE
Van Nest, NY, 159, NW
Van Nest Shops, NY, 159, NW
Van Nostrand Ave., NJ, 158A
Van Wagner's, NY, 142, SW
Van Wagners, NY, 142, SW
Vanceboro, ME, 15, SW
Vandalia, NY, 112, SE
Vandeveer Park, NY, 159, SW
Vanity Fair, RI, 147, NW
Varna, NY, 117, NW
Varysburg, NY, 94, NW

Vassalboro, ME, 56, NE
Veazie, ME, 39, NE
Veneer, NB, 3, NE
Verbank, NY, 142, SE
Verbank Village, NY, 142, SE
Vergennes, VT, 49, SE
Vermontville, NY, 47, NE
Vernon, CT, 145, NW
Vernon, NJ, 151, SW
Vernon, NY, 80, SE
Vernon, VT, 105, NE
Verona, NJ, 158, NW
Verona, NY, 80, SE
Versailles, CT, 145, SE
Versailles Road, NY, 76, SE
Vestal, NY, 117, SE
Victor, NY, 96, NW
Victoria Park, ON, 73, SE, 73A
Victory, VT, 33, SW
Victory Mills, NY, 84, SE
Vienna, NY, 80, SE
Vineland, ON, 73, SW
Violette, ME, 3, SW
Violette Brook, NB, 3, NE
Voorheesville, NY, 103, SW
Vosburg, PA, 137, SE
Vreeland Ave., NJ, 158, NE

Waban, MA, 127D
Waddington, NY, 26, NE
Wade Road, ME, 5, NE
Wadhams, NY, 49, SW
Wading River, NY, 161, NW
Wadsworth, NY, 95, NE
Wadsworth Jct., NY, 95, NE
Wainfleet, ON, 92, NW
Wainscott, NY, 162, NW
Waites Crossing, NY, 92, SW
Wakefield, NH, 69, SE
Wakefield, NY, 159, NW
Wakefield, RI, 156, NE
Wakefield Jct., MA, 127C
Walden, NY, 141, SE
Walden, VT, 51, NW
Waldo, ME, 38, SE
Waldoboro, ME, 57, SW
Waldwick, NJ, 151, SE
Wales, ME, 55, SE
Wales, NY, 93, SE
Walker, ME, 5, SE
Walker, NY, 76, NW
Walkill, NY, 141, SE
Wall St., CT, 153, SW

Wallace (DL&W; Erie), NY, 115, NW
Wallagrass, ME, 1, SE
Wallingford, CT, 154, NW
Wallingford, VT, 86, NW
Wallington, NY, 77, SE
Walloomsac, NY, 104, NW
Wallum Lake, RI, 126, SW
Walmore, NY, 74, SW
Walnut Hill, MA, 127B
Walnut St., CT, 144A
Walnut St., NJ, 158, NE
Walpole, MA, 127, SW
Walpole, NH, 87, SW
Walpole Heights, MA, 127, SW
Waltham, Bleachery, MA, 127D
Waltham, MA, 127D
Waltham (North), MA, 127B
Waltham Highlands, MA, 127B
Waltham, Newton Street, MA, 127D
Walton, CT, 123, SE
Walton, NY, 119, SE
Walton, PA, 134, NW
Walton-Bridge St., NY, 119, SE
Walworth, NY, 77, SW
Wamesit, MA, 108, SW
Wampsville, NY, 80, SE
Wanakena, NY, 46, SW
Wanaque-Midvale, NJ, 151, SW
Wantage, NJ, 150, NE
Wantagh, NY, 159, SE
Wapping, CT, 144, NE
Ward, ME, 41, SW
Ward Hill, MA, 108, SE
Wardsboro, VT, 86, SE
Ware, MA, 125, NE
Wareham, MA, 148, NE
Warehouse Point, CT, 144, NE
Warner, NH, 88, NW
Warners, NJ, 158, SE
Warners, NY, 79, SW
Warren, MA, 125, SE
Warren, ME, 57, SW
Warren, NH, 68, NW
Warren (NYC; PRR), PA, 131, NE
Warren, RI, 147, SW
Warrensburg, NY, 84, NW
Warsaw, NY, 94, SE
Warwick, NY, 151, NW
Washburn (AV; BAR), ME, 5, NE
Washburn Jct., ME, 6, SW
Washington, CT, 143, SW

Washington, MA, 123, NE
Washington, RI, 146, SE
Washington Hunt, NY, 95, SW
Washington Jct., ME, 40, SE
Washington Mills, NY, 81, SW
Washington St., MA, 128, SW
Washington St., NY, 115, NW
Washingtonville, NY, 151, NE
Wassaic, NY, 142, NE
Watchung Ave., NJ, 158, NE
Watchusett, MA, 107, SW
Waterboro, ME, 70, SE
Waterboro, NY, 111, SE
Waterbury, CT, 143, SE
Waterbury, VT, 50, NW
Waterford, PA, 130, NW
Waterloo, NH, 88, NW
Waterloo, NY, 97, NW
Watermill, NY, 162, NW
Waterport, NY, 75, NE
Waters River, MA, 109, SW
Watertown, MA, 127D
Watertown, Main St., NY, 60, NW
Watertown, NY, 60, NW
Watertown Jct., NY, 60, NW
Waterville, CT, 143, SE
Waterville, MA, 106, SE
Waterville, ME, 37, SE
Waterville, NY, 100, NE
Watervliet, NY, 103, SE, 103A
Watessing Ave., NJ, 158, NE
Watkins Glen (NYC; PRR), NY, 116, NW
Watrous, PA, 134, SE
Watson (B&O; TIV), PA, 131, SE
Watson, ME, 2, SE
Watson, NY, 117, SE
Watt, NB, 24, NE
Watts Flats, NY, 111, SW
Watuppa, MA, 147, SE
Waukeag, ME, 40, SE
Waumbec Tank, NH, 53, NW
Waumbek Jct., NH, 52, NW
Wauregan, CT, 146, SW
Wave Crest, NY, 159, SW
Waveland, MA, 128, NW
Waverly (Erie), NY, 116, SE
Waverly, MA, 127B
Waverly, PA, 136, NE
Wawarsing, NY, 141, NW
Waweig, NB, 24, SE
Wayland, MA, 127, NW
Wayland, NY, 95, SE

Waymart, PA, 139, SW
Wayne, NJ, 158, NW
Wayneport, NY, 77, SW
Wayneport Icing Station, NY, 77, SW
Wayville, NY, 84, SE
Weatogue, CT, 144, NW
Webb, NH, 106, NE
Webb Farms, NY, 142, NW
Webbs, NY, 114, NE
Webster, MA, 126, SW
Webster, ME, 19, NE
Webster, NY, 77, SW
Webster Ave., NY, 159, NW
Webster Jct., MA, 126, SW
Webster Lake, NH, 88, NE
Webster Mills, MA, 126, SW
Webster Mills, NH, 89, NW
Websters, NY, 95, SE
Websterville, VT, 51, SW
Wedgemere, MA, 127B
Wedgewood, NY, 116, NW
Weedsport, NY, 78, SE
Weehawken, NJ, 158A
Weeks Mills, ME, 56, NE
Weeksboro, ME, 7, NE
Weir Jct., MA, 147, NE, 147C
Weir Village, MA, 147, NE, 147C
Weirs, NH, 69, SW
Welland (CN), ON, 92, NW
Welland (NYC), ON, 92, NW
Welland Jct., ON, 92, NE
Wellesley, MA, 127, NW
Wellesley Farms, MA, 127, NW
Wellesley Hills, MA, 127, NW
Wellfleet, MA, 149, NE
Wellington, MA, 127C
Wells Beach, ME, 90, NE
Wells Bridge, NY, 119, NE
Wells River, VT, 51, SE
Wellsboro, PA, 135, NW
Wellsboro Jct., PA, 135, NW
Wellsburg, NY, 116, SE
Wellsville (B&O; Erie; W&B), NY, 114, SW
Wemple, NY, 103, SW
Wende, NY, 93, NE
Wendell, MA, 106, SW
Wenlock, VT, 33, NW
Wentworth, NH, 68, NW
Wentworth Ave., NY, 158, SE
Wescott, ME, 70, SE
Wesley, NY, 113, NE

West, ME, 21, SW
West Acton, MA, 127, NW
West Albany, NY, 103, SW, 103A
West Alton, NH, 69, SW
West Anburn, MA, 126, SW
West Andover, MA, 108, SE
West Arlington, NJ, 158, NE
West Athens, NY, 122, NW
West Baldwin, ME, 70, NW
West Barrington, RI, 147, SW
West Bedford, MA, 127, NW
West Bergen, NY, 75, SE
West Berlin Jct., MA, 126, NE
West Bingham, PA, 134, NW
West Bloomfield, NY, 95, NE
West Boyleston, MA, 126, NW
West Bradford, PA, 132, NE
West Bridgewater, MA, 127, SE
West Brighton, NY, 158, SE
West Brimfield, MA, 125, SW
West Brookfield, MA, 125, SE
West Burke, VT, 33, SW
West Cambridge, MA, 127B
West Cambridge, NY, 104, NW
West Camp, NY, 122, SW
West Candor, NY, 117, SW
West Caribou (AV), ME, 5, NE
West Carteret, NJ, 158, SE
West Chazy, NY, 29, NE
West Chelmsford, MA, 108, SE
West Cheshire, CT, 144, SW
West Concord, MA, 127, NW
West Cornwall, CT, 143, NW
West Cornwall, NY, 151, NE
West Danby, NY, 116, NE
West Danville, VT, 51, NE
West Davenport, NY, 120, NW
West Deerfield, MA, 105, SE
West Deering, NH, 88, SW
West Dover, ME, 20, SE
West Dudley, MA, 126, SW
West Dummerston, VT, 105, NE
West Edmeston, NY, 100, NW
West Eldred, PA, 133, NW
West End, NY, 29, SE
West Englewood, NJ, 158, NE
West Epping, NH, 89, SE
West Everett, MA, 127C
West Falls, NY, 93, SE
West Falmouth, ME, 71, SW
West Farmington, ME, 36, SE
West Farms, NY, 159, NW
West Fitchburg, MA, 107, SW

West Gloucester, MA, 109, SE
West Gonic, NH, 90, NW
West Gorham, NY, 96, NE
West Graniteville, MA, 108, SW
West Greenfield East, MA, 105, SE
West Greenfield West, MA, 105, SE
West Groton, MA, 107, SE
West Hanover, CT, 144, SW
West Hanover, MA, 128, SW
West Hartford, VT, 67, SW
West Haven, CT, 154, NW
West Haverstraw, NY, 152, SW
West Hempstead, NY, 159, SE
West Henrietta, NY, 76, SE
West Hickory, PA, 131, SW
West Hingham, MA, 128, SW
West Hollis, NH, 107, SE
West Hopkinton, NH, 88, SE
West Hurley, NY, 141, NE
West Kennebunk, ME, 90, NE
West Line, PA, 132, NW
West Lockport, NY, 74, SE
West Lynn, MA, 128, NW
West Manchester, MA, 109, SW
West Manchester, NH, 108A
West Mansfield, MA, 147, NW
West Medford, MA, 127B
West Milan, NH, 34, SW
West Minot, ME, 55, SW
West Monroe, NY, 79, NE
West Mystic, CT, 156, NW
West Newton, MA, 127D
West Norwood, NJ, 152, SW
West Norwood, NJ, 159, NW
West Notch, NY, 113, SE
West Nyack, NY, 152, SW
West Oakland, NJ, 151, SW
West Orange, NJ, 158, NE
West Oxford, MA, 126, SW
West Palmyra, ME, 38, NW
West Park, NY, 142, NW
West Patterson, NY, 142, SE
West Pawlet, VT, 85, NE
West Pawling, NY, 142, SE
West Peabody, MA, 109, SW
West Perrysburg, NY, 111, NE
West Pine Plains, NY, 142, NE
West Pittsfield, MA, 123, NW
West Point, NY, 152, NW
West Portal, MA, 104, SE

West Presque Isle (AV), ME, 5, SE
West Rindge, NH, 106, NE
West River, NY, 96, SW
West River, VT, 105, NE
West Roxbury, MA, 127D
West Rupert, VT, 85, SE
West Rush, NY, 95, NE
West Rutland, MA, 126, NW
West Rutland, VT, 65, SE
West Scarboro, ME, 71, SW
West Sebago, ME, 70, NE
West Seboois, ME, 12, SW
West Shore, NY, 78, SE
West Shore Line Switch, NY, 158, SE
West Side Ave., NJ, 158A
West Somerset, NY, 74, NE
West Sparta, NY, 95, SW
West Springfield, MA, 124, SE
West Springfield Yard, MA, 124, SE, 124A
West Stewartstown, NH, 33, NE
West Stockbridge, MA, 123, NW
West St., NY, 159, NE
West Suffield, CT, 144, NE
West Swanzey, NH, 106, NW
West Thompson, CT, 146, NW
West Thornton, NH, 68, NE
West Townsend, MA, 107, SE
West Townshend, VT, 86, SE
West Upton, MA, 126, SE
West Utica, NY, 81, SW, 81A
West Valley, NY, 112, NE
West Ware, MA, 125, NW
West Warren, MA, 125, SE
West Warwick, RI, 146, SE
West Waterford, NY, 103, NE
West Watertown, MA, 127D
West Willington, CT, 145, NW
West Winfield, NY, 100, NE
West Wrentham, MA, 127, SW
West Yard, NY, 78, NE, 78A
Westboro, MA, 126, NE
Westboro, NH, 67, SW
Westbrook, CT, 155, NW
Westbrook, ME, 71, SW
Westbrookville, NY, 140, SE
Westbury, NY, 159, NE
Westchester, CT, 145, NW
Westchester, NY, 159, NW
Westchester Ave., NY, 152, SW
Westcolang Park, PA, 139, SE

Westdale, MA, 128, SW
Westdale, NY, 80, NW
Westerly, RI, 156, NW
Westfield, CT, 144, SE
Westfield (JW&NW), NY, 110, NE
Westfield, MA, 124, SE
Westfield, ME, 6, SW
Westfield, NJ, 158, SW
Westfield (NKP), NY, 110, NE
Westfield (NYC), NY, 110, NE
Westfield, PA, 134, NE
Westfield Pit, ME, 6, SW
Westford, MA, 108, SE
Westhampton, NY, 161, NE
Westminster, MA, 107, SW
Westminster, VT, 87, SW
Westminster St., RI, 147A
Westmoreland, NH, 87, SW
Westmoreland, NY, 81, SW
Weston, MA, 127, NW
Weston, NY, 113, SW
Weston, PA, 136, SE
Westons, NY, 113, SW
Westport, CT, 153, SW
Westport, NH, 106, NW
Westport, NY, 49, SW
Westport Factory, MA, 147, SE
Westtown, NY, 150, NE
Westview, NJ, 158, NE
Westville, NH, 108, NE
Westway, CT, 145, NW
Westwood, NJ, 158, NE
Westwood, NY, 159, SE
Wethersfield, CT, 144, SE
Wetmore, NH, 106, SE
Wetmore, PA, 132, SW
Weymouth, MA, 128, SW
Weymouth Heights, MA, 128, SW
Wharton, PA, 133, SE
Whately, MA, 124, NE
Wheatland, NY, 76, SW
Wheatville, NY, 75, SW
Wheeler, NY, 115, NW
Wheelers, NY, 96, NW
Wheelerville, PA, 136, SW
Wheelock, ME, 1, SE
Wheelwright, MA, 125, NE
Whippany, NJ, 158, NW
Whipple, RI, 146, NE
Whipples, MA, 125, SW
Whippleville, NY, 28, NE
White Bridge, NY, 95, SE
White Creek, NY, 104, NW
White Gravel, PA, 132, NW
White Mills, PA, 139, SE
White Mountain House, NH, 53, NW
White Mountains Transfer, NH, 51, SE
White Plains, NY, 152, SW
White Plains North, NY, 152, SW
White River Jct., B&M Yard, VT, 67, SW
White River Jct., VT, 67, SW
Whitefield, ME, 56, SE
Whitefield, NH, 52, NE
Whitefield Jct., NH, 52, NE
Whitehall, NY, 65, SW
Whitehead, MA, 128, NW
Whites Corners, NY, 99, NE
Whitesboro, NY, 81, SW
Whitestone, NY, 159, NW
Whitestone Landing, NY, 159, NW
Whitesville, NY, 114, SW
Whiting, VT, 65, NE
Whiting River, CT, 123, SW
Whitins, MA, 126, SE
Whitman, MA, 128, SW
Whitney, MA, 128, SW
Whitney Brook, ME, 55, NW
Whitney Point, NY, 118, NW
Whitneys, CT, 144, NW
Whitneys, MA, 127, SW
Whitneyville, ME, 41, SE
Whittenton, MA, 147, NE, 147C
Wickford, RI, 147, SW
Wickford Jct., RI, 147, SW
Wickford Landing, RI, 147, SW
Wickham, NB, 8, SE
Wilawanna, PA, 136, NE
Wilbers, NY, 99, SE
Wilcox, NY, 151, NW
Wilcox, PA, 132, SE
Wild Goose Club, ME, 38, NW
Wilder, VT, 67, SW
Wiley Road, ME, 8, SW
Wilkes-Barre Pier, RI, 147A
Wilkinsonville, MA, 126, SE
Willard, ME, 7, NW
Willard, NY, 97, SW
Willets, NY, 97, SE
Willey, NH, 108, NW
Willey House, NH, 53, SW
Williams, MA, 148, SE
Williams, ME, 6, NW
Williams & Clark, NJ, 158, SE
Williams Ave., NJ, 158, NE
Williams Bridge, NY, 159, NW
Williamsburg, MA, 124, NE
Williamsburg, ME, 20, NE
Williamsburg Jct., MA, 124, NE
Williamson, NY, 77, SE
Williamstown, MA, 104, SE
Williamstown, NY, 80, NW
Williamstown, VT, 50, SE
Williamsville, MA, 125, NE
Williamsville, VT, 105, NE
Willimansett, MA, 124, SE
Willimantic, CT, 145, SE
Williston, VT, 49, NE
Willow Ave., NJ, 158A
Willow Creek, NY, 97, SE
Willows, MA, 107, SE
Willows, NY, 142, NE
Willsboro, NY, 49, NW
Willseyville (DL&W; LV), NY, 117, NW
Wilmington, MA, 108, SE
Wilmington, VT, 105, NW
Wilna Siding, NY, 60, NE
Wilson, NH, 108, NW
Wilson, NY, 74, NW
Wilson Point, CT, 153, SW
Wilton, CT, 153, SW
Wilton, ME, 36, SE
Wilton, NH, 107, NE
Winchell, NY, 142, NE
Winchendon, MA, 106, SE
Winchester, MA, 127B
Winchester, NH, 106, NW
Winchester Highlands, MA, 127B
Windermere, MA, 128, NW
Windham, NH, 108, NW
Windmill Point, ON, 92, NE
Windsor, CT, 144, NE
Windsor, ME, 56, NE
Windsor, NY, 118, SE
Windsor, VT, 87, NW
Windsor Beach, NY, 76, SE
Windsor Locks, CT, 144, NE
Wing Road, NH, 52, NE
Wingdale, NY, 142, SE
Wings, ME, 41, SW
Winhall, VT, 86, SW
Winn, ME, 22, NW
Winnecook, ME, 38, SW
Winnipauk, CT, 153, SW
Winnisquam, NH, 68, SE
Winona, NH, 68, SE
Winooski, VT, 49, NE
Winslow, ME, 37, SE
Winslows Crossing, MA, 128, SW
Winslows Falls, ME, 57, SW
Winsted, CT, 143, NE
Winter Hill, MA, 127A
Winterport, ME, 39, SW
Winterport Ferry, ME, 39, SW
Winterton, NY, 141, SW
Winterville, ME, 4, NE
Winthrop, ME, 56, NW
Winthrop, NY, 27, NW
Winthrop Beach, MA, 128, NW
Winthrop Center, MA, 128, NW
Wiscasset, ME, 56, SE
Wisner, NY, 151, NW
Woburn, Central Square, MA, 127B
Woburn, MA, 127B
Woburn Highlands, MA, 127B
Wolcott, NY, 78, SW
Wolcott, VT, 32, SW
Wolf Hollow, NY, 120, SW
Wolf Pond, NY, 28, SE
Wolf Run, NY, 112, SW
Wolfboro, NH, 69, SE
Wolfboro Falls, NH, 69, SE
Wollaston, MA, 127E
Wood River Jct. (NH; WRB), RI, 156, NE
Woodard, ME, 21, NE
Woodard, NY, 79, SE
Woodbine, NY, 151, SE
Woodbridge, NJ, 158, SW
Woodbury, NY, 151, NE
Woodbury, VT, 51, NW
Woodburys, MA, 109, SW
Woodcliff Lake, NJ, 151, SE
Woodfords, ME, 71, SW, 71A
Woodgate, NY, 61, SE
Woodhaven Jct., NY, 159, SW
Woodland, MA, 127D
Woodland, ME, 24, SW
Woodland Center, ME, 5, NE
Woodland Jct., ME, 24, SW
Woodlands, NH, 69, SW
Woodlands, NY, 152, SW
Woodlawn, NY, 159, NW
Woodmere, NH, 106, NE
Woodmere, NY, 159, SE
Woodridge, NY, 140, SE
Woodrow, CT, 143, SW
Woodruff's Gap, NJ, 150, SE

Woods, ME, 7, NW
Woods, NY, 62, NW
Woods Falls, NY, 29, NE
Woods Hole, MA, 148, SE
Woodside, MA, 127, SW
Woodside, NJ, 158, NE
Woodside, NY, 159, SW
Woodstock, NB, 8, SE
Woodstock, NH, 68, NE
Woodstock, VT, 66, SE
Woodsville, NH, 51, SE
Woodvale, PA, 132, SE
Woodville, PA, 134, NW
Woodville, RI, 156, NE
Woodway, CT, 152, SE

Woolwich, ME, 72, NW
Woonsocket, RI, 126, SE
Worcester, MA, 126, NW, 126A
Worcester, NY, 101, SE
World's Fair, NY, 159, NW
Woronoco, MA, 124, SW
Wortendyke, NJ, 158, NE
Worthington, NY, 152, SW
Worthley, ME, 55, NW
Wrentham, MA, 127, SW
Wrights, MA, 106, SE
Wrights, ME, 72, NW
Wrights, PA, 133, SE
Wurlitzer, NY, 74, SW

Wurtzboro, NY, 141, SW
Wyalusing, PA, 137, SW
Wyandanch, NY, 160, NW
Wycoff, NJ, 151, SE
Wykagyl, NY, 159, NW
Wyoming, MA, 127C
Wyoming, NY, 94, NE
Wysox, PA, 137, NW
Wytopitlock, ME, 13, SE

Yalesville, CT, 144, SW
Yantic, CT, 145, SE
Yaphank, NY, 161, NW
Yarmouth, MA, 149, SW
Yarmouth, ME, 71, NE

Yarmouth Jct. (CN; MEC), ME, 71, NE
Yonge's Mills, ON, 25, SW
Yonkers, NY, 159, NW
York Beach, ME, 90, SE
York Harbor, ME, 90, SE
Yorkshire, NY, 94, SW
Yorktown Heights, NY, 152, NW
Yosts, NY, 102, NW
Youngsville (NYC; PRR), PA, 131, NW

Zoar, MA, 105, SW
Zurich, NY, 77, SE
Zylonite, MA, 104, SE

INDEX OF TRACK PANS

On main lines of certain railroads, especially the New York Central and the Pennsylvania, steam locomotives could be resupplied with water without stopping, from water troughs or track pans that were placed between the rails. A scoop would be lowered from the underside of the tender, while over the track pan, and lifted before reaching the end. The railroad operating each track pan is signified by its reporting marks, shown in parentheses. The Putnam and Rowayton track pans were no longer in service in 1946.

Churchville (NYC), NY, 76, SW
Clinton Point (NYC), NY, 142, SW

Dunellen (CNJ), NJ, 158, SW

Linden (PRR), NJ, 158, SW

Palmyra (NYC), NY, 77, SE

Putnam (NH), CT, 146, NW

Rome (NYC), NY, 81, SW
Rowayton (NH), CT, 153, SW

Schodack Landing (NYC), NY, 122, NW

Scotia (NYC), NY, 103, NW

Silver Creek (NYC), NY, 92, SE

Tivoli (NYC), NY, 122, SW

Wende (NYC), NY, 93, NE

Westfield (NYC), NY, 110, NE

Yosts (NYC), NY, 102, NW

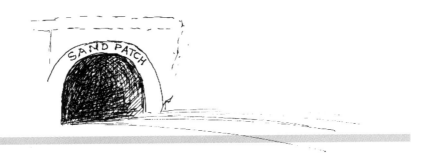

INDEX OF TUNNELS

The tunnels listed in this index are those that have been identified by name on USGS maps, railroad track charts, or other reliable sources. The reporting marks of the owning railroad are indicated in parentheses.

Anthony's Nose (NYC), NY, 152, NW

Belden (D&H), NY, 118, SE
Bergen (DL&W), NJ, 158A
Bergen Arches (ERIE), NJ, 158A
Breakneck (NYC), NY, 152, NW

Cazenovia (NYC), NY, 99, NW

Fallsburgh (NYO&W), NY, 140, SE
Fort Montgomery (NYC), NY, 152, NW

Garrison (NYC), NY, 152, NW

Haverstraw (NYC), NY, 152, SW
Hawk's Mountain (NYO&W), NY, 139, NE
High View (NYO&W), NY, 141, SW
Hoosac (B&M), MA, 104, SE

Laurel Hill (NH), CT, 145, SE
Little (NYC), NY, 152, NW

Middle (NYC), NY, 152, NW

Nicholson (DL&W), PA, 138, SW
North River (PRR), NJ, 158A
Northfield (NYO&W), NY, 119, NE
Norwich (CV), CT, 145, SE

Oscawanna (NYC), NY, 152, SW
Oswego (NYC), NY, 78A
Otisville (ERIE), NY, 150, NE

Park Ave. (NYC), NY, 159, NW

State Line (B&A), NY, 123, NW

Tafts (NH), CT, 145, SE

Vosburg (LV), PA, 137, SE

West Point (NYC), NY, 152, NW

INDEX OF VIADUCTS

Rivers and valleys were often spanned by large bridges or viaducts, many of which were named. The railroad owning the viaduct is signified by its reporting marks, shown in parentheses.

Alfred H. Smith Bridge (NYC), NY, 103, SW

Belfast (ERIE), NY, 113, NE

Fillmore (ERIE), NY, 113, NE
Frankenstein (MEC), NH, 53, SW

Harpursville (D&H), NY, 118, SE
Hell Gate (NYCN), NY, 159, NW, 159A

Kingsley (DL&W), PA, 138, NW
Kinzua (ERIE), PA, 132, NE

Lyman (NH), CT, 145, SW

McIntyre (P&C), NY, 142, NE
Moodna (ERIE), NY, 151, NE

Oswego (NYC), NY, 78A

Portage (ERIE), NY, 94, SE
Poughkeepsie (NH), NY, 142, SW

Rapello (NH), CT, 145, SW
Rosendale (NYC), NY, 141, NE

Salmon River (CN), NB, 3, SE
Salt Point (P&C), NY, 142, NW

Starrucca (ERIE), PA, 138, NE
Suspension Bridge (MC), NY, 73, SE, 73A

Tunkhannock (DL&W), PA, 138, SW

Wappingers Creek (P&C), NY, 142, NW
Whirlpool Rapids Bridge (CN), NY, 73A
Willey House (MEC), NH, 53, SW
Woodbury (ERIE), NY, 151, NE

RICHARD C. CARPENTER was born in Hartford, Connecticut, in 1933 and was raised there and in Wethersfield, Connecticut. He received his bachelor of science degree in history from Boston College, served as a first lieutenant in the U.S. Army artillery, and then received his master's in city and regional planning at the University of Pennsylvania. He was town planner for Wilton, Connecticut, and then, in 1966, joined the South Western Regional Planning Agency of Connecticut as its executive director, retiring in 1999. He continues to serve as a member of the Connecticut Public Transportation Commission. For more than fifty years, Dick Carpenter has closely observed and studied railroad track and signal operating characteristics, collecting in the process invaluable charts, maps, timetables, and other extensive railroad materials. He is a member of the National Railway Historical Society and the Irish Railway Record Society, a corresponding member of the Signalling Record Society, and a member of the Historical and Technical Organizations of nine railroads. He resides in East Norwalk, Connecticut, where he keeps his sailboat, *Phoebe Snow*, named for the symbol of the Delaware, Lackawanna & Western Railroad.